STUDY GUIDE AND SOLUTIONS MANUAL
TO ACCOMPANY

ORGANIC CHEMISTRY

SEVENTH EDITION

STUDY GUIDE
AND
SOLUTIONS MANUAL
TO ACCOMPANY
ORGANIC
CHEMISTRY
SEVENTH EDITION

T. W. GRAHAM SOLOMONS

University of South Florida

CRAIG B. FRYHLE

Pacific Lutheran University

ROBERT G. JOHNSON

Xavier University

JOHN WILEY & SONS, INC.

New York   Chichester   Weinheim   Brisbane   Toronto   Singapore

Acquisitions Editor	David Harris
Associate Editor	Jennifer Yee
Marketing Manager	Charity Robey
Senior Production Editor	Deborah Herbert
Cover Designer	Karin Gerdes Kincheloe

This book was set in 10/12 Times Roman by York Production Services and printed and bound by Hamilton Printing. The cover was printed by The Lehigh Press.

Recognizing the importance of preserving what has been written, it is a policy of John Wiley & Sons, Inc., to have books of enduring value published in the United States printed on acid-free paper, and we exert out best efforts to that end.

ISBN 0-471-35196-2

Printed in the United States of America

10 9 8 7 6 5 4 3

TO THE STUDENT

Contrary to what you may have heard, organic chemistry does not have to be a difficult course. It will be a rigorous course, and it will offer a challenge. But you will learn more in it than in almost any course you will take—and what you learn will have a special relevance to life and the world around you. However, because organic chemistry can be approached in a logical and systematic way, you will find that with the right study habits, mastering organic chemistry can be a deeply satisfying experience. Here, then, are some suggestions about how to study:

1. **Keep up with your work from day to day— never let yourself get behind.** Organic chemistry is a course in which one idea almost always builds on another that has gone before. It is essential, therefore, that you keep up with, or better yet, be a little ahead of your instructor. Ideally, you should try to stay one day ahead of your instructor's lectures in your own class preparations. The lecture, then, will be much more helpful because you will already have some understanding of the assigned material. Your time in class will clarify and expand ideas that are already familiar ones.

2. **Study material in small units, and be sure that you understand each new section before you go on to the next.** Again, because of the cumulative nature of organic chemistry, your studying will be much more effective if you take each new idea as it comes and try to understand it completely before you move on to the next concept. Many key concepts are emphasized in the main text by *Toolbox icons* in the margin and their accompanying captions. These concepts, once you learn them, will be part of your toolbox for success in organic chemistry. Similarly, tips or suggestions for studying and thinking about organic chemistry are highlighted in the text by *Study Tip icons* and captions. Whether or not a concept is highlighted by a Toolbox or Study Tip icon, be sure you understand it before moving on.

3. **Work all of the in-chapter and assigned problems.** One way to check your progress is to work each of the in-chapter problems when you come to it. These problems have been written just for this purpose and are designed to help you decide whether or not you understand the material that has just been explained. If you can work the in-chapter problem, then you should go on; if you cannot, then you should go back and study the preceding material again. Work all of the problems assigned by your instructor from the end of the chapter, as well. Do all of your problems in a notebook and bring this book with you when you go to see your instructor for extra help.

4. **Write when you study.** Write the reactions, mechanisms, structures, and so on, over and over again. Organic chemistry is best assimilated through the fingertips by writing, and not through the eyes by simply looking, or by highlighting material in the text, or by referring to flash cards. There is a good reason for this. Organic structures, mechanisms, and reactions are complex. If you simply examine them, you may think you understand them thoroughly, but that will be a misperception. The reaction mechanism may make sense to you in a certain way, but you need a deeper understanding than this. You need to know the material so thoroughly that you can explain it to someone else. This level of understanding comes to most of us (those of us without photographic memories) through writing. Only by writing the reaction mechanisms do we pay sufficient attention to their details, such as which atoms are connected to which atoms, which bonds break in a reaction and which bonds form, and the three-dimensional aspects of the structures. When we write reactions and mechanisms, connections are made in our brains that provide the long-term memory needed for success in organic chemistry. We virtually guarantee that your grade in the course will be directly proportional to the number of pages of paper that

you fill with your own writing in studying during the term.

5. **Learn by teaching and explaining.** Study with your student peers and practice explaining concepts and mechanisms to each other. Use the *Learning Group Problems* and other exercises your instructor may assign as vehicles for teaching and learning interactively with your peers.

6. **Use the answers to the problems in the *Study Guide* in the proper way.** Refer to the answers only in two circumstances: (1) When you have finished a problem, use the Study Guide to check your answer. (2) When, after making a real effort to solve the problem, you find that you are completely stuck, then look at the answer for a clue and go back to work out the problem on your own. The value of a problem is in solving it. If you simply read the problem and look up the answer, you will deprive yourself of an important way to learn.

7. **Use the introductory material in the *Study Guide* entitled "Solving the Puzzle—or—Structure Is Everything (Almost)"** as a bridge from general chemistry to your beginning study of organic chemistry. You might find this section helpful as a way to see the relevance to organic chemistry of some of the concepts you learned in general chemistry, while at the same time refreshing these ideas in your mind and gearing you up to study organic chemistry. It is also meant to help you see that an understanding of certain fundamental principles, largely having to do with structure, can help immensely to reduce the complexity of the puzzle you may feel lies ahead of you. Indeed, once you have a firm understanding of structure, the puzzle of organic chemistry can become one of very manageable size and comprehensible pieces.

8. **Use molecular models when you study.** Because of the three-dimensional nature of most organic molecules, molecular models can be an invaluable aid to your understanding of them. Buy yourself an inexpensive molecular model set and use it when you need to see the three-dimensional aspect of the particular topic. An appendix in the *Study Guide* provides a set of highly useful molecular model exercises.

INTRODUCTION

"Solving the Puzzle"
or
"Structure Is Everything (Almost)"

As you begin your study of organic chemistry it may seem like a puzzling subject. In fact, in many ways organic chemistry is like a puzzle—a jigsaw puzzle. But it is a jigsaw puzzle with useful pieces, and a puzzle with fewer pieces than perhaps you first thought. In order to put a jigsaw puzzle together you must consider the shape of the pieces and how one piece fits together with another. In other words, solving a jigsaw puzzle is about **structure**. In organic chemistry, molecules are the pieces of the puzzle. Much of organic chemistry, indeed life itself, depends upon the fit of one molecular puzzle piece with another. For example, when an antibody of our immune system acts upon a foreign substance, it is the puzzle-piece-like fit of the antibody with the invading molecule that allows "capture" of the foreign substance. When we smell the sweet scent of a rose, some of the neural impulses are initiated by the fit of a molecule called geraniol in an olfactory receptor site in our nose. When an adhesive binds two surfaces together, it does so by billions of interactions between the molecules of the two materials. Chemistry is truly a captivating subject.

As you make the transition from your study of general to organic chemistry, it is important that you solidify those concepts that will help you understand the structure of organic molecules. A number of concepts are discussed below using several examples. It is suggested that you consider the examples and the explanations given, and refer to information from your general chemistry studies when you need more elaborate information. There are also occasional references below to sections in your text, Solomons's and Fryhle's *Organic Chemistry*, because some of what follows foreshadows what you will learn in the course.

SOME FUNDAMENTAL PRINCIPLES WE NEED TO CONSIDER

What do we need to know to understand the structure of organic molecules? First, we need to know where electrons are located around a given atom. To understand this we need to recall from general chemistry the ideas of **electron configuration** and **valence shell electron orbitals,** especially in the case of atoms such as carbon, hydrogen, oxygen, and nitrogen. We also need to use **Lewis valence shell electron structures.** These concepts are useful because the shape of a molecule is defined by its constituent atoms, and the placement of the atoms follows from the location of the electrons that bond the atoms. Once we have a Lewis structure for a molecule, we can consider **orbital hybridization** and **valence shell electron pair repulsion (VSEPR) theory** in order to generate a three-dimensional image of the molecule.

Secondly, in order to understand why specific organic molecular puzzle pieces fit together we need to consider the attractive and repulsive forces between them. To understand this we need to know how electronic charge is distributed in a molecule. We must use tools such as **formal charge** and **electronegativity.** That is, we need to know which parts of a molecule are relatively positive and which are relatively negative—in other words, their **polarity.** Associations between molecules strongly depend on both shape and the complementarity of their electrostatic charges (polarity).

When it comes to organic chemistry it will be much easier for you to understand why organic molecules have certain properties and react the way they do if you have an appreciation for the structure of the molecules involved. Structure is, in fact, almost everything, in that whenever we want to know why or how something works we look ever more deeply into its structure. This is true whether we are considering a toaster, jet engine, or an organic reaction. If you can visualize the shape of the puzzle pieces in organic chemistry (molecules), you will see more easily how they fit together (react).

SOME EXAMPLES

In order to review some of the concepts that will help us understand the structure of organic molecules, let's consider three very important molecules—water, methane, and methanol (methyl alcohol). These three are small and relatively simple molecules that have certain similarities among them, yet distinct differences that can be understood on the basis of their structures. Water is a liquid with a moderately high boiling point that does not dissolve organic compounds well. Methanol is also a liquid, with a lower boiling point than water, but one that dissolves many organic compounds easily. Methane is a gas, having a boiling point well below room temperature. Water and methanol will dissolve in each other, that is, they are miscible. We shall study the structures of water, methanol, and methane because the principles we learn with these compounds can be extended to much larger molecules.

Water

HOH

Let's consider the structure of water, beginning with the central oxygen atom. Recall that the atomic number (the number of protons) for oxygen is eight. Therefore, an oxygen atom also has eight electrons. (An ion may have more or less electrons than the atomic number for the element, depending on the charge of the ion.) Only the valence (outermost) shell electrons are involved in bonding. Oxygen has six valence electrons—that is, six electrons in the second principal shell. (Recall that the number of valence electrons is apparent from the group number of the element in the periodic table, and the row number for the element is the principal shell number for its valence electrons.) Now, let's consider the electron configuration for oxygen. The sequence of atomic orbitals for the first three shells of any atom is shown below. Oxygen uses only the first two shells in its lowest energy state.

$$1s, 2s, 2p_x, 2p_y, 2p_z, 3s, 3p_x, 3p_y, 3p_z$$

The p orbitals of any given principal shell (second, third, etc.) are of equal energy. Recall also that each orbital can hold a maximum of two electrons and that each orbital must accept one electron before a second can reside there (Hund's rule). So, for oxygen we place two electrons in the $1s$ orbital, two in the $2s$ orbital, and one in each of the $2p$ orbitals, for a subtotal of seven electrons. The final eighth electron is paired with another in one of the $2p$ orbitals. The configuration for the eight electrons of oxygen is, therefore

$$1s^2 \, 2s^2 \, 2p_x^2 \, 2p_y^1 \, 2p_z^1$$

where the superscript numbers indicate how many electrons are in each orbital. In terms of relative energy of these orbitals, the following diagram can be drawn. Note that the three $2p$ orbitals are depicted at the same relative energy level.

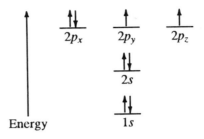

Now, let's consider the shape of these orbitals. The shape of an *s* **orbital** is that of a sphere with the nucleus at the center. The shape of each *p* **orbital** is approximately that of a dumbbell or lobe-shaped object, with the nucleus directly between the two lobes. There is one pair of lobes for each of the three p orbitals (p_x, p_y, p_z), and they are aligned along the x, y, and z coordinate axes, with the nucleus at the origin. Note that this implies that the three p orbitals are at 90° angles to each other.

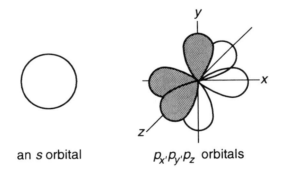

an *s* orbital p_x, p_y, p_z orbitals

Now, when oxygen is bonded to two hydrogens, bonding is accomplished by the sharing of an electron from each of the hydrogens with an unpaired electron from the oxygen. This type of bond, involving the sharing of electrons between atoms, is called a **covalent bond.** The formation of covalent bonds between the oxygen atom and the two hydrogen atoms is advantageous because each atom achieves a full valence shell by the sharing of these electrons. For the oxygen in a water molecule, this amounts to satisfying the octet rule.

A **Lewis structure** for the water molecule (which shows only the valence shell electrons) is depicted in the following structure. There are two nonbonding pairs of electrons around the oxygen as well as two bonding pairs.

In the left-hand structure the six valence electrons contributed by the oxygen are shown as dots, while those from the hydrogens are shown as x's. This is done strictly for bookkeeping purposes. All electrons are, of course, identical. The right-hand structure uses the convention that a bonding pair of electrons can be shown by a single line between the bonded atoms.

This structural model for water is only a *first approximation,* however. While it is a proper Lewis structure for water, it is not an entirely correct three-dimensional structure. It might appear that the angle between the hydrogen atoms (or between any two pairs of electrons in a water molecule) would be 90°, but this is not what the true angles are in a water molecule. The angle between the two hydrogens is in fact about 105°, and the nonbonding electron pairs are in a different plane than the hydrogen atoms. The reason for this arrangement is that groups of bonding and nonbonding electrons tend to repel each other due to the negative charge of the electrons. Thus, the ideal angles between bonding and nonbonding groups of electrons are those angles that allow maximum separation in three-dimensional space. This principle and the theory built around it are called the **valence shell electron pair repulsion (VSEPR) theory.**

VSEPR theory predicts that the ideal separation between four groups of electrons around an atom is 109.5°, the so-called **tetrahedral angle.** At an angle of 109.5° all four electron groups are separated equally from each other, being oriented toward the corners of a regular tetrahedron. The exact tetrahedral angle of 109.5° is found in structures where the four groups of electrons and bonded groups are identical.

In water, there are two different types of electron groups—pairs bonding the hydrogens with the oxygen and nonbonding pairs. Nonbonding electron pairs repel each other with greater force than bonding pairs, so the separation between them is greater. Consequently, the angle between the pairs bonding the hydrogens to the oxygen in a water molecule is compressed slightly from 109.5°, being actually about 105°. As we shall see shortly, the angle between the four groups of bonding electrons in methane (CH_4) *is* the ideal tetrahedral angle of 109.5°. This is because the four groups of electrons and bound atoms are identical in a methane molecule.

Orbital hybridization is the reason that 109.5° is the ideal tetrahedral angle. As noted earlier, an *s* orbital is spherical, and each *p* orbital is shaped like two symmetrical lobes aligned along the *x, y,* and *z* coordinate axes. Orbital hybridization involves taking a weighted average of the valence electron orbitals of the atom, resulting in the same number of new hybridized orbitals. With four groups of valence electrons, as in the structure of water, one *s* orbital and three *p* orbitals from the second principal shell in oxygen are hybridized (the $2s$ and $2p_x$, $2p_y$, and $2p_z$ orbitals). The result is four new hybrid orbitals of equal energy designated as sp^3 orbitals (instead of the original three *p* orbitals and one *s* orbital). Each of the four sp^3 orbitals has roughly 25% *s* character and 75% *p* character. The geometric result is that the major lobes of the four sp^3 orbitals are oriented toward the corners of a tetrahedron with an angle of 109.5° between them.

sp³ hybrid orbitals
(109.5° angle between lobes)

In the case of the oxygen in a water molecule, where two of the four sp^3 orbitals are occupied by nonbonding pairs, the angle of separation between them is larger than 109.5° due to additional electrostatic repulsion of the nonbonding pairs. Consequently, the angle between the bonding electrons is slightly smaller, about 105°.

More detail about orbital hybridization than provided above is given in Sections 1.10–1.14 of *Organic Chemistry*. With that greater detail it will be apparent from consideration of orbital hybridization that for three groups of valence electrons the ideal separation is 120° (trigonal planar), and for two groups of valence electrons the ideal separation is 180° (linear). VSEPR theory allows us to come to essentially the same conclusion as by the mathematical hybridization of orbitals, and it will serve us for the moment in predicting the three-dimensional shape of molecules.

Methane

$$CH_4$$

Now let's consider the structure of methane (CH_4). In methane there is a central carbon atom bearing four bonded hydrogens. Carbon has six electrons in total, with four of them being valence electrons. (Carbon is in Group 4 in the periodic table.) In methane each valence electron is shared with an electron from a hydrogen atom to form four covalent bonds. This information allows us to draw a Lewis structure for methane (see below). With four groups of valence electrons the VSEPR theory allows us to predict that the three-dimensional shape of a methane molecule should be tetrahedral, with an angle of 109.5° between each of the bonded hydrogens. This is indeed the case. Orbital hybridization arguments can also be used to show that there are four equivalent sp^3 hybrid orbitals around the carbon atom, separated by an angle of 109.5°.

All H-C-H angles are 109.5°.

The structure at the far right above uses the dash-wedge notation to indicate three dimensions. A solid wedge indicates that a bond projects out of the paper toward the reader. A dashed bond indicates that it projects behind the paper away from the viewer. Ordinary lines represent bonds in the plane of the paper. The dash-wedge notation is an important and widely used tool for depicting the three-dimensional structure of molecules.

Methanol

$$CH_3OH$$

Now let's consider a molecule that incorporates structural aspects of both water and methane. Methanol (CH_3OH), or methyl alcohol, is such a molecule. In methanol, a central carbon atom has three hydrogens and an O–H group bonded to it. Three of the four valence electrons of the carbon atom are shared with a valence electron from the hydrogen atoms, forming three C–H bonds. The fourth valence electron of the carbon is shared with a valence electron from the oxygen atom, forming a C–O bond. The carbon atom now has an octet of valence electrons through the formation of four covalent bonds. The angles between these four covalent bonds is very near the ideal tetrahedral angle of 109.5°, allowing maximum separation between them. (The valence orbitals of the carbon are sp^3 hybridized.) At the oxygen atom, the situation is very similar to that in water. The oxygen uses its two unpaired valence electrons to form covalent bonds. One valence electron is used in the bond with the carbon atom, and the other is paired with an electon from the hydrogen to form the O—H bond. The remaining valence electrons of the oxygen are present as two nonbonding pairs, just as in water. The angles separating the four groups of electrons around the oxygen are thus near the ideal angle of 109.5°, but reduced slightly in the C–O–H angle due to repulsion by the two nonbonding pairs on the oxygen. (The valence orbitals of the oxygen are also sp^3 hybridized since there are four groups of valence electrons.) A Lewis structure for methanol is shown below, along with a three-dimensional perspective drawing.

THE "CHARACTER" OF THE PUZZLE PIECES

With a mental image of the three-dimensional structures of water, methane, and methanol, we can ask how the structure of each, as a "puzzle piece," influences the interaction of each molecule with identical and different molecules. In order to answer this question we have to move one step beyond the three-dimensional shape of these molecules. We need to consider not only the location of the electron groups (bonding and nonbonding) but also the distribution of electronic charge in the molecules.

First, we note that nonbonding electrons represent a locus of negative charge, more so than electrons involved in bonding. Thus, water would be expected to have some partial negative charge localized in the region of the nonbonding electron pairs of the oxygen. The same would be true for a methanol molecule.

Secondly, the phenomenon of electronegativity influences the distribution of electrons, and hence the charge in a molecule, especially with respect to electrons in covalent bonds. **Electronegativity** is the propensity of an element to draw electrons toward it in a covalent bond. The trend among elements is that of increasing electronegativity toward the upper right corner of the periodic table. (Fluorine is the most electronegative element.) By observing the relative locations of carbon, oxygen, and hydrogen in the periodic table, we can see that oxygen is the most electronegative of these three elements. Carbon is more electronegative than hydrogen, although only slightly. Oxygen is significantly more electronegative than hydrogen. Thus, there is substantial separation of charge in a water molecule, due not only to the nonbonding electron pairs on the oxygen but also to the greater electronegativity of the oxygen with respect to the hydrogens. The oxygen tends to draw electron density toward itself in the bonds with the hydrogens, leaving the hydrogens partially positive. The resulting separation of charge is called **polarity.** The oxygen–hydrogen bonds are called **polar covalent bonds** due to this separation of charge. If one considers the net effect of the two nonbonding electron pairs in a water molecule as being a region of negative charge, and the hydrogens as being a region of relative positive charge, it is clear that a water molecule has substantial separation of charge, or polarity.

An analysis of polarity for a methanol molecule would proceed similarly to that for water. Methanol, however, is less polar than water because only one O–H bond is present. Nevertheless, the region of the molecule around the two nonbonding electron pairs of the oxygen is relatively negative, and the region near the hydrogen is relatively positive. The electronegativity difference between the oxygen and the carbon is not as large as that between oxygen and hydrogen, however, so there is less polarity associated with the C–O bond. Since there is even less difference in electronegativity between hydrogen and carbon in the three C–H bonds, these bonds contribute essentially no polarity to the molecule. The net effect for methanol is to make it a polar molecule, but less so than water due to the nonpolar character of the CH_3 region of the molecule.

Now let's consider methane. Methane is a nonpolar molecule. This is evident first because there are no nonbonding electron pairs, and secondly because there is relatively little electronegativity difference between the hydrogens and the central carbon. Furthermore, what little electronegativity difference there is between the hydrogens and the central carbon atom is negated by the symmetrical distribution of the C–H bonds in the tetrahedral shape of methane. The slight polarity of each C–H bond is canceled by the symmetrical orientation of the four C–H bonds. If considered as vectors, the vector sum of the four slightly polar covalent bonds oriented at 109.5° to each other would be zero.

 Net dipole is zero.

The same analysis would hold true for a molecule with identical bonded groups, but groups having electronegativity significantly different from carbon, so long as there were symmetrical distribution of the bonded groups. Tetrachloromethane (carbon tetrachloride) is such a molecule. It has no net polarity.

 Net dipole is zero.

INTERACTIONS OF THE PUZZLE PIECES

Now that you have an appreciation for the polarity and shape of these molecules it is possible to see how molecules might interact with each other. The presence of polarity in a molecule bestows upon it attractive or repulsive forces in relation to other molecules. The negative part of one molecule is attracted to the positive region of another. Conversely, if there is little polarity in a molecule, the attractive forces it can exert are very small [though not completely nonexistent, due to van der Waals forces (Section 2.14D in *Organic Chemistry*)]. Such effects are called **intermolecular forces** (forces between molecules), and strongly depend on the polarity of a molecule or certain bonds within it (especially O–H, N–H, and other bonds between hydrogen and more electronegative atoms with nonbonding pairs). Intermolecular forces have profound effects on physical properties such as **boiling point, solubility,** and **reactivity.** An important manifestation of these properties is that the ability to isolate a pure compound after a reaction often depends on differences in boiling point, solubility, and sometimes reactivity among the compounds present.

Boiling Point

An intuitive understanding of boiling points will serve you well when working in the laboratory. The polarity of water molecules leads to relatively strong intermolecular attraction between water molecules. One result is the moderately high boiling point of water (100°C, as compared to 65°C for methanol and −162°C for methane, which we will discuss shortly). Water has the highest boiling point of these three example molecules because it will strongly associate with itself by attraction of the partially positive hydrogens of one molecule (from the electronegativity difference between the O and H) to the negatively charged region in another water molecule (where the nonbonding pairs are located).

The specific attraction between a partially positive hydrogen atom attached to a heteroatom (an atom with both nonbonding and bonding valence electrons, e.g., oxygen or nitrogen) and the nonbonding electrons of another heteroatom is called **hydrogen bonding.** It is a form of **dipole-dipole attraction** due to the polar nature of the hydrogen–heteroatom bond. A given water molecule can associate by hydrogen bonding with several other water molecules, as shown above. Each water molecule has two hydrogens that can associate with the nonbonding pairs of other water molecules, and two nonbonding pairs that can associate with the hydrogens of other water molecules. Thus, several hydrogen bonds are possible for each water molecule. It takes a significant amount of energy (provided by heat, for example) to give the molecules enough kinetic energy (motion) for them to overcome the polarity-induced attractive forces between them and escape into the vapor phase (evaporation or boiling).

Methanol, on the other hand, has a lower boiling point than water (65°C), in large part due to the decreased hydrogen bonding ability of methanol in comparison with water. Each methanol molecule has only one hydrogen atom that can participate in a hydrogen bond with the nonbonding electron pairs of another methanol molecule (as compared with two for each water molecule). The result is reduced intermolecular attraction between methanol molecules and a lower boiling point since less energy is required to overcome the lesser intermolecular attractive forces.

The CH_3 group of methanol does not participate in dipole–dipole attractions between molecules because there is not sufficient polarity in any of its bonds to lead to significant partial positive or negative charges. This is due to the small electronegativity difference between the carbon and hydrogen in each of the C–H bonds.

Now, on to methane. Methane has no hydrogens that are eligible for hydrogen bonding, since none is attached to a heteroatom such as oxygen. Due to the small difference in electronegativity between carbon and hydrogen there are no bonds with any significant polarity. Furthermore, what slight polarity there is in each C–H bond is canceled due to the tetrahedral symmetry of the molecule. [The minute attraction that is present between methane molecules is due to van der Waals forces, but these are negligible in comparison to dipole–dipole interactions that exist when significant differences in electronegativity are present in molecules such as water and methanol.]. Thus, because there is only a very weak attractive force between methane molecules, the boiling point of methane is very low ($-162°C$) and it is a gas at ambient temperature and pressure.

Solubility

An appreciation for trends in solubility is very useful in gaining a general understanding of many practical aspects of chemistry. The ability of molecules to dissolve other molecules or solutes is strongly affected by polarity. The polarity of water is frequently exploited during the isolation of an organic reaction product because water will not dissolve most organic compounds but will dissolve salts, many inorganic materials, and other polar byproducts that may be present in a reaction mixture.

As to our example molecules, water and methanol are miscible with each other because each is polar and can interact with the other by dipole–dipole hydrogen bonding interactions. Since methane is a gas under ordinary conditions, for the purposes of this discussion let's consider a close relative of methane – hexane. Hexane (C_6H_{14}) is a liquid having only carbon – carbon and carbon – hydrogen bonds. It belongs to the same chemical family as methane. Hexane is *not* soluble in water due to the essential absence of polarity in its bonds. Hexane is slightly soluble in methanol due to the compatibility of the nonpolar CH_3 region of methanol with hexane. The old saying "like dissolves like" definitely holds true. This can be extended to solutes, as well. Very polar substances, such as ionic compounds, are usually freely soluble in water. The high polarity of salts generally prevents most of them from being soluble in methanol, however. And, of course, there is absolutely no solubility of ionic substances in hexane. On the other hand, very nonpolar substances, such as oils, would be soluble in hexane.

Thus, the structure of each of these molecules we've used for examples (water, methanol, and methane) has a profound effect on their respective physical properties. The presence of nonbonding electron pairs and polar covalent bonds in water and methanol versus the complete absence of these features in the structure of methane imparts markedly different physical properties to these three compounds. Water, a small molecule with strong intermolecular forces, is a moderately high boiling liquid. Methane, a small molecule with only very weak intermolecular forces, is a gas. Methanol, a molecule combining structural aspects of both water and methane, is a relatively low boiling liquid, having sufficient intermolecular forces to keep the molecules associated as a liquid, but not so strong that mild heat can't disrupt their association.

Reactivity

While the practical importance of the physical properties of organic compounds may only be starting to become apparent, one strong influence of polarity is on the **reactivity** of molecules. It is often possible to understand the basis for a given reaction in organic chemistry by considering the relative polarity of molecules and the propensity, or lack thereof, for them to interact with each other.

Let us consider one example of reactivity that can be understood at the initial level by considering structure and polarity. When chloromethane (CH_3Cl) is exposed to hydroxide ions (OH^-) in water a reaction occurs that produces methanol. This reaction is shown below.

$$CH_3Cl \ + \ HO^- \ (\text{as NaOH dissolved in water}) \ \rightarrow \ HOCH_3 \ + \ Cl^-$$

This reaction is called a substitution reaction, and it is of a general type that you will spend considerable time studying in organic chemistry. The reason this reaction occurs readily can be understood by considering the principles of structure and polarity that we have been discussing. The hydroxide ion has a negative charge associated with it, and thus should be attracted to a species that has positive charge. Now recall our discussion of electronegativity and polar covalent bonds, and apply these ideas to the structure of chloromethane. The chlorine atom is significantly more electronegative than carbon (note its position in the periodic table). Thus, the covalent bond between the carbon and the chlorine is polarized such that there is partial negative charge on the chlorine and partial positive charge on the carbon. This provides the positive site that attracts the hydroxide anion!

The intimate details of this reaction will be studied in Chapter 6 of your text. Suffice it to say for the moment that the hydroxide ion attacks the carbon atom using one of its nonbonding electron pairs to form a bond with the carbon. At the same time, the chlorine atom is pushed away from the carbon and takes with it the pair of electrons that used to bond it to the carbon. The result is substitution of OH for Cl at the carbon atom and the synthesis of methanol. By calculating **formal charges** (Section 1.7 in the text) one can show that the oxygen of the hydroxide anion goes from having a formal negative charge in hydroxide to zero formal charge in the methanol molecule. Similarly, the chlorine atom goes from having zero formal charge in chloromethane to a formal negative charge as a chloride ion after the reaction. *The fact that the reaction takes place at all rests largely upon the complementary polarity of the interacting species. This is a pervasive theme in organic chemistry.*

   Acid–base reactions are also very important in organic chemistry. Many organic reactions involve at least one step in the overall process that is fundamentally an acid–base reaction. Both Bronsted-Lowry acid–base reactions (those involving proton donors and acceptors) and Lewis acid–base reactions (those involving electron pair acceptors and donors, respectively) are important. In fact, the reaction above can be classified as a Lewis acid–base reaction in that the hydroxide ion acts as a Lewis base to attack the partially positive carbon as a Lewis acid. It is strongly recommended that you review concepts you have learned previously regarding acid–base reactions. Chapter 3 in *Organic Chemistry* will help in this regard, but it is advisable that you begin some early review about acids and bases based on your previous studies. Acid–base chemistry is widely applicable to understanding organic reactions.

JOINING THE PIECES

   Finally, while what we have said above has largely been in reference to three specific compounds, water, methanol, and methane, the principles involved find exceptionally broad application in understanding the structure, and hence reactivity, of organic molecules in general. You will find it constantly useful in your study of organic chemistry to consider the electronic structure of the molecules with which you are presented, the shape caused by the distribution of electrons in a molecule, the ensuing polarity, and the resulting potential for that molecule's reactivity. What we have said about these very small molecules of water, methanol, and methane can be extended to consideration of molecules with 10 to 100 times as many atoms. You would simply apply these principles to small sections of the larger molecule one part at a time. The following structure of Streptogramin A provides an example.

A region with trigonal planar bonding

A few of the partially positive and partially negative regions are shown, as well as regions of tetrahedral and trigonal planar geometry. See if you can identify more of each type.

A region with tetrahedral bonding

Streptogramin A

A natural antibacterial compound that blocks protein
synthesis at the 70S ribosomes of Gram-positive bacteria.

   We have not said much about how overall shape influences the ability of one molecule to interact with another, in the sense that a key fits in a lock or a hand fits in a glove.

This type of consideration is also extremely important, and will follow with relative ease if you have worked hard to understand the general principles of structure outlined here and expanded upon in the early chapters of *Organic Chemistry*. An example would be the following. Streptogramin A, shown above, interacts in a hand-in-glove fashion with the 70S ribosome in bacteria to inhibit binding of transfer RNA at the ribosome. The result of this interaction is the prevention of protein synthesis in the bacterium, which thus accounts for the antibacterial effect of Streptogramin A. Other examples of hand-in-glove interactions include the olfactory response to geraniol mentioned earlier, and the action of enzymes to speed up the rate of reactions in biochemical systems.

FINISHING THE PUZZLE

In conclusion, if you pay attention to learning aspects of structure during this initial period of "getting your feet wet" in organic chemistry, much of the three-dimensional aspects of molecules will become second nature to you. You will immediately recognize when a molecule is tetrahedral, trigonal planar, or linear in one region or another. You will see the potential for interaction between a given section of a molecule and that of another molecule based on their shape and polarity, and you will understand why many reactions take place. Ultimately, you will find that there is much less to memorize in organic chemistry than you first thought. You will learn how to put the pieces of the organic puzzle together, and see that structure is indeed almost everything, just applied in different situations!

ACKNOWLEDGMENTS

We are especially grateful to Keith Buszek of Kansas State University for reviewing all of the answers to the new problems contained in the seventh edition.

We also thank those who have made many helpful suggestions for earlier versions of this study guide. These include: George R. Jurch, George R. Wenzinger, and J. E. Fernandez at the University of South Florida; Darell Berlin, Oklahoma State University; John Mangravite, West Chester State College; J. G. Traynham, Louisiana State University; and Desmond M. S. Wheeler, University of Nebraska.

We are very grateful to Professor John Thompson of Texas A&M University—Kingsville for completing the arduous task of creating almost all of the structures found in this Study Guide.

T. W. Graham Solomons

C. B. Fryhle

R. G. Johnson

CONTENTS

1 CARBON COMPOUNDS AND CHEMICAL BONDS

SOLUTIONS TO PROBLEMS

Another Approach to Writing Lewis Structures

When we write Lewis structures using this method, we assemble the molecule or ion from the constituent atoms showing only the valence electrons (i.e., the electrons of the outermost shell). By having the atoms share electrons, we try to give each atom the electronic structure of a noble gas. For example, we give hydrogen atoms two electrons because this gives them the structure of helium. We give carbon, nitrogen, oxygen, and fluorine atoms eight electrons because this gives them the electronic structure of neon. The number of valence electrons of an atom can be obtained from the periodic table because it is equal to the group number of the atom. Carbon, for example, is in group 4A and has four valence electrons; fluorine, in group 7A, has seven; hydrogen, in group 1A, has one. As an illustration, let us write the Lewis structure for CH_3F. In the example below, we will at first show a hydrogen's electron as x, carbon's electrons as o's, and fluorine's electrons as dots.

Example A

$$3 \; H^\times, \; \overset{\circ}{\underset{\circ}{\text{o}}}C\text{o}, \; \text{and} \; \cdot\overset{\cdot\cdot}{\underset{\cdot\cdot}{F}}\text{: are assembled as}$$

$$
\begin{array}{cc}
\text{H} & \text{H} \\
\overset{\times\text{o}}{\underset{\times\text{o}}{\text{H}\,\overset{\circ}{\underset{\circ}{\text{C}}}\,\overset{\cdot\cdot}{\underset{\cdot\cdot}{\text{F}}}}} \quad \text{or} & \text{H}\,\overset{\cdot\cdot}{\underset{\cdot\cdot}{\text{C}}}\,\overset{\cdot\cdot}{\underset{\cdot\cdot}{\text{F}}}\,\text{:} \\
\text{H} & \text{H}
\end{array}
$$

If the structure is an ion, we add or subtract electrons to give it the proper charge. As an example, consider the chlorate ion, ClO_3^-.

Example B

$$\text{:}\overset{\cdot\cdot}{Cl}\cdot, \; \text{and} \; \overset{\circ\circ}{\underset{\circ\circ}{O}} \; \text{and an extra electron} \times \text{are assembled as}$$

$$
\left[\; \overset{\overset{\circ\circ}{O}}{\underset{\circ\circ}{\underset{\circ\circ}{O}\text{:}Cl\text{:}\overset{\circ\circ}{O}}} \; \right]^{-} \quad \text{or} \quad \left[\; \overset{\text{:}\ddot{O}\text{:}}{\text{:}\ddot{O}\text{:}\overset{\cdot\cdot}{Cl}\text{:}\ddot{O}\text{:}} \; \right]^{-}
$$

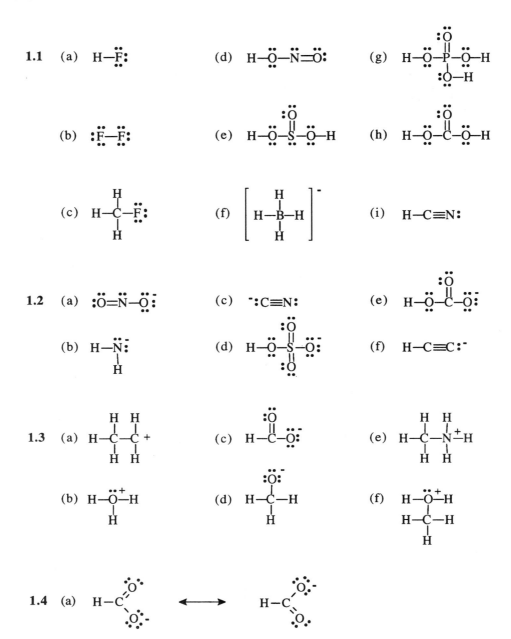

(b) and (c).   Since the two resonance structures are equivalent, each should make an equal contribution to the overall hybrid. The C—O bonds should therefore be of equal length (they should be of bond order 1.5), and each oxygen atom should bear a 0.5 negative charge.

1.5 (a) In its ground state, the valence electrons of carbon might be disposed as shown in the following figure.

The electronic configuration of a ground state carbon atom. The p orbitals are designated $2p_x$, $2p_y$, and $2p_z$ to indicate their respective orientations along the x, y, and z axes. The assignment of the unpaired electrons to the $2p_y$ and $2p_x$ orbitals is arbitrary. They could also have been placed in the $2p_x$ and $2p_z$ or $2p_y$ and $2p_z$ orbitals. (To have placed them both in the same orbital would not have been correct, however, for this would have violated Hund's rule.) (Section 1.10.)

The formation of the covalent bonds of methane *from individual atoms* requires that the carbon atom overlap its orbitals containing *single electrons* with 1s orbitals of hydrogen atoms (which also contain a single electron). If a ground state carbon atom were to combine with hydrogen atoms in this way, the result would be that depicted below. *Only two carbon-hydrogen bonds would be formed, and these would be at right angles to each other.*

The hypothetical formation of CH_2 from a carbon atom in its ground state.

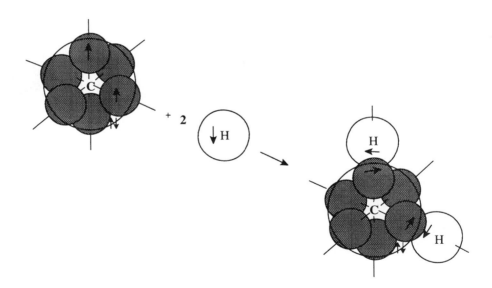

(b) An excited-state carbon atom might combine with four hydrogen atoms as shown in the figure above.

The promotion of an electron from the 2s orbital to the $2p_z$ orbital requires energy. The amount of energy required has been determined and is equal to 400 kJ mol^{-1}. This expenditure of energy can be rationalized by arguing that the energy released when two additional covalent bonds form would more than compensate for that required to excite the electron. No doubt this is true, but it solves only one problem. The problems that cannot be solved by using an excited-state carbon as a basis for a model of methane are the problems of the carbon-hydrogen bond angles and the apparent equivalence of all four carbon-hydrogen bonds. Three of the hydrogens—those overlapping their $1s$ orbitals with the three p orbitals—would, in this model, be at angles of 90° with respect to each other; the fourth hydrogen, the one overlapping its $1s$ orbital with the $2s$ orbital of carbon, would be at some other angle, probably as far from the other bonds as the confines of the molecule would allow. Basing our model of methane on this excited state of carbon gives us a carbon that is tetravalent *but one that is not tetrahedral*, and it predicts a structure for methane in which one carbon-hydrogen bond differs from the other three.

The hypothetical formation of CH_4 from an excited-state carbon atom.

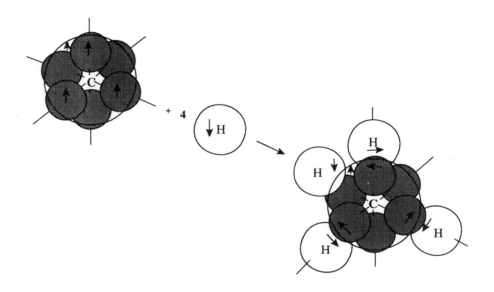

1.6   (a)   Cis-trans isomers are not possible.

(b)   and

(c)   Cis-trans isomers are not possible.

(d)   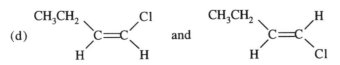   and

1.7   If the geometry of the carbon atom were square planar, two isomeric compounds with the formula CH_2Cl_2 should exist.

   and

The fact that only one form of CH_2Cl_2 has ever been detected supports a tetrahedral geometry for carbon, because with a tetrahedral geometry only one compound with the formula CH_2Cl_2 is possible. All of the following structures are just different orientations of the same molecule.

Make models to convince yourself of this fact.

1.8 (a) There are four bonding pairs. The geometry is tetrahedral.

(b) There are two bonding pairs about the central atom. The geometry is linear.

$$:\!F\!\!-\!\!Be\!\!-\!\!F\!:$$

(c) There are four bonding pairs. The geometry is tetrahedral.

(d) There are two bonding pairs and two nonbonding pairs. The geometry is angular.

(e) There are three bonding pairs. The geometry is trigonal planar.

(f) There are four bonding pairs around the central atom. The geometry is tetrahedral.

(g) There are four bonding pairs around the central atom. The geometry is tetrahedral.

(h) There are three bonding pairs and one nonbonding pair around the central atom. The geometry is trigonal pyramidal.

1.9   (a)   F, F ... C=C 120°, 120° , trigonal planar at each carbon atom.

(b) CH_3—C≡C—CH_3 180°   linear

(c)   H—C≡N: 180°   linear

1.10   H—C—O—C—C—H (with H atoms)

1.11   (a) CH_3 / H_3C—CH—CH_2—CH_3 =

(b) CH_3 / H_3C—CH—CH_2—CH_2—OH   =   OH

(c) H_3C, H_3C C=CH—CH_2—CH_3 =

(d) H_3C—CH_2—CH_2—CH_2—CH_3 =

(e) H_3C—CH_2—CH—CH_2—CH_3 | OH   =

(f) CH_2=C (CH_2—CH_3) (CH_2—CH_3)   =

(g)   =

(h)   =

1.12   (a) and (d) are constitutional isomers with the molecular formula C_5H_{12}.

(b) and (e) are constitutional isomers with the molecular formula $C_5H_{12}O$.

(c) and (f) are constitutional isomers with the molecular formula C_6H_{12}.

1.13   (a)

(b)

(c)

1.14   (a)   (Note that the **Cl** atom and the three **H** atoms
                                  may be written at any of the four positions.)

(b)   or   and so on

(c)   and others

(d)   and others

1.15 (a) Na^+ has the electronic configuration, $1s^22s^22p^6$, of Ne.

(b) Cl^- has the electronic configuration, $1s^22s^22p^63s^23p^6$, of Ar.

(c) F^+ and (h) Br^+ do not have the electronic configuration of a noble gas.

(d) H^- has the electronic configuration, $1s^2$, of He.

(e) Ca^{2+} has the electronic configuration, $1s^22s^22p^63s^23p^6$ of Ar.

(f) S^{2-} has the electronic configuration, $1s^22s^22p^63s^23p^6$ of Ar.

(g) O^{2-} has the electronic configuration, $1s^22s^22p^6$ of Ne.

1.16 (a)   (b)   (c)   (d)

1.17 (a)   (c)

(b)   (d)

1.18 (a) $(CH_3)_2CHCH_2OH$ (c) $HC—CH_2$ / $HC—CH_2$

(b) $(CH_3)_2CHCCH(CH_3)_2$ (with O double bond)   (d) $(CH_3)_2CHCH_2CH_2OH$

1.19 (a) $C_4H_{10}O$ (c) C_4H_6

(b) $C_7H_{14}O$ (d) $C_5H_{12}O$

1.20 (a) Different compounds, not isomeric   (i) Different compounds, not isomeric

(b) Constitutional isomers   (j) Same compound

(c) Same compound   (k) Constitutional isomers

(d) Same compound   (l) Different compounds, not isomeric

(e) Same compound   (m) Same compound

(f) Constitutional isomers   (n) Same compound

(g) Different compounds, not isomeric   (o) Same compound

(h) Same compound   (p) Constitutional isomers

1.23   CH$_2$=CHCH$_2$CH$_3$ CH$_3$CH=CHCH$_3$ CH$_2$=CCH$_3$
                                                                                      CH$_3$

1.25 (a) While the structures differ in the position of their electrons, they also differ in the positions of their nuclei and thus *they are not resonance structures.* (In cyanic acid the hydrogen nucleus is bonded to oxygen; in isocyanic acid it is bonded to nitrogen.)

(b) The anion obtained from either acid is a resonance hybrid of the following structures: :O̤—C≡N: ⟷ :O̤=C=N̈:

1.26

$$\begin{array}{c} H \\ | \\ H-C \\ | \\ H \end{array}$$

(a) A + charge. ($F = 4 - \% = + 1$)

(b) A + charge. (It is called a methyl cation.)

(c) Trigonal planar, that is,

(d) sp^2

1.27

$$\begin{array}{c} H \\ | \\ H-\overset{..}{C}: \\ | \\ H \end{array}$$

(a) A − charge. ($F = 4 - \% - 2 = - 1$)

(b) A − charge. (It is called a methyl anion.)

(c) Trigonal pyramidal, that is

(d) sp^3

1.28

$$\begin{array}{c} H \\ | \\ H-C \cdot \\ | \\ H \end{array}$$

(a) No formal charge. ($F = 4 - \% - 1 = 0$)

(b) No charge.

(c) sp^2, that is,

1.29   (a) and (b)

(c) Because the two resonance structures are equivalent, they should make equal contributions to the hybrid and, therefore, the bonds should be the same length.

(d) Yes. We consider the central atom to have two groups or units of bonding electrons and one unshared pair.

1.30

  A   B   C

Structures **A** and **C** are equivalent and, therefore, make equal contributions to the hybrid. The bonds of the hybrid, therefore, have the same length.

*1.31   (a)

(b) $(CH_3)_2NH$ $CH_3CH_2NH_2$

(c) $(CH_3)_3N$ $CH_3CH_2NHCH_3$ $CH_3CH_2CH_2NH_2$ CH_3CHCH_3
                                                                              NH_2

(d)

*1.32   (a) constitutional isomers   (b) the same
        (c) resonance forms   (d) constitutional isomers
        (e) resonance forms   (f) the same

*1.33   (a) $\overset{..}{O}=\overset{+}{N}=\overset{..}{O}$

        (b) Linear

        (c) Carbon dioxide

*1.34   Set A:

        Set B:

Set C:

a   b

[and unstable enol forms of a, b, and c]

c

Set D:

Set E: (i.e., $CH_3CH_2\overset{..}{C}H_2^-$ and $CH_3\overset{..}{C}HCH_3$)

QUIZ

1.1   Which of the following is a valid Lewis dot formula for the nitrite ion (NO_2^-)?

(a) $\overset{..}{\overset{.}{O}}$—$\overset{..}{N}$=$\overset{..}{O}$:   (b) :$\overset{..}{O}$=$\overset{..}{N}$—$\overset{..}{O}$:   (c) :$\overset{..}{O}$—$\overset{..}{N}$=$\overset{..}{O}$:   (d) Two of these

(e) None of the above

1.2   What is the hybridization state of the boron atom in BF_3?

(a) s (b) p (c) sp (d) sp^2 (e) sp^3

1.3 BF_3 reacts with NH_3 to produce a compound, F—$\overset{\overset{F}{|}}{\underset{\underset{F}{|}}{B}}$—$\overset{\overset{H}{|}}{\underset{\underset{H}{|}}{N}}$—$H$. The hybridization state of **B** is

(a) s (b) p (c) sp (d) sp^2 (e) sp^3

1.4   The formal charge on N in the compound given in Problem 1.3 is

(a) -2 (b) -1 (c) 0   (d) $+1$ (e) $+2$

1.5   The correct bond-line formula of the compound whose condensed formula is $CH_3CHClCH_2CH(CH_3)CH(CH_3)_2$ is

(a)   (b)   (c)

(d)   (e)

1.6   Write another resonance structure for the formate ion.

Formate ion

1.7   In the boxes below write condensed structural formulas for constitutional isomers of $CH_3(CH_2)_3CH_3$.

1.8   Write a three-dimensional formula for a constitutional isomer of compound **A** given below. Complete the partial structure shown.

A

Constitutional isomer of A

1.9   Consider the molecule $(CH_3)_3B$ and give the following:

(a) Hybridization state of boron

(b) Hybridization state of carbon atoms

(c) Formal charge on boron

(d) Orientation of groups around boron

(e) Dipole moment of $(CH_3)_3B$

1.10   Give the formal charge on oxygen in each compound.

(a) CH₃–Ö–CH₃
         |
         CH₃

(c)   H
         |
    CH₃–C–Ö:
         |
         H

(b) CH₃–C=Ö:
         |
         CH₃

1.11   Write another resonance structure in which all of the atoms have a formal charge of zero.

:Ö:⁻
 |
H–C=N⁺–H   ⟷
    |
    H

1.12   Indicate the direction of the net dipole moment of the following molecule.

H₃C⟋   ⟍Cl
       C
H₃C⟋   ⟍F

2 REPRESENTATIVE CARBON COMPOUNDS: FUNCTIONAL GROUPS, INTERMOLECULAR FORCES, AND INFRARED (IR) SPECTROSCOPY

SOLUTIONS TO PROBLEMS

2.1 (a) H—F or $\overset{\delta+ \quad \delta-}{H—F}$ (c) Br—Br $\mu = 0$

(b) I—Br or $\overset{\delta+ \quad \delta-}{I—Br}$ (d) F—F $\mu = 0$

2.2 VSEPR theory predicts a planar structure for BF_3.

The vector sum of the bond moments of a trigonal planar structure would be zero, resulting in a prediction of $\mu = 0$ for BF_3. This correlates with the experimental observation and confirms the prediction of VSEPR theory.

2.3 The shape of $CCl_2{=}CCl_2$ (below) is such that the vector sum of all of the C—Cl bond moments is zero.

2.4 The fact that SO_2 has a dipole moment indicates that the molecule is angular, not linear.

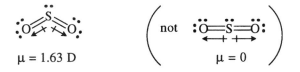

An angular shape is what we would expect from VSEPR theory, too.

The fact that CO_2 has no dipole moment indicates that its shape is linear, not angular.

$$\mu = 0 \qquad \qquad \left(\text{not} \qquad \mu \neq 0 \right)$$

2.5   Again, this is what VSEPR theory predicts.

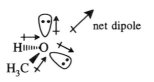

net dipole

2.6   In $CFCl_3$ the large C—F bond moment opposes the C—Cl moments, leading to a net dipole moment in the direction of the fluorine. Because hydrogen is much less electronegative than fluorine no such opposing effect occurs in $CHCl_3$; therefore, it has a net dipole moment that is larger and in the direction of the chlorine atoms.

2.9 (a) $CH_3CH_2CH_2CH_2Br$ and CH_3CHCH_2Br
$|$
CH_3

(b) $CH_3CHCH_2CH_3$
$|$
Br

(c) $CH_3{-}\overset{\displaystyle CH_3}{\underset{\displaystyle CH_3}{\overset{|}{\underset{|}{C}}}}{-}Br$

2.10 (a) CH_3CH_2F

(b) CH_3CHCH_3 or $(CH_3)_2CHCl$
$|$
Cl

(c) Propyl bromide

(d) Isopropyl fluoride

(e) Phenyl iodide

2.11 (a) $CH_3CH_2CH_2CH_2OH$ and CH_3CHCH_2OH
$|$
CH_3

(b) $CH_3CHCH_2CH_3$
$|$
OH

(c) $CH_3{-}\overset{\displaystyle CH_3}{\underset{\displaystyle CH_3}{\overset{|}{\underset{|}{C}}}}{-}OH$

2.12 (a) $CH_3CH_2CH_2OH$

(b) CH_3CHCH_3 or $CH_3CH{-}OH$
$\quad\quad |$ $\quad\quad\quad\quad |$
$\quad\quad OH$ $\quad\quad\quad\quad CH_3$

2.13 (a) $CH_3CH_2{-}O{-}CH_2CH_3$

(b) $CH_3CH_2{-}O{-}CH_2CH_2CH_3$

(c) $CH_3CH_2{-}O{-}CHCH_3$
$|$
CH_3

(d) Methyl propyl ether

(e) Diisopropyl ether

(f) Methyl phenyl ether

2.14 (a) $CH_3CH_2CH_2NH_2$

(b) $CH_3{-}N{-}CH_3$ or $(CH_3)_3N$
$|$
CH_3

(c) $CH_3{-}N{-}CH_2CH_3$
$|$
$\quad\quad CH$
$\quad\quad H_3C\quad CH_3$

(d) Isopropylpropylamine

(e) Tripropylamine

(f) Methylphenylamine

(g) Dimethylphenylamine

2.15 (a) (a) only   (b) (d,f)   (c) (b, c, e, g)

2.16 (a) $CH_3{-}\overset{\displaystyle CH_3}{\underset{\displaystyle CH_3}{\overset{|}{\underset{|}{N}}}}{:}$ $+$ $H{-}\ddot{\underset{\cdot\cdot}{C}l}{:}$ $\longrightarrow$ $CH_3{-}\overset{\displaystyle CH_3}{\underset{\displaystyle CH_3}{\overset{|}{\underset{|}{\overset{+}{N}}}}}{-}H$ $+$ $:\ddot{\underset{\cdot\cdot}{C}l}{:}^{-}$

(b) sp^3

2.17   (a) $CH_3CH_2CH_2CH_2OH$ would boil higher because its molecules can form hydrogen bonds to each other through the $-\ddot{O}-H$ group.

   (b) $CH_3CH_2NHCH_3$ would boil higher because its molecules can form hydrogen bonds to each other through the $-\ddot{N}-H$ group.

   (c) $HOCH_2CH_2CH_2OH$ because by having two $-\ddot{O}-H$ groups, it can form more hydrogen bonds.

2.18   Cyclopropane would have the higher melting point because its cyclic structure gives it a rigid compact shape that would permit stronger crystal lattice forces.

2.19   (a) Ketone   (b) Alkyne   (c) Alcohol   (d) Aldehyde
   (e) Alcohol   (f) Alkene

2.20   (a) Three carbon-carbon double bonds (alkene) and a 2° alcohol
   (b) Phenyl, carboxylic acid, amide, ester, and a 1° amine
   (c) Phenyl and a 1° amine
   (d) Carbon-carbon double bond and a 2° alcohol
   (e) Phenyl, ester, and a 3° amine
   (f) Carbon-carbon double bond and an aldehyde
   (g) Carbon-carbon double bond and 2 ester groups

2.21 $CH_3CH_2CH_2CH_2Br$ $CH_3CH_2CHCH_3$ CH_3CHCH_2Br $CH_3-\overset{\overset{\displaystyle CH_3}{|}}{\underset{\underset{\displaystyle Br}{|}}{C}}-CH_3$
                                                    $\overset{}{\underset{\underset{\displaystyle Br}{|}}{}}$ $\overset{}{\underset{\underset{\displaystyle CH_3}{|}}{}}$

        1° Alkyl   2° Alkyl   1° Alkyl   3° Alkyl
        halide   halide   halide   halide

2.22 $CH_3CH_2CH_2CH_2OH$ $CH_3CH_2CHCH_3$ CH_3CHCH_2OH $CH_3-\overset{\overset{\displaystyle CH_3}{|}}{\underset{\underset{\displaystyle OH}{|}}{C}}-CH_3$
                                                    $\overset{}{\underset{\underset{\displaystyle OH}{|}}{}}$ $\overset{}{\underset{\underset{\displaystyle CH_3}{|}}{}}$

        1° Alcohol   2° Alcohol   1° Alcohol   3° Alcohol

        $CH_3OCH_2CH_2CH_3$ CH_3OCHCH_3 $CH_3CH_2OCH_2CH_3$
             Ether $\underset{\underset{\displaystyle CH_3}{|}}{}$ Ether
                               Ether

2.23 Any four of the following:

CH_3CCH_3
Ketone

CH_3CH_2CH
Aldehyde

H_2C-CH_2 / H_2C-O
Ether

$H_3C-HC\!\!-\!\!CH_2$ (epoxide, O)
Ether

$CH_2{=}CHCH_2OH$
Alkene, alcohol

$CH_2{=}CH-O-CH_3$
Alkene, ether

$\underset{H_2C}{\overset{H_2C}{>}}CHOH$
Alcohol

2.24 (a) 1°   (b) 2°   (c) 3°   (d) 3°   (e) 2°

2.25 (a) 2°   (b) 1°   (c) 3°   (d) 2°   (e) 2°   (f) 3°

2.26 (a) $CH_3OCH_2CH_2CH_3$ CH_3OCHCH_3 (CH_3) $CH_3CH_2OCH_2CH_3$

(b) $\underset{H_2C}{\overset{H_2C}{>}}CH\text{-}CH_2OH$ $CH_2{=}CHCH_2CH_2OH$ $CH_3CH{=}CHCH_2OH$

(c) $\underset{H_2C}{\overset{H_2C}{>}}CH\text{-}OH$ (d) $\underset{H_2C}{\overset{H_2C}{>}}\overset{OH}{\underset{}{C}}\text{-}CH_3$

(e) $H\text{-}\overset{O}{\overset{\|}{C}}\text{-}O\text{-}CH_2CH_3$ $CH_3\text{-}\overset{O}{\overset{\|}{C}}\text{-}O\text{-}CH_3$

(f) $CH_3CH_2CH_2CH_2CH_2Br$ $CH_3CHCH_2CH_2Br$ (CH_3) $CH_3CH_2CHCH_2Br$ (CH_3)

$CH_3\text{-}\overset{CH_3}{\underset{CH_3}{C}}\text{-}CH_2\text{-}Br$

(g) $CH_3CHCH_2CH_2CH_3$ (Br) $CH_3CH_2CHCH_2CH_3$ (Br) $CH_3CHCHCH_3$ (CH_3, Br)

(h) $CH_3CH_2\text{-}\overset{CH_3}{\underset{CH_3}{C}}\text{-}Br$

(i) $CH_3CH_2CH_2CH_2CH$ (O) CH_3CHCH_2CH (O, CH_3) CH_3CH_2CHCH (O, CH_3)

2.27 (a) CH₃CH₂CH₂OH because its molecules can form hydrogen bonds to each other through its —Ö—H group.

(b) HOCH₂CH₂OH because with two—Ö—H groups, its molecules can form more hydrogen bonds with each other.

(c) ⟍⟋⟍⟍OH because its molecules can form hydrogen bonds to each other.

(d) ▷—OH [Same reason as (c)].

(e) ⬡NH because its molecules can form hydrogen bonds to each other through its —N̈—H group.

(f) [F₂C=CF₂ structure] because its molecules will have a larger dipole moment. (The trans compound will have μ = 0.)

(g) ⟍⟋⟍⟍(C=O)OH [Same reason as (c)].

(h) Nonane, because of its larger molecular weight and larger size, will have larger van der Waals attractions.

(i) ⟩=O because its carbonyl group is far more polar than the double bond of ⟩= .

2.28 (a) The alcohol would have a broad absorption from the O—H group in the 3200 to 3500 cm⁻¹ region of its IR spectrum. The ether would have no such absorption.

(c) The ketone would have a strong absorption from its carbonyl group near 1700 cm⁻¹ in its IR spectrum. The alcohol would have a broad absorption due to its hydroxyl group in the 3200 to 3500 cm⁻¹ region of its IR spectrum.

(d) Same rationale as for (a)

(e) The secondary amine would have an absorption near 3300 to 3500 cm^{-1} arising from N—H stretching. The tertiary amine would have no such absorption in this region since there is no N—H group present.

(g) Both compounds would exhibit absorptions near 1710 to 1780 cm^{-1} due to carbonyl stretching vibrations. The carboxylic acid would also have a broad absorption somewhere between 2500 and 3500 cm^{-1} due to its hydroxyl group. The ester would not have a hydroxyl absorption.

(i) The ketone would have a strong absorption from its carbonyl group near 1700 cm^{-1} in its IR spectrum. The alkene would have no such absorption but would have an absorption between 1620 and 1680 cm^{-1} due to C=C stretching.

2.29 (a)

(b) The last one given above [i.e., HCN(CH$_3$)$_3$] because it does not have a hydrogen that is covalently bonded to nitrogen and, therefore, its molecules cannot form hydrogen bonds to each other. The other molecules all have a hydrogen covalently bonded to nitrogen and, therefore, hydrogen-bond formation is possible. With the first molecule, for example, hydrogen bonds could form in the following way:

2.30 An ester group, $-\overset{|}{\underset{|}{C}}-\overset{O}{\overset{||}{C}}-O-\overset{|}{\underset{|}{C}}-$ (CH$_2$)$_n$

2.31 The attractive forces between hydrogen fluoride molecules are the very strong dipole–dipole attractions that we call *hydrogen bonds*. (The partial positive charge of a hydrogen fluoride molecule is relatively exposed because it resides on the hydrogen nucleus. By contrast, the positive charge of an ethyl fluoride molecule is buried in the ethyl group and is shielded by the surrounding electrons. Thus the positive end of one hydrogen fluoride molecule can approach the negative end of another hydrogen fluoride molecule much more closely, with the result that the attractive force between them is much stronger.)

2.32 (a) and (b) are polar and hence are able to dissolve ionic compounds. (c) and (d) are nonpolar and will not dissolve ionic compounds.

2.33 (a) (b) (c) (d) F—C⁗F No dipole moment (e) (f) Cl—B—Cl No dipole moment (g) F—Be—F No dipole moment (h) (i) (j)

2.34 (a) Dimethyl ether: There are four electron pairs around the central oxygen: two bonding pairs and two nonbonding pairs. We would expect sp^3 hybridization of the oxygen with a bond angle of approximately 109.5° between the methyl groups.

(b) Trimethylamine: There are four electron pairs around the central nitrogen: three bonding pairs and one nonbonding pair. We would expect sp^3 hybridization of the nitrogen with a bond angle of approximately 109.5° between the methyl groups.

(c) Trimethylboron: There are only three bonding electron pairs around the central boron. We would expect sp^2 hybridization of the boron with a bond angle of 120° between the methyl groups.

(d) Dimethylberyllium: There are only two bonding electron pairs around the central beryllium atom. We would expect sp hybridization of the beryllium atom with a bond angle of 180° between the methyl groups.

$$H_3C—Be—CH_3$$

2.35 Without one (or more) polar bonds, a molecule cannot possess a dipole moment and, therefore, it cannot be polar. If the bonds are directed so that the bond moments cancel, however, the molecule will not be polar even though it has polar bonds.

2.36 Crixivan has the following functional groups:

2.37

Hydrogen bond

*2.38

The 1780-cm^{-1} band is in the general range for C=O stretching so structure B′ is considered one of the possible answers, but only B would have its C=O stretch at this high frequency (B′ would be at about 1730 cm^{-1}).

*2.39 (a)

(b) The *cis* isomer will have the 3572-cm^{-1} band because only in it are the two hydroxyl groups close enough to permit intramolecular hydrogen-bonding. (Intermolecular hydrogen-bonding is not possible at high dilution in a non-polar solvent like CCl$_4$.)

*2.40   (a)

(b)

(Enantiomers and diastereomers are defined in Chapter 5.)

QUIZ

2.1   Which of the following pairs of compounds is *not* a pair of constitutional isomers?

(a) $CH_3-O-CH=CH_2$ and $CH_3CH_2\overset{\overset{O}{\|}}{C}H$

(b) ⬠   and $CH_3CH=CHCH_2CH_3$

(c) $CH_3\overset{\overset{O}{\|}}{C}-OH$ and $HO-CH_2\overset{\overset{O}{\|}}{C}H$

(d) $CH_3CH_2C\equiv CH$ and $CH_3CH=C=CH_2$

(e) $CH_3\underset{\underset{CH_3}{|}}{C}HCH(CH_3)_2$ and $(CH_3)_2CHCH(CH_3)_2$

2.2   Which of the answers to Problem 2.1 contains an ether group?

2.3   Which of the following pairs of structures represents a pair of isomers?

(a)

(b) $CH_3C\equiv CCH_3$ and $CH_3CH_2C\equiv CH$

(c)   and

(d) $CH_3CH_2\overset{\underset{\displaystyle CH_3}{|}}{C}HCH_2CH_3$ and $CH_3CH_2\overset{\underset{\displaystyle CH_2CH_3}{|}}{C}HCH_3$

(e)   More than one of these pairs are isomers.

2.4   Give a structural formula for each of the following:

(a) A tertiary alcohol with the formula $C_5H_{12}O$

(b) An *N,N*–disubstituted amide with the formula C_4H_9NO

(c) The alkene isomer of $C_2H_2Cl_2$ that has no dipole moment

(d) An ester with the formula $C_2H_4O_2$

(e) The isomer of $C_2H_2Cl_2$ that cannot show cis-trans isomerism

(f) The isomer of C_3H_8O that would have the lowest boiling point

(g) The isomer of $C_4H_{11}N$ that would have the lowest boiling point

2.5   Write the bond-line formula for a constitutional isomer of the compound shown below that does not contain a double bond.

$CH_3CH_2CH{=}CH_2$

2.6   Circle the compound in each pair that would have the higher boiling point.

(a) $CH_3CH_2CH_2OH$ or CH_3CH_2CHO

(b)

(c) $CH_3\overset{\overset{\displaystyle O}{\|}}{C}OCH_3$ or $CH_3CH_2\overset{\overset{\displaystyle O}{\|}}{C}OH$

(d) $CH_3OCH_2CH_2OH$ or $CH_3OCH_2OCH_3$

(e) $CH_3CH_2\overset{\overset{\displaystyle O}{\|}}{C}NHCH_3$ or $CH_3\overset{\overset{\displaystyle O}{\|}}{C}N(CH_3)_2$

2.7   Give an acceptable name for each of the following:

(a)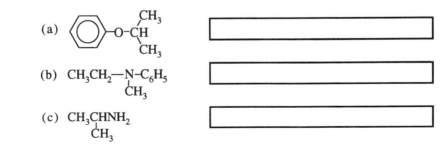

(b) $CH_3CH_2{-}\underset{\underset{\displaystyle CH_3}{|}}{N}{-}C_6H_5$

(c) $CH_3\underset{\underset{\displaystyle CH_3}{|}}{C}HNH_2$

3

AN INTRODUCTION TO
ORGANIC REACTIONS: ACIDS AND BASES

SOLUTIONS TO PROBLEMS

3.1 (a)

3.2 (a) Lewis base   (b) Lewis acid

(c) Lewis base   (d) Lewis base

(e) Lewis acid   (f) Lewis base

3.3

Lewis base   Lewis acid

3.4 (a) $K_a = \dfrac{[H_3O^+][HCO_2^-]}{[HCO_2H]} = 1.77 \times 10^{-4}$

Let $x = [H_3O^+] = [HCO_2^-]$ at equilibrium
then, $0.1 - x = [HCO_2H]$ at equilibrium
but, since the K_a is very small, x will be very small and $0.1 - x \simeq 0.1$
Therefore,

$$\frac{(x)(x)}{0.1} = 1.77 \times 10^{-4}$$
$$x^2 = 1.77 \times 10^{-5}$$
$$x = 0.0042 = [H_3O^+] = [HCO_2^-]$$

(b)   % Ionized $= \dfrac{[H_3O^+]}{0.1} \times 100$ or $\dfrac{[HCO_2{}^-]}{0.1} \times 100$

$= \dfrac{.0042}{0.1} \times 100 = 4.2\%$

3.5   (a) $pK_a = -\log 10^{-7} = -(-7) = 7$

(b) $pK_a = -\log 5.0 = -0.699$

(c)   Since the acid with a $K_a = 5$ has a larger K_a, it is the stronger acid.

3.6   When H_3O^+ acts as an acid in aqueous solution, the equation is

$$H_3O^+ + H_2O \rightleftharpoons H_2O + H_3O^+$$

and K_a is

$$K_a = \dfrac{[H_2O][H_3O^+]}{[H_3O^+]} = [H_2O]$$

The molar concentration of H_2O in pure H_2O, that is, $[H_2O] = 55.5$; therefore, $K_a = 55.5$

The pK_a is

$$pK_a = -\log 55.5 = -1.74$$

3.7   The pK_a of the methylaminium ion is equal to 10.6 (Section 3.5C). Since the pK_a of the anilinium ion is equal to 4.6, the anilinium ion is a stronger acid than the methylaminium ion, and aniline ($C_6H_5NH_2$) is a weaker base than methylamine (CH_3NH_2).

3.8   (a)   Negative. Because the atoms are constrained to one molecule in the product, they have to become more ordered.

(b)   Approximately zero.

(c)   Positive. Because the atoms are in two separate product molecules, they become more disordered.

3.9   (a)   If $K_{eq} = 1$
then,
$\log K_{eq} = 0 = \dfrac{-\Delta G^\circ}{2.303RT}$
$\Delta G^\circ = 0$

(b)   If $K_{eq} = 10$
then,
$\log K_{eq} = 1 = \dfrac{-\Delta G^\circ}{2.303RT}$
$\Delta G^\circ = -(2.303)(0.008314 \text{ kJ mol}^{-1}\text{ K}^{-1})(298 \text{ K}) = -5.71 \text{ kJ mol}^{-1}$

(c) $\Delta G^\circ = \Delta H^\circ - T\Delta S^\circ$
$\Delta G^\circ = \Delta H^\circ = -5.71 \text{ kJ mol}^{-1}$ if $\Delta S^\circ = 0$

3.10   Structures **A** and **B** make equal contributions to the overall hybrid. This means that the carbon-oxygen bonds should be the same length and that the oxygens should bear equal positive charges.

A   **B**

3.11   (a) $CHCl_2CO_2H$ would be the stronger acid because the electron-withdrawing inductive effect of *two chlorine atoms* would make its hydroxyl proton more positive. The electron-withdrawing effect of the two chlorine atoms would also stabilize the dichloroacetate ion more effectively by dispersing its negative charge more extensively.

(b) CCl_3CO_2H would be the stronger acid for reasons similar to those given in (a), except here there are three versus two electron-withdrawing chlorine atoms involved.

(c) CH_2FCO_2H would be the stronger acid because the electron-withdrawing effect of a fluorine atom is greater than that of a bromine atom (fluorine is more electronegative).

(d) CH_2FCO_2H is the stronger acid because the fluorine atom is nearer the carboxyl group and is, therefore, better able to exert its electron-withdrawing inductive effect. (*Remember:* Inductive effects weaken steadily as the distance between the substituent and the group increases.)

3.12   All compounds containing oxygen and most compounds containing nitrogen will have an unshared electron pair on their oxygen or nitrogen atom. These compounds can, therefore, act as bases and accept a proton from concentrated sulfuric acid. When they accept a proton, these compounds become either oxonium ions or ammonium ions, and having become ionic, they are soluble in the polar medium of sulfuric acid. The only nitrogen compounds that do not have an electron pair on their nitrogen atom are quaternary ammonium compounds, and these, already being ionic, also dissolve in the polar medium of concentrated sulfuric acid.

3.13   (a) $CH_3\ddot{O}-H$ + $:H^-$ $\xrightarrow{\text{methanol}}$ $CH_3\ddot{O}:^-$ + H_2

Stronger acid   Stronger   Weaker   Weaker
$pK_a = \sim 16$ base   base   acid
                    (from NaH) $pK_a = 35$

(b) $CH_3CH_2\ddot{O}-H$ + $:\ddot{N}H_2^-$ $\xrightarrow{\text{ethanol}}$ $CH_3CH_2\ddot{O}:^-$ + $:NH_3$

Stronger acid   Stronger   Weaker   Weaker
$pK_a = 16$ base   base   acid
                  (from NaNH$_2$) $pK_a = 38$

(c) $H-\overset{..}{N}-H$ + $^{-}:CH_2CH_3$ $\xrightarrow{\text{hexane}}$ $:\overset{..}{N}H_2^{-}$ + CH_3CH_3

Stronger acid   Stronger base   Weaker base   Weaker acid

$pK_a = 38$ (from CH_3CH_2Li) $pK_a = 50$

(d) $H-\overset{+}{\underset{}{N}}-H$ + $:\overset{..}{N}H_2^{-}$ $\xrightarrow{\text{liq. NH}_3}$ $:NH_3$ + $:NH_3$

Stronger acid   Stronger base   Weaker base   Weaker acid

$pK_a = 9.2$ (from $NaNH_2$) $pK_a = 38$

(from NH_4Cl)

(e) $H-\overset{..}{\underset{..}{O}}-H$ + $^{-}:\overset{..}{\underset{..}{O}}C(CH_3)_3$ $\xrightarrow{\text{H}_2\text{O}}$ $H-\overset{..}{\underset{..}{O}}:^{-}$ + $HOC(CH_3)_3$

Stronger acid   Stronger base   Weaker base   Weaker acid

$pK_a = 15.7$ $pK_a = 18$

[from $(CH_3)_3CONa$]

(f)   No appreciable acid-base reaction would occur because NaOH is not a strong enough base to remove a proton from $(CH_3)_3COH$.

3.14 (a) $HC\equiv CH$ +   NaH $\xrightarrow{\text{hexane}}$ $HC\equiv CNa$ + H_2

(b) $HC\equiv CNa$ + D_2O $\xrightarrow{\text{hexane}}$ $HC\equiv CD$ +   NaOD

(c) CH_3CH_2Li + D_2O $\xrightarrow{\text{hexane}}$ CH_3CH_2D +   LiOD

(d) CH_3CH_2OH +   NaH $\xrightarrow{\text{hexane}}$ CH_3CH_2ONa + H_2

(e) CH_3CH_2ONa + T_2O $\xrightarrow{\text{hexane}}$ CH_3CH_2OT +   NaOT

(f) $CH_3CH_2CH_2Li$ + D_2O $\xrightarrow{\text{hexane}}$ $CH_3CH_2CH_2D$ +   LiOD

3.15 (a) $:\overset{..}{N}H_2^{-}$ (the amide ion)   (d) $H-C\equiv C:^{-}$ (the ethynide ion)

(b) $H-\overset{..}{\underset{..}{O}}:^{-}$ (the hydroxide ion)   (e) $CH_3\overset{..}{\underset{..}{O}}:^{-}$ (the methoxide ion)

(c) $:H^{-}$ (the hydride ion)   (f) H_2O (water)

3.16 $:\overset{..}{N}H_2^{-} > :H^{-} > H-C\equiv C:^{-} > CH_3\overset{..}{\underset{..}{O}}:^{-} \approx H-\overset{..}{\underset{..}{O}}:^{-} > H_2O$

3.17 (a) H_2SO_4 (d) NH_3

(b) H_3O^{+} (e) CH_3CH_3

(c) $CH_3NH_3^{+}$ (f) CH_3CO_2H

3.18 $H_2SO_4 > H_3O^+ > CH_3CO_2H > CH_3NH_3^+ > NH_3 > CH_3CH_3$

3.19 (a) $CH_3CH_2-\overset{..}{\underset{..}{Cl}}:$ + $AlCl_3$ $\longrightarrow$ $CH_3CH_2-\overset{+}{\underset{..}{Cl}}-\overset{:\overset{..}{Cl}:}{\underset{:\underset{..}{Cl}:}{\overset{|}{\underset{|}{Al}}}}-\overset{..}{\underset{..}{Cl}}:$

Lewis base   Lewis acid

(b) $CH_3-\overset{..}{O}H$ + BF_3 $\longrightarrow$ $CH_3-\overset{+}{\underset{H}{O}}-\overset{:\overset{..}{F}:}{\underset{:\overset{..}{F}:}{B}}-\overset{..}{\underset{..}{F}}:$

Lewis base   Lewis acid

(c) $CH_3-\overset{CH_3}{\underset{CH_3}{\overset{|}{\underset{|}{C}}}}{}^+$ + $H_2\overset{..}{O}:$ $\longrightarrow$ $CH_3-\overset{CH_3}{\underset{CH_3}{\overset{|}{\underset{|}{C}}}}-\overset{..}{O}H_2{}^+$

Lewis acid   Lewis base

3.20 (a) $CH_3-\overset{..}{O}H$ + $H-\overset{..}{\underset{..}{I}}:$ $\longrightarrow$ $CH_3-\overset{+}{\underset{H}{O}}-H$ + $:\overset{..}{\underset{..}{I}}:^-$

(b) $CH_3-\overset{..}{N}H_2$ + $H-\overset{..}{\underset{..}{Cl}}:$ $\longrightarrow$ $CH_3-\overset{H}{\underset{H}{\overset{|}{\underset{|}{N}}}}{}^+-H$ + $:\overset{..}{\underset{..}{Cl}}:^-$

(c) $\overset{H}{\underset{H}{}}C=C\overset{H}{\underset{H}{}}$ + $H-\overset{..}{\underset{..}{F}}:$ $\longrightarrow$ $H-\overset{H}{\underset{H}{\overset{|}{\underset{|}{C}}}}-\overset{H}{\underset{H}{\overset{|}{\underset{|}{C}}}}{}^+-H$ + $:\overset{..}{\underset{..}{F}}:^-$

3.21 Because the proton attached to the highly electronegative oxygen atom of CH_3OH is much more acidic than the protons attached to the much less electronegative carbon atom.

3.22 $CH_3CH_2-\overset{..}{\underset{..}{O}}H$ + $^-:C\equiv C-H$ $\xrightarrow{\text{liq. } NH_3}$ $CH_3CH_2-\overset{..}{\underset{..}{O}}:^-$ + $H-C\equiv C-H$

3.23 (a) $pK_a = -\log 1.77 \times 10^{-4} = 4 -0.248 = 3.752$

(b) $K_a = 10^{-13}$

3.24 (a) HB is the stronger acid because it has the smaller pK_a.

(b) Yes. Since A^- is the stronger base and HB is the stronger acid, the following acid-base reaction will take place.

A:$^-$ +	H—B	$\longrightarrow$	A—H	+	B:$^-$
Stronger base	Stronger acid		Weaker acid		Weaker base
	$pK_a = 10$		$pK_a = 20$		

3.25 (a) $CH_3CH_2-\overset{\overset{\displaystyle O}{\|}}{C}-\ddot{\overset{..}{O}}-H + :\overset{..}{\underset{..}{O}}-H \rightleftharpoons CH_3CH_2-\overset{\overset{\displaystyle O}{\|}}{C}-\overset{..}{\underset{..}{O}}:^- + H-\overset{..}{\underset{..}{O}}-H$

(b) $C_6H_5-\overset{\overset{\displaystyle :\overset{..}{O}}{\|}}{\underset{\underset{\displaystyle :O}{\|}}{S}}-\overset{..}{O}-H + :\overset{..}{\underset{..}{O}}-H \longrightarrow C_6H_5-\overset{\overset{\displaystyle :\overset{..}{O}}{\|}}{\underset{\underset{\displaystyle :O}{\|}}{S}}-\overset{..}{\underset{..}{O}}:^- + H-\overset{..}{\underset{..}{O}}-H$

(c) No appreciable acid-base reaction takes place because CH_3CH_2ONa is too weak a base to remove a proton from ethyne.

(d) $H-C\equiv C-H + {}^-:CH_2CH_3 \xrightarrow{\text{hexane}} H-C\equiv C:^- + CH_3CH_3$
    (from LiCH$_2$CH$_3$)

(e) $CH_3-CH_2-\overset{..}{\underset{..}{O}}-H + {}^-:CH_2CH_3 \xrightarrow{\text{hexane}} CH_3-CH_2-\overset{..}{\underset{..}{O}}:^- + CH_3CH_3$
    (from LiCH$_2$CH$_3$)

3.26 (a) $C_6H_5-C\equiv C-H + NaNH_2 \xrightarrow{\text{ether}} C_6H_5-C\equiv C:^- Na^+ + NH_3$
    then

$C_6H_5-C\equiv C:^- Na^+ + T_2O \longrightarrow C_6H_5-C\equiv C-T + NaOT$

(b) $CH_3-\underset{\underset{\displaystyle CH_3}{|}}{CH}-O-H + NaH \longrightarrow CH_3-\underset{\underset{\displaystyle CH_3}{|}}{CH}-ONa + H_2$

then

$CH_3-\underset{\underset{\displaystyle CH_3}{|}}{CH}-ONa + D_2O \longrightarrow CH_3-\underset{\underset{\displaystyle CH_3}{|}}{CH}-O-D + NaOD$

(c) $CH_3CH_2CH_2OH + NaH \longrightarrow CH_3CH_2CH_2ONa + H_2$

then

$CH_3CH_2CH_2ONa + D_2O \longrightarrow CH_3CH_2CH_2OD + NaOD$

3.27 (a) $CH_3CH_2OH > CH_3CH_2NH_2 > CH_3CH_2CH_3$

Oxygen is more electronegative than nitrogen, which is more electronegative than carbon. The O-H bond is most polarized, the N-H bond is next, and the C-H bond is least polarized.

(b) $CH_3CH_2O^- < CH_3CH_2NH^- < CH_3CH_2CH_2{}^-$

The weaker the acid, the stronger the conjugate base.

3.28 (a) $CH_3C{\equiv}CH$ > $CH_3CH{=}CH_2$ > $CH_3CH_2CH_3$

(b) $CH_3CHClCO_2H$ > $CH_3CH_2CO_2H$ > $CH_3CH_2CH_2OH$

(c) $CH_3CH_2OH_2^+$ > CH_3CH_2OH > CH_3OCH_3

3.29 (a) $CH_3NH_3^+$ < CH_3NH_2 < CH_3NH^-

(b) CH_3O^- < CH_3NH^- < $CH_3CH_2^-$

(c) $CH_3C{\equiv}C^-$ < $CH_3CH{=}CH^-$ < $CH_3CH_2CH_2^-$

3.30 The acidic hydrogens must be attached to oxygen atoms. In H_3PO_3, one hydrogen is bonded to a phosphorus atom:

3.31 (a)

3.32   (a)   Assume that the acidic and basic groups of glycine in its two forms have acidities and basicities similar to those of acetic acid and methylamine. Then consider the equilibrium between the two forms:

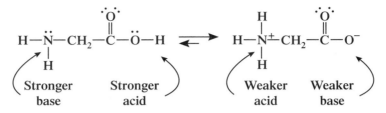

We see that the ionic form contains the groups that are the weaker acid and weaker base. The equilibrium, therefore, will favor this form.

(b)   The high melting point shows that the ionic structure better represents glycine.

3.33   (a)   The second carboxyl group of malonic acid acts as an electron-withdrawing group and stabilizes the conjugate base formed (i.e., $HO_2CCH_2CO_2^-$) when malonic acid loses a proton. [Any factor that stabilizes the conjugate base of an acid always increases the strength of the acid (Section 3.10C).] An important factor here may be an entropy effect as explained in Section 3.9.

(b)   When $^-O_2CCH_2CO_2H$ loses a proton, it forms a dianion, $^-O_2CCH_2CO_2^-$. This dianion is destabilized by having two negative charges in close proximity.

3.34   HB is the stronger acid.

3.35 $\Delta G° = \Delta H° - T\Delta S°$

$\quad\quad = 6.3 \text{ kJ mol}^{-1} - (298 \text{ K})(0.0084 \text{ kJ mol}^{-1}\text{K}^{-1})$

$\quad\quad = 3.8 \text{ kJ mol}^{-1}$

$\log K_{eq} = \log K_a = -p K_a = -\dfrac{\Delta G°}{2.303RT}$

$pK_a = \dfrac{\Delta G°}{2.303RT}$

$pK_a = \dfrac{3.8 \text{ kJ mol}^{-1}}{(2.303)(0.008314 \text{ kJ mol}^{-1}\text{K}^{-1})(298 \text{ K})}$

$pK_a = 0.66$

3.36   The dianion is a hybrid of the following resonance structures:

If we mentally fashion a hybrid of these structures, we see that each carbon-carbon bond is a single bond in three structures and a double bond in one. Each carbon-oxygen bond is a double bond in two structures and a single bond in two structures. Therefore, we would expect all of the carbon-carbon bonds to be equivalent and of the same length, and exactly the same can be said for the carbon-oxygen bonds.

*3.37 (a) A is $CH_3CH_2S^-$ B is CH_3OH

C is $CH_3CH_2SCH_2CH_2O^-$ D is $CH_3CH_2SCH_2CH_2OH$

E is OH^-

(b) $CH_3CH_2-\overset{..}{\underset{..}{S}}-H + CH_3-\overset{..}{\underset{..}{O}}:^- \longrightarrow CH_3CH_2-\overset{..}{\underset{..}{S}}:^- + CH_3-\overset{..}{\underset{..}{O}}-H$

$CH_3CH_2-\overset{..}{\underset{..}{S}}:^- + \overset{}{CH_2}-CH_2 \longrightarrow CH_3CH_2-\overset{..}{\underset{..}{S}}-CH_2CH_2-\overset{..}{\underset{..}{O}}:^-$

$CH_3CH_2-\overset{..}{\underset{..}{S}}-CH_2CH_2-\overset{..}{\underset{..}{O}}:^- + H-\overset{..}{\underset{..}{O}}-H \longrightarrow$

$CH_3CH_2-\overset{..}{\underset{..}{S}}-CH_2CH_2-\overset{..}{\underset{..}{O}}-H + H-\overset{..}{\underset{..}{O}}:^-$

*3.38 (a) $CH_3(CH_2)_8OD + CH_3(CH_2)_8 Li \longrightarrow CH_3(CH_2)_8O^- Li^+ + CH_3(CH_2)_8D$

Hexane could be used as solvent. Liquid ammonia and ethanol could not because they would compete with $CH_3(CH_2)_8OD$ and generate mostly non-deuterio-labelled $CH_3(CH_2)_7CH_3$.

(b) $NH_2^- + CH_3C\equiv CH \longrightarrow NH_3 + CH_3C\equiv C:^-$

Hexane or liquid ammonia could be used; ethanol is too acidic and would lead to $CH_3CH_2O^-$ (ethoxide ion) instead of the desired alkynide ion.

(c) $HCl + $

Hexane or ethanol could be used; liquid ammonia is too strong a base and would lead to NH_4^+ instead of the desired anilinium ion.

*3.39 (a,b)

$$\underset{H}{\overset{\overset{\displaystyle:\overset{..}{\overset{..}{O}}}{\|}}{C}}\!\!-\!\!\underset{CH_3}{\overset{..}{N}} \longleftrightarrow \underset{H}{\overset{\overset{\displaystyle:\overset{..}{\underset{..}{O}}:^-}{|}}{C}}\!\!=\!\!\underset{CH_3}{\overset{+}{N}}-CH_3$$

(c)   Since DMF does not bind with (solvate) anions, their electron density remains high and their size small, both of which make nucleophiles more reactive.

*3.40 (a)

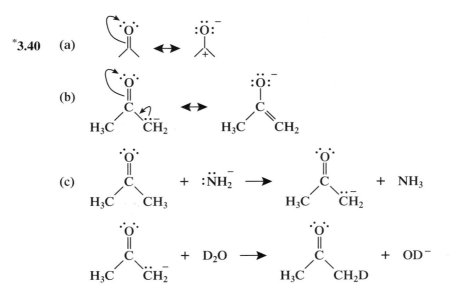

QUIZ

3.1 Which of the following is the strongest acid?

(a) $CH_3CH_2CO_2H$ (b) CH_3CH_3 (c) CH_3CH_2OH (d) $CH_2{=}CH_2$

3.2 Which of the following is the strongest base?

(a) CH_3ONa (b) $NaNH_2$ (c) CH_3CH_2Li (d) $NaOH$ (e) CH_3CO_2Na

3.3 Dissolving $NaNH_2$ in water will give:

(a) A solution containing solvated Na^+ and $NH_2{}^-$ ions.

(b) A solution containing solvated Na^+ ions, OH^- ions, and NH_3.

(c) NH_3 and metallic Na.

(d) Solvated Na^+ ions and hydrogen gas.

(e) None of the above.

3.4 Which base is strong enough to convert $(CH_3)_3COH$ into $(CH_3)_3CONa$ in a reaction that goes to completion?

(a) $NaNH_2$ (b) CH_3CH_2Na (c) $NaOH$ (d) CH_3CO_2Na

(e) More than one of the above.

3.5 Which would be the strongest acid?

(a) $CH_3CH_2CH_2CO_2H$ (b) $CH_3CH_2CHFCO_2H$ (c) $CH_3CHFCH_2CO_2H$

(d) $CH_2FCH_2CH_2CO_2H$ (e) $CH_3CH_2CH_2CH_2OH$

3.6   Which would be the weakest base?

(a) CH_3CO_2Na (b) CF_3CO_2Na (c) CHF_2CO_2Na (d) CH_2FCO_2Na

3.7   What acid-base reaction (if any) would occur when NaF is dissolved in H_2SO_4?

3.8   The pK_a of $CH_3NH_3^+$ equals 10.6; the pK_a of $(CH_3)_2NH_2^+$ equals 10.7. Which is the stronger base, CH_3NH_2 or $(CH_3)_2NH$?

3.9   Supply the missing reagents.

3.10   Supply the missing intermediates and reagents.

4 ALKANES: NOMENCLATURE, CONFORMATIONAL ANALYSIS, AND AN INTRODUCTION TO SYNTHESIS

SOLUTIONS TO PROBLEMS

4.1

Heptane

2-Methylhexane

3-Methylhexane

2,2-Dimethylpentane

3,3-Dimethylpentane

2,3-Dimethylpentane

2,4-Dimethylpentane

3-Ethylpentane

2,2,3-Trimethylbutane

4.2 (a) $CH_3CH_2CH_2CH_2CH_2$—

Pentyl

$CH_3CH_2CH_2\overset{|}{C}HCH_3$

1-Methylbutyl

$CH_3CH_2\overset{|}{C}HCH_2CH_3$

1-Ethylpropyl

$CH_3CH_2CH(CH_3)CH_2$—

2-Methylbutyl

$CH_3CH(CH_3)CH_2CH_2$—

3-Methylbutyl

$CH_3\overset{|}{C}HCH(CH_3)CH_3$

1,2-Dimethylpropyl

$CH_3\overset{|}{C}(CH_3)CH_2CH_3$—

1,1-Dimethylpropyl

(b) See the answer to Problem 4.1 for the names of C_7H_{16} isomers.

4.3   (a) $CH_3CH_2CH_2CH_2Cl$
1-Chlorobutane

CH_3CHCH_2Cl
$\quad\ \ |$
$\quad\ \ CH_3$
1-Chloro-2-methylpropane

$CH_3CH_2CHCH_3$
$\qquad\quad |$
$\qquad\quad Cl$
2-Chlorobutane

$\qquad\ \ CH_3$
$\qquad\ \ |$
$CH_3{-}C{-}CH_3$
$\qquad\ \ |$
$\qquad\ \ Cl$
2-Chloro-2-methylpropane

(b) $CH_3CH_2CH_2CH_2CH_2Br$

1-Bromopentane

$CH_3CHCH_2CH_2Br$
$\quad\ \ |$
$\quad\ \ CH_3$
1-Bromo-3-methylbutane

$CH_3CH_2CH_2CHCH_3$
$\qquad\qquad\ |$
$\qquad\qquad\ Br$
2-Bromopentane

$CH_3CH_2CHCH_2Br$
$\qquad\quad |$
$\qquad\quad CH_3$
1-Bromo-2-methylbutane

$CH_3CH_2CHCH_2CH_3$
$\qquad\quad |$
$\qquad\quad Br$
3-Bromopentane

$\qquad\ \ CH_3$
$\qquad\ \ |$
$CH_3CHCHCH_3$
$\qquad\quad |$
$\qquad\quad Br$
2-Bromo-3-methylbutane

$\quad\ \ CH_3$
$\quad\ \ |$
CH_3CCH_2Br
$\quad\ \ |$
$\quad\ \ CH_3$
1-Bromo-2,2-dimethylpropane

$\qquad\ \ CH_3$
$\qquad\ \ |$
$CH_3CH_2CCH_3$
$\qquad\qquad |$
$\qquad\qquad Br$
2-Bromo-2-methylbutane

4.4   (a) $CH_3CH_2CH_2CH_2OH$
1-Butanol

CH_3CHCH_2OH
$\quad\ \ |$
$\quad\ \ CH_3$
2-Methyl-1-propanol

$CH_3CH_2CHCH_3$
$\qquad\qquad |$
$\qquad\qquad OH$
2-Butanol

$\qquad\ \ CH_3$
$\qquad\ \ |$
CH_3COH
$\qquad\ \ |$
$\qquad\ \ CH_3$
2-Methyl-2-propanol

(b) $CH_3CH_2CH_2CH_2CH_2OH$

1-Pentanol

$CH_3CHCH_2CH_2OH$
　　|
　　CH_3

3-Methyl-1-butanol

$CH_3CH_2CH_2CHCH_3$
　　　　　|
　　　　　OH

2-Pentanol

$CH_3CH_2CHCH_2OH$
　　　　|
　　　　CH_3

2-Methyl-1-butanol

$CH_3CH_2CHCH_2CH_3$
　　　　|
　　　　OH

3-Pentanol

　　　CH_3
　　　|
$CH_3CHCHCH_3$
　　　　|
　　　　OH

3-Methyl-2-butanol

　　　CH_3
　　　|
CH_3CCH_2OH
　　　|
　　　CH_3

2,2-Dimethyl-1-propanol

　　　　CH_3
　　　　|
$CH_3CH_2CCH_3$
　　　　|
　　　　OH

2-Methyl-2-butanol

4.5 (a) 1-(1-Methylethyl)-2-(1,1-dimethylethyl)cyclopentane or
1-*tert*-butyl-2-isopropylcyclopentane

(b) 1-Methyl-2-(2-methylpropyl)cyclohexane or
1-isobutyl-2-methylcyclohexane

(c) Butylcyclohexane

(d) 1-Chloro-2,4-dimethylcyclohexane

(e) 2-Chlorocyclopentanol

(f) 3-(1,1-Dimethylethyl)cyclohexanol or 3-*tert*-butylcyclohexanol

4.6 (a) 2-Chlorobicyclo[1.1.0]butane

(b) Bicyclo[3.2.1]octane

(c) Bicyclo[2.1.1]hexane

(d) 9-Chlorobicyclo[3.3.1]nonane

(e) 2-Methylbicyclo[2.2.2]octane

(f) Bicyclo[3.1.0]hexane or bicyclo[2.1.1]hexane

4.7 (a) *trans*-3-Heptene

(b) 2,5-Dimethyl-2-octene

(c) 4-Ethyl-2-methyl-1-hexene

(d) 3,5-Dimethylcyclohexene

(e) 4-Methyl-4-penten-2-ol

(f) 2-Chloro-3-methyl-3-cyclohexen-1-ol

4.8

4.9

1-Hexyne

2-Hexyne

3-Hexyne

3-Methyl-1-pentyne

4-Methyl-1-pentyne

4-Methyl-2-pentyne

4.10

4.11

$$\log K_{eq} = \frac{\Delta G^\circ}{-2.303RT} = \frac{-7600 \text{ J}}{(-2.303)(8.314 \text{ J K}^{-1})(298 \text{ K})} = 1.32$$

$$K_{eq} = 21.38$$

Let e = amount of equatorial form

and a = amount of axial form

then, $K_{eq} = \dfrac{e}{a} = 21.38$

$$e = 21.38a$$

$$\%e = \frac{21.38a}{a + 21.38a} \times 100 = 95.5\%$$

4.12 **(a)**

(cis)   (trans)

(b)

(cis)   (trans)

4.13 **(a-d)**

More stable because larger group is equatorial

Less stable because larger group is axial

4.14 **(a)**

(b) Yes

(c)

Less stable because the large *tert*-butyl group is axial

More stable because the large *tert*-butyl group is equatorial

(d)

More stable because both methyl groups are equatorial

Less stable because both methyl groups are axial

4.18   (a) Nonane

3-ethylheptadecane

(after hydrogenation of the alkyne from one of the possible retrosynthetic disconnections)

(or homologous pairs)

(Note that [structure] X + [structure] is not a good choice because the alkyl halide is branched at the carbon adjacent to the halogen.

Neither would [structure] + [structure] work because the alkyl halide is secondary.)

(b) For any pair of reactants above that is a feasible retrosynthetic disconnection, the steps for the synthesis would be

$$RC{\equiv}C{-}H \xrightarrow[(-NH_3)]{NaNH_2} R{-}C{\equiv}C{:}^{-} \xrightarrow[\substack{(R' \text{ is} \\ \text{primary} \\ \text{and} \\ \text{unbranched} \\ \text{at the} \\ \text{second} \\ \text{carbon})}]{R'{-}X} R{-}C{\equiv}C{-}R'$$

(a terminal alkyne; R = alkyl, H)

(an alkynide anion)

$\downarrow$ H₂ Pd, Pt, or Ni

$$R{-}CH_2CH_2{-}R'$$

4.19   (a) ClCH₂CH₂CH₂CHCH₃
                          |
                          Cl

(b) CH₃CH₂CHCH₃
                   |
                   Br

(c) CH₃CH₂CH₂CHCH₂CH₂CH₃
                    |
                    CHCH₃
                    |
                    CH₃

(d) CH₃C—CHCH₂CH₃
         | |
         CH₃   CH₃
       (CH₃ above C)

(e) CH₃CHCHCH₂CH₂CH₃
         |
         CH₂
         |
         CH₃
       (CH₃ above first CH)

(f)

(g)

(h)

(i) CH₃CHCH₂CHCH₃
         | |
         OH   CH₃

(j) HO⟍⟍⟍⟍ᴵᴵᴵᴵCH₂CHCH₃
                              |
                              CH₃

or

HO⟍⟍⟍⟍CH₂CH(CH₃)₂

(k) ▷—CH₂CH₂CH₂CHCH₂CH₃
                              |
                              △

(l) CH₃—C—CH₂OH
              |
              CH₃
       (CH₃ above C)

(m)

(n)

(o)

4.20   (a) 3,3,4-Trimethylhexane

(b) 2,2-Dimethyl-1-butanol

(c) 3,5,7-Trimethylnonane

(d) 3-Methyl-4-heptanol

(e) 2-Bromobicyclo[3.3.1]nonane

(f) 2,5-Dibromo-4-ethyloctane

(g) Cyclobutylcyclopentane

(h) 7-Chlorobicyclo[2.2.1]heptane

4.21   The two secondary carbon atoms in *sec*-butyl alcohol are equivalent; however, there are three five-carbon alcohols (pentyl alcohols) that contain a secondary carbon atom.

4.22 (a) 2,2,3,3-Tetramethylbutane   (b) Cyclohexane

(c) 1,1-Dimethylcyclobutane   (d) Bicyclo [2.2.2] octane

4.23   Each of the desired alkenes must have the same carbon skeleton as 2-methylbutane,

C—C—C—C; they are therefore

$$CH_2{=}\overset{CH_3}{\underset{}{C}}CH_2CH_3$$

$$CH_3\overset{CH_3}{\underset{}{C}}{=}CHCH_3 \Bigg\} + H_2 \xrightarrow[\substack{C_2H_5OH\\pressure}]{Pd,\ Pt,\ or\ Ni} CH_3\overset{CH_3}{\underset{}{CH}}CH_2CH_3$$

$$CH_3\overset{CH_3}{\underset{}{CH}}CH{=}CH_2$$

4.24   Only one isomer of C_6H_{14} can be produced from five isomeric hexyl chlorides ($C_6H_{13}Cl$).

The alkane is 2-methylpentane, $CH_3\overset{CH_3}{\underset{}{CH}}CH_2CH_2CH_3$. The five alkyl chlorides are

$ClCH_2\overset{CH_3}{\underset{}{CH}}CH_2CH_2CH_3$ $CH_3\overset{CH_3}{\underset{}{C}}ClCH_2CH_2CH_3$ $CH_3\overset{CH_3}{\underset{}{CH}}CHClCH_2CH_3$

$CH_3\overset{CH_3}{\underset{}{CH}}CH_2CHClCH_3$ and $CH_3\overset{CH_3}{\underset{}{CH}}CH_2CH_2CH_2Cl$

4.25 $CH_3\overset{CH_3}{\underset{}{CH}}{-}\overset{CH_3}{\underset{}{CH}}CH_3$ 2,3-Dimethylbutane

From two alkyl chlorides

$$CH_3\overset{CH_3}{\underset{Cl}{C}}{-\!\!-}\overset{CH_3}{\underset{H}{C}}CH_3 \Bigg\} \xrightarrow[H_3O^+]{Zn} CH_3\overset{CH_3}{\underset{}{CH}}{-}\overset{CH_3}{\underset{}{CH}}CH_3$$

$$ClCH_2\overset{CH_3}{\underset{}{CH}}{-}\overset{CH_3}{\underset{}{CH}}CH_3$$

From two alkenes

4.26

4.27 $(CH_3)_3CCH_3$ is the most stable isomer (i.e., it is the isomer with the lowest potential energy) because it evolves the least amount of heat on a molar basis when subjected to complete combustion.

$CH_3(CH_2)_3CH_3 + 8 O_2$

$CH_3CH(CH_3)CH_2CH_3 + 8 O_2$

$(CH_3)_3CCH_3 + 8 O_2$

$\Delta H°$

$\Delta H° = -3536$ kJ mol^{-1}

$\Delta H° = -3529$ kJ mol^{-1}

$\Delta H° = -3515$ kJ mol^{-1}

$5 CO_2 + 6 H_2O$

4.28   A homologous series is one in which each member of the series differs from the one preceding it by a constant amount, usually a CH_2 group. A homologous series of alkyl halides would be the following:

$$CH_3X$$
$$CH_3CH_2X$$
$$CH_3(CH_2)_2X$$
$$CH_3(CH_2)_3X$$
$$CH_3(CH_2)_4X$$
etc.

4.29

This conformation is *less stable* because 1,3-diaxial interactions with the large *tert*-butyl group cause considerable repulsion.

This conformation is *more stable* because 1,3-diaxial interactions with the smaller methyl group are less repulsive.

4.30

Cyclopentane   Methylcyclobutane   *cis*-1,2-Dimethylcyclopropane

trans-1,2-Dimethylcyclopropane 1,1-Dimethylcyclopropane Ethylcyclopropane

4.31   (a)   (b)   (c)   (d)

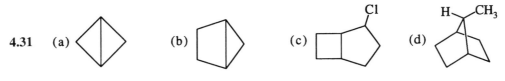

4.32   S − A + 1 = N

For cubane, S = 12 and A = 8. Thus 12 − 8 + 1 = N; N = 5 rings in cubane.

4.33 (a)

Smallest heat
of combustion   <   Largest heat
of combustion

(b)

Least stable   <   Most stable

4.34

4.35 (a) Pentane would boil higher because its chain is unbranched. Chain-branching lowers the boiling point.

(b) Heptane would boil higher because it has the larger molecular weight and would, because of its larger surface area, have larger van der Waals attractions.

(c) 2-Chloropropane because it is more polar and because it has a larger molecular weight.

(d) 1-Propanol would boil higher because its molecules would be associated with each other through hydrogen-bond formation.

(e) Propanone (CH_3COCH_3) would boil higher because its molecules are more polar.

4.36 C_4H_6

Bicyclo[1.1.0]butane   1-Butyne

The IR stretch at ~2250 cm^{-1} for the alkyne C≡C bond distinguishes these two compounds.

4.37 *trans*-1,2-Dimethylcyclopropane would be more stable because there is less crowding between its methyl groups.

Less stable   More stable

4.38 (a)

                    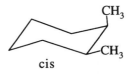

trans   cis

More stable because both methyl groups are equatorial in the most stable conformation

Less stable because one methyl group must be axial and so has the larger heat of combustion

(b)

cis   trans

More stable because both methyl groups are equatorial in the most stable conformation

Less stable because one methyl group must be axial and so has the larger heat of combustion

(c)

trans

More stable because both methyl
groups are equatorial in the most
stable conformation

cis

Less stable because one methyl
group must be axial and so has
the larger heat of combustion

4.39   (a)

More stable conformation because
both alkyl groups are equatorial

(b)

More stable because larger group
is equatorial

(c)

More stable conformation because
both alkyl groups are equatorial

(d)

More stable because larger group
is equatorial

4.40   If the cyclobutane ring were planar, the C−Br bond moments would exactly
cancel in the trans isomer. The fact that *trans*-1,3-dibromocyclobutane has a
dipole moment shows the ring is not planar.

Planar form
$\mu = 0$

Bent form
$\mu \neq 0$

4.41   (a) H_2; Pd, Pt, or Ni catalyst, pressure

(b) (1) $NaNH_2$ (2) $BrCH_2CH_2CH_2CH_3$

(c) (1)

(2) H_2; Pd, Pt, or Ni catalyst, pressure

(d)

(e) H_2; Pd, Pt, or Ni catalyst, pressure

(f) $X-CH_2\overset{\overset{\displaystyle CH_3}{|}}{\underset{\underset{\displaystyle CH_3}{|}}{C}}CH_3$

*4.42   (a)   (b) From Table 4.8 we find that this is
cis-1,2-dimethylcyclohexane.

(c) Since catalytic hydrogenation produces the cis isomer, both hydrogen atoms must
have added from the same side of the double bond. (As we shall see in Section 7.14,
this type of addition is called a syn addition.

cis-1,2-Dimethylcyclohexane

The cis isomer is produced
when both hydrogen atoms
add from the same side.

*4.43   (a) From Table 4.8 we find that this is *trans*-1,2-dichlorocyclohexane.

(b) Since the product is the trans isomer, we can conclude that the chlorine atoms
have added from opposite sides of the double bond.

trans-1,2-Dichlorocyclohexane

The trans isomer is produced
when the chlorine atoms
add from opposite sides of
the double bond.

4.44 If *trans*-1,3-di-*tert*-butylcyclohexane were to adopt a chair conformation, one *tert*-butyl group would have to be axial. It is, therefore, more energetically favorable for the molecule to adopt a twist boat conformation.

*4.45 (a) More rules are needed (see Chapter 7) to indicate relative stereochemistry for these 1-bromo-2-chloro-1-fluoroethenes.

(b) Bicyclo[4.4.0]decane (or decalin)

(c) Bicyclo[4.4.0]dec-2-ene (or Δ^1-octalin)

(d) Bicyclo[4.4.0]dec-1-ene (or $\Delta^{1(8a)}$-octalin)

NOTE: The common name decalin comes from decahydronaphthalene, the deriva-

tive of naphthalene that has had all of its five double bonds con-

verted to single bonds by addition of 10 atoms of hydrogen. Octalin similarly comes from octahydronaphthalene and contains one surviving double bond. When using these common names derived from naphthalene, their skeletons are usually numbered like that of naphthalene. When, as in case (d), a double bond does not lie between the indicated carbon and the next higher numbered carbon, its location is specified as shown.

Also, the symbol Δ is one that has been used with commom names to indicate the presence of a double bond at the position specified by the accompanying superscript number(s).

*4.46 D-Glucose has all of its attached groups equatorial in the lowest energy chair conformation. Other sugars have at least one group axial in their lowest energy chair conformation.

Because if the configuration of stereocenters is changed, not all of the ring substituents can simultaneously be in the preferred equatorial position, as they are in glucose.

*4.47

trans   *cis*

The *trans* isomer would be more stable because it can have both of its rings in the favored chair conformation, avoiding the many eclipsing interactions found in the doubly boat conformation of the *cis* isomer (which would actually increase its stability slightly by twisting to avoid some eclipsing interactions).

*4.48 $H—C\equiv C—C\equiv C—H$ $\xrightarrow[\text{2 mol equiv. } CH_3(CH_2)_3Br]{\text{2 mol equiv. LiNH}_2}$

$CH_3(CH_2)_3—C\equiv C—C\equiv C—(CH_2)_3CH_3$ $\xrightarrow[\text{Pt catalyst}]{H_2}$ $CH_3(CH_2)_{10}CH_3$

QUIZ

4.1 Consider the properties of the following compounds:

NAME	FORMULA	BOILING POINT (°C)	MOLECULAR WEIGHT
Ethane	CH_3CH_3	−88.2	30
Fluoromethane	CH_3F	−78.6	34
Methanol	CH_3OH	+64.7	32

Select the answer that explains why methanol boils so much higher than ethane or fluoromethane, even though they all have nearly equal molecular weights.

(a)   Ion-ion forces between molecules.

(b)   Weak dipole-dipole forces between molecules.

(c)   Hydrogen bonding between molecules.

(d)   van der Waals forces between molecules.

(e)   Covalent bonding between molecules.

4.2 Select the correct name of the compound whose structure is

$$\overset{\displaystyle CH_3}{\underset{\displaystyle \underset{CH_2CH_3}{|}\quad\underset{CH_2CH_3}{|}}{CH_3CH_2CHCHCH_2CH_2CHCH_3}}$$

(a)   2,5-Diethyl-6-methyloctane

(b)   4,7-Diethyl-3-methyloctane

(c)   4-Ethyl-3,7-dimethylnonane

(d)   6-Ethyl-3,7-dimethylnonane

(e)   More than one of the above

4.3 Select the correct name of the compound whose structure is $CH_3\overset{\displaystyle CH_3}{\overset{|}{C}}HCH_2Cl$

(a)   Butyl chloride

(b)   Isobutyl chloride

(c)   *sec*-Butyl chloride

(d)   *tert*-Butyl chloride

(e)   More than one of the above

4.4   The structure shown in Problem 4.2 has:

(a)   1°, 2°, and 3° carbon atoms

(b)   1° and 2° carbon atoms only

(c)   1° and 3° carbon atoms only

(d)   2° and 3° carbon atoms only

(e)   None of the above

4.5   How many isomers are possible for C_3H_7Br?

(a) 1 (b) 2 (c) 3 (d) 4 (e) 5

4.6   Which isomer of 1,3-dimethylcyclohexane is more stable?

(a)   cis (b)   trans   (c)   Both are equally stable

(d)   Impossible to tell

4.7   Which is the lowest energy conformation   of *trans*-1,4-dimethylcyclohexane?

(e)   More than one of the above

4.8   Supply the missing structures

2-Bromobicyclo[2.2.1]heptane

(c)   Newman projection for a gauche form of 1,2-dibromoethane

4.9   Supply the missing reagents:

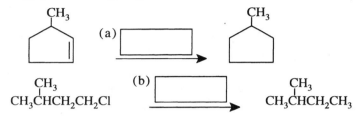

4.10   The most stable conformation of *trans*-1-isopropyl-3-methylcyclohexane:

5

STEREOCHEMISTRY: CHIRAL MOLECULES

SOLUTIONS TO PROBLEMS

5.1 (a) Achiral (c) Chiral (e) Chiral (g) Chiral

 (b) Achiral (d) Chiral (f) Chiral (h) Achiral

5.2 (a) Yes (b) No (c) No

5.3 (a) They are the same molecule. (b) They are enantiomers.

5.4 (a), (b), and (f) do not have stereocenters.

(g) (1) / (2)

(h) (1) / (2)

5.5   (a) Limonene

(b) Thalidomide

(R)   (S)

(S)   (R)

5.6   (a) Two in each instance.

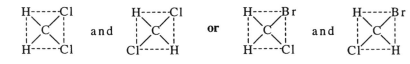

(b)   Only one in each instance.

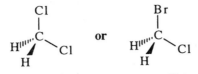

Any other tetrahedral arrangements will be superposable on one or the other of the above.

(c)   Three:

(d)   Two:

5.7   The following items possess a plane of symmetry, and are, therefore, achiral.

(a)   A screwdriver

(b)   A baseball bat (ignoring any writing on it)

(h)   A hammer

5.8   In each instance below, the plane defined by the page is a plane of symmetry.

(a)   (b)   (f)

5.9

(S)   (R)

5.10   (c)   (1) is (S)
              (2) is (R)

(d)   (1) is (S)
       (2) is (R)

(e)   (1) is (S)
       (2) is (R)

(g)   (1) is (S)
       (2) is (R)

(h)   (1) is (S)
       (2) is (R)

5.11 (a) $-Cl$ > $-SH$ > $-OH$ > $-H$

(b) $-CH_2Br$ > $-CH_2Cl$ > $-CH_2OH$ > $-CH_3$

(c) $-OH$ > $-CHO$ > $-CH_3$ > $-H$

(d) $-C(CH_3)_3$ > $-CH=CH_2$ > $-CH(CH_3)_2$ > $-H$

(e) $-OCH_3$ > $-N(CH_3)_2$ > $-CH_3$ > $-H$

5.12 (a) (*S*)   (b) (*R*)   (c) (*S*)

5.13 (a) Enantiomers

(b) Two molecules of the same compound

(c) Enantiomers

5.14

(*S*)-(+)-Carvone   (*R*)-(−)-Carvone

5.15 (a) Enant. Excess $= \dfrac{\text{observed rotation}}{\text{specific rotation of pure enantiomer}} \times 100$

$= \dfrac{+1.151°}{+5.756°} \times 100$

$= 20.00\%$

(b) since the (*R*) enantiomer (see Section 5.7C) is +, the (*R*) enantiomer is present in excess.

5.16 (a)

(*S*)-Ibuprofen

(b)

(*S*)-Methyldopa

(c)

(*S*)-Penicillamine

5.17 (a) Diastereomers.

(b) Diastereomers in each instance.

(c) No, diastereomers have different melting points.

(d) No, diastereomers have different boiling points.

(e) No, diastereomers have different vapor pressures.

STUDY AID

An Approach to the Classification of Isomers

We can classify isomers by asking and answering a series of questions:

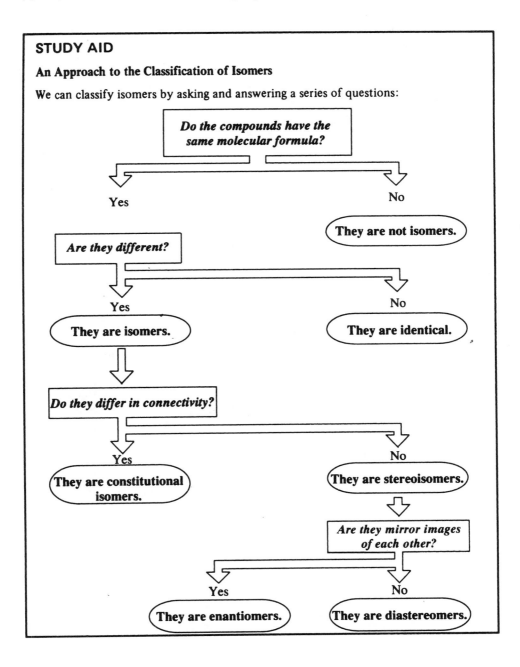

5.18 (a) It would be optically active.

(b) It would be optically active.

(c) No, because it is a meso compound.

(d) No, because it would be a racemic form.

5.19 (a) Represents **A**

(b) Represents **C**

(c) Represents **B**

5.20 (a)

5.21   **B** is (2S,3S)-2,3-Dibromobutane
  C is (2R,3S)-2,3-Dibromobutane

5.22   (a) (**1**) is (2S,3S)-2,3-Dichlorobutane
    (**2**) is (2R,3R)-2,3-Dichlorobutane
    (**3**) is (2R,3S)-2,3-Dichlorobutane

  (b) (**1**) is (2S,4S)-2,4-Pentanediol
    (**2**) is (2R,4R)-2,4-Pentanediol
    (**3**) is (2R,4S)-2,4-Pentanediol

  (c) (**1**) is (2R,3R)-1,4-Dichloro-2,3-difluorobutane
    (**2**) is (2S,3S)-1,4-Dichloro-2,3-difluorobutane
    (**3**) is (2R,3S)-1,4-Dichloro-2,3-difluorobutane

  (d) (**1**) is (2S,4S)-4-Chloro-2-pentanol
    (**2**) is (2R,4R)-4-Chloro-2-pentanol
    (**3**) is (2S,4R)-4-Chloro-2-pentanol
    (**4**) is (2R,4S)-4-Chloro-2-pentanol

  (e) (**1**) is (2S,3S)-2-Bromo-3-fluorobutane
    (**2**) is (2R,3R)-2-Bromo-3-fluorobutane
    (**3**) is (2S,3R)-2-Bromo-3-fluorobutane
    (**4**) is (2R,3S)-2-Bromo-3-fluorobutane

5.23

Chloramphenicol

5.24   (a) No   (b) Yes   (c) No   (d) No   (e) Diastereomers

(f) Diastereomers

5.25

Meso compound   Enantiomers

5.26   (a)

(b)

(1S,3R)

Enantiomers
(both cis)

(1R,3S)

(1S,3S)

Enantiomers
(both trans)

(1R,3R)

(c)

Achiral (trans)

Achiral (cis)

5.27   See Problem 5.26. The molecules in (c) are achiral, so they have no (R-S) designation.

5.28

(S)-(−)-Glyceraldehyde   (S)-(+)-Glyceric acid

$\xrightarrow{\text{HgO}}$ $\xleftarrow[\text{H}_2\text{O}]{\text{HNO}_2}$ (S)-(−)-Isoserine
(See the following
reaction, also.)

(S)-(−)-Isoserine   (R)-(+)-3-Bromo-
2-hydroxypro-
panoic acid   (S)-(+)-Lactic acid

$\xrightarrow[\text{HBr}]{\text{HNO}_2}$ $\xrightarrow{\text{Zn, H}_3\text{O}^+}$

5.29

(S)-(+)-1-Chloro-2-methylbutane   (R)-1-Deuterio-2-methylbutane

5.30   (a), (b), (f) and (g) only

5.31   (a)   Seven.

(b)   (R)- and (S)-3-Methylhexane and (R)- and (S)-2,3-dimethylpentane.

5.32   (a) and (b)

(c)   Four

(d)   Because a trans arrangement of the one carbon bridge is structurally impossible. Such a molecule would have too much strain.

5.33   (a)   **A** is (R,S)-2,3-dichlorobutane; **B** is (S,S)-2,3-dichlorobutane; **C** is (R,R)-2,3-dichlorobutane.

(b)   **A**

5.34   (a)

(e)

(other answers are possible)

5.35 (a) Same   (j) Enantiomers that are interconvertible by a ring flip

     (b) Enantiomers   (k) Diastereomers

     (c) Diastereomers   (l) Same

     (d) Same   (m) Diastereomers

     (e) Same   (n) Constitutional isomers

     (f) Constitutional isomers   (o) Diastereomers

     (g) Diastereomers   (p) Same

     (h) Enantiomers   (q) Same

     (i) Same

5.36 All of these molecules are expected to be planar. Their stereochemistry is identical to that of the corresponding chloroethenes. (a) can exist as cis and trans isomers. Only one compound exists in the case of (b) and (c).

5.37 (a)

     (b) (**3**) and (**4**) are chiral and are enantiomers of each other.

     (c) Three fractions: a fraction containing (**1**), a fraction containing (**2**), and a fraction containing (**3**) and (**4**) [because, being enantiomers, (**3**) and (**4**) would have the same vapor pressure].

     (d) None

5.38 (a)

     (b) No, they are not superposable.

     (c) No, and they are, therefore, enantiomers of each other.

     (d)

(e)   No, they are not superposable.

(f)   Yes, and they are, therefore, just different conformations of the same molecule.

5.39   (a)

(b)   Yes, and therefore *trans*-1.4-diethylcyclohexane is achiral.
(c)   No, they are different orientations of the same molecule.
(d)   Yes, *cis*-1,4-diethylcyclohexane is a stereoisomer (a diastereomer) of *trans*-1.4-diethylcyclohexane.

cis-1,4-Diethylcyclohexane

(e)   No, it, too, is superposable on its mirror image. (Notice, too, that the plane of the page constitutes a plane of symmetry for both *cis*-1,4-diethylcyclohexane and for *trans*-1,4-diethylcyclohexane as we have drawn them.)

5.40   *trans*-1,3-Diethylcyclohexane can exist in the following enantiomeric forms.

trans-1,3-Diethylcyclohexane enantiomers

cis-1,3-Diethylcyclohexane consists of achiral molecules because they have a plane of symmetry. [The plane of the page (below) is a plane of symmetry.]

cis-1,3-Diethylcyclohexane

*5.41   (a)   Since it is optically active and not resolvable, it must be the meso form:

(c)   No   (d)   A racemic modification

*5.42 (a) $[\alpha]_D = \dfrac{-30°}{(0.10 \text{ g/mL})(1.0 \text{ dm})} = -300°$

(b) $[\alpha]_D = \dfrac{+165°}{(0.05 \text{ g/mL})(1.0 \text{ dm})} = +3300°$

The two rotation values can be explained by recognizing that this is a powerfully optically active substance and that the first reading, assumed to be −30°, was really +330°. Making this change the $[\alpha]_D$ becomes +3300° in both cases.

(c) No, the apparent 0° rotation could actually be + or −360° (or an integral multiple of these values).

*5.43 Yes, it could be a *meso* form or an enantiomer whose stereocenters, by rare coincidence, happen to cancel each others activities.

*5.44 A compound $C_3H_6O_2$ has an index of hydrogen deficiency of 1. Thus, it could possess a carbon-carbon double bond, a carbon-oxygen double bond, or a ring.

   The IR spectral data rule out a carbonyl group but indicate the presence of an —OH group.

   No stable structure having molecular formula $C_3H_6O_2$ with a C=C bond can exist in stereoisomeric forms and 1,2-cyclopropanediol can exist in three stereoisomeric forms.

   Only ethylene oxide (oxirane) derivatives are possible for Y:

QUIZ

5.1 Describe the relationship between the two structures shown.

   (a) Enantiomers   (b) Diastereomers   (c) Constitutional isomers

   (d) Conformations   (e) Two molecules of the same compound

5.2 Which of the following molecule(s) possess(es) a plane of symmetry?

   (a)   (b)   (c)

   (d) More than one of these   (e) None of these

5.3   Give the (*R-S*) designation of the structure shown.

(a)   (*R*)   (b)   (*S*)   (c)   Neither, because this molecule has no stereocenter

(d)   Impossible to tell

5.4   Select the words that best describe the following structure:

(a)   Chiral   (b)   Meso form   (c)   Achiral   (d)   Has a plane of symmetry

(e)   More than one of these

5.5   Select the words that best describe what happens to the optical rotation of the alkene shown when it is hydrogenated to the alkane according to the following equation:

(a)   Increases   (b)   Drops to zero   (c)   Changes sign

(d)   Stays the same   (e)   Impossible to predict

5.6   There are two compounds with the formula C_7H_{16} that are capable of existing as enantiomers. Write three-dimensional formulas for the (*S*) isomer of each.

5.7   Compound A is optically active and is the (*S*) isomer.

A

5.8   Compound B is a hydrocarbon with the minimum number of carbon atoms necessary for it to possess a stereocenter and, as well, alternative stereochemistries about a double bond.

B

6

IONIC REACTIONS— NUCLEOPHILIC SUBSTITUTION AND ELIMINATION REACTIONS OF ALKYL HALIDES

SOLUTIONS TO PROBLEMS

6.1 (a) $CH_3—\ddot{\underset{..}{I}}:$ + $CH_3CH_2—\ddot{\underset{..}{O}}:^-$ $\longrightarrow$ $CH_3—\ddot{\underset{..}{O}}—CH_2CH_3$ + $:\ddot{\underset{..}{I}}:^-$

       Substrate   Nucleophile   Leaving group

(b) $:\ddot{\underset{..}{I}}:^-$ + $CH_3CH_2—\ddot{\underset{..}{Br}}:$ $\longrightarrow$ $CH_3CH_2—\ddot{\underset{..}{I}}:$ + $:\ddot{\underset{..}{Br}}:^-$

     Nucleo-   Substrate   Leaving
     phile   group

(c) $CH_3\ddot{\underset{..}{O}}H$ + $(CH_3)_3C—\ddot{\underset{..}{C}}l:$ $\longrightarrow$ $(CH_3)_3C—\ddot{\underset{..}{O}}—CH_3$ + $:\ddot{\underset{..}{C}}l:^- + H^+$

     Nucleo-   Substrate   Leaving
     phile   group

(d) $CH_3CH_2CH_2—\ddot{\underset{..}{Br}}:$ + $^-:C≡N:$ $\longrightarrow$ $CH_3CH_2CH_2—CN$ + $:\ddot{\underset{..}{Br}}:^-$

       Substrate   Nucleo-   Leaving
                               phile   group

(e) $C_6H_5CH_2—\ddot{\underset{..}{Br}}:$ + $:NH_3$ $\longrightarrow$ $C_6H_5CH_2—\ddot{\underset{..}{N}}H_2$ + $:\ddot{\underset{..}{Br}}:^- + H^+$

      Substrate   Nucleo-   Leaving
                           phile   group

6.2

6.3 (a)   We know that when a secondary alkyl halide reacts with hydroxide ion by substitution, the reaction occurs with *inversion of configuration* because the reaction is S_N2. If we know that the configuration of (−)-2-butanol (from Section 5.7C) is that shown here, then we can conclude that (+)-2-chlorobutane has the opposite configuration.

(R)-(-)-2-Butanol
$[\alpha]\,^{25°}_D$ = -13.52°

(S)-(+)-2-Chlorobutane
$[\alpha]\,^{25°}_D$ = +36.00°

(b)   Again the reaction is S_N2. Because we now know the configuration of (+)-2-chloro-butane to be (S) [cf., part (a)], we can conclude that the configuration of (−)-2-iodobutane is (R).

(S)-(+)-2-Chlorobutane   (R)-(-)-2-Iodobutane

(+)-2-Iodobutane has the (S) configuration

6.4 (a) (b)

By path (a)   By path (b)

6.5

6.6 (a) Being primary halides, the reactions are most likely to be S_N2, with the nucleophile in each instance being a molecule of the solvent (i.e., a molecule of ethanol).

(b) Steric hindrance is provided by the substituent or substituents on the carbon β to the carbon bearing the leaving group. With each addition of a methyl group at the β carbon (below), the number of pathways open to the attacking nucleophile becomes fewer.

6.7 **Protic solvents** are those that have an H bonded to an oxygen or nitrogen (or to another

strongly electronegative atom). Therefore, the protic solvents are formic acid, $HCOH$;

$$O \\ \parallel$$

formamide, $HCNH_2$: ammonia, NH_3; and ethylene glycol, $HOCH_2CH_2OH$

Aprotic solvents lack an H bonded to a strongly electronegative element. Aprotic solvents in this list are acetone, CH_3CCH_3; acetonitrile, $CH_3C\equiv N$; sulfur dioxide, SO_2; and trimethylamine, $N(CH_3)_3$.

6.8 The reaction is an S_N2 reaction. In the polar aprotic solvent (DMF), the nucleophile (CN^-) will be relatively unencumbered by solvent molecules, and, therefore, it will be more reactive than in ethanol. As a result, the reaction will occur faster in *N,N*-dimethylformamide.

6.9 (a) CH_3O^-

(b) H_2S

(c) $(CH_3)_3P$

6.10 (a) Increasing the percentage of water in the mixture increases the polarity of the solvent. (Water is more polar than methanol.) Increasing the polarity of the solvent increases the rate of the solvolysis because separated charges develop in the transition state. The more polar the solvent, the more the transition state is stabilized (Section 6.14D).

(b) In an S_N2 reaction of this type, the charge becomes dispersed in the transition state:

Increasing the polarity of the solvent increases the stabilization of the reactant I⁻ more than the stabilization of the transition state, and thereby increases the free energy of activation, thus decreasing the rate of reaction.

6.11 $CH_3OSO_2CF_3$ > CH_3I > CH_3Br > CH_3Cl > CH_3F > $^{14}CH_3OH$

(Most reactive)   (Least reactive)

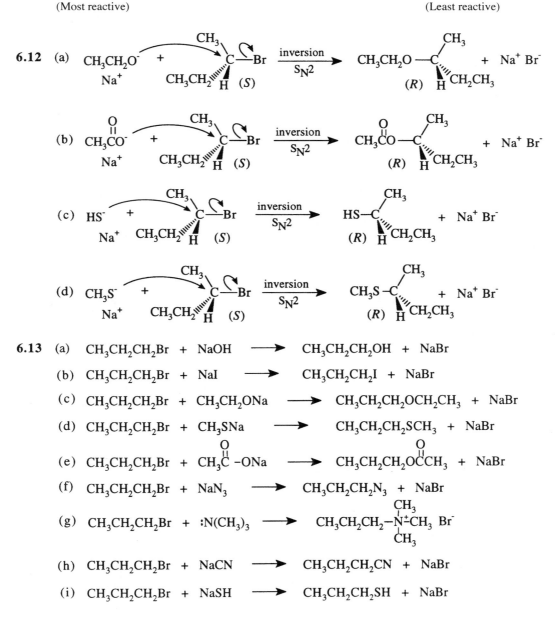

6.13   (a) $CH_3CH_2CH_2Br$ + $NaOH$ ⟶ $CH_3CH_2CH_2OH$ + $NaBr$

(b) $CH_3CH_2CH_2Br$ + NaI ⟶ $CH_3CH_2CH_2I$ + $NaBr$

(c) $CH_3CH_2CH_2Br$ + CH_3CH_2ONa ⟶ $CH_3CH_2CH_2OCH_2CH_3$ + $NaBr$

(d) $CH_3CH_2CH_2Br$ + CH_3SNa ⟶ $CH_3CH_2CH_2SCH_3$ + $NaBr$

(e) $CH_3CH_2CH_2Br$ + $CH_3\overset{O}{\overset{\|}{C}}-ONa$ ⟶ $CH_3CH_2CH_2O\overset{O}{\overset{\|}{C}}CH_3$ + $NaBr$

(f) $CH_3CH_2CH_2Br$ + NaN_3 ⟶ $CH_3CH_2CH_2N_3$ + $NaBr$

(g) $CH_3CH_2CH_2Br$ + $:N(CH_3)_3$ ⟶ $CH_3CH_2CH_2-\overset{CH_3}{\underset{CH_3}{N^{+}CH_3}}$ Br^-

(h) $CH_3CH_2CH_2Br$ + $NaCN$ ⟶ $CH_3CH_2CH_2CN$ + $NaBr$

(i) $CH_3CH_2CH_2Br$ + $NaSH$ ⟶ $CH_3CH_2CH_2SH$ + $NaBr$

6.14   (a)   1-Bromopropane would react more rapidly because, being a primary halide, it is less hindered.

(b) 1-Iodobutane, because iodide ion is a better leaving group than chloride ion.

(c) 1-Chlorobutane, because the carbon bearing the leaving group is less hindered than in 1-chloro-2-methylpropane.

(d) 1-Chloro-3-methylbutane, because the carbon bearing the leaving group is less hindered than in 1-chloro-2-methylbutane.

(e) 1-Chlorohexane because it is a primary halide. Phenyl halides are unreactive in S_N2 reactions.

6.15 (a) Reaction (1) because ethoxide ion is a stronger nucleophile than ethanol.

(b) Reaction (2) because the ethyl sulfide ion is a stronger nucleophile than the ethoxide ion in a protic solvent. (Because sulfur is larger than oxygen, the ethyl sulfide ion is less solvated and it is more polarizable.)

(c) Reaction (2) because triphenylphosphine $[(C_6H_5)_3P]$ is a stronger nucleophile than triphenylamine. (Phosphorus atoms are larger than nitrogen atoms.)

(d) Reaction (2) because in an S_N2 reaction the rate depends on the concentration of the substrate and the nucleophile. In reaction (2) the concentration of the nucleophile is twice that of the reaction (1).

6.16 (a) Reaction (2) because bromide ion is a better leaving group than chloride ion.

(b) Reaction (1) because water is a more polar solvent than methanol, and S_N1 reactions take place faster in more polar solvents.

(c) Reaction (2) because the concentration of the substrate is twice that of reaction (1). (The major reaction would be E2. However, the problem asks us to consider that small portion of the overall reaction that proceeds by an S_N1 pathway.)

(d) Both reactions would take place at the same rate because, being S_N1 reactions, they are independent of the concentration of the nucleophile.
(The major reaction would be E2. However, the problem asks us to consider that small portion of the overall reaction that proceeds by an S_N1 pathway.)

(e) Reaction (1) because the substrate is a tertiary halide. Phenyl halides are unreactive in S_N1 reactions.

6.17 Possible methods are given here.

(f) CH_3CH_2Cl $\xrightarrow[\substack{CH_3OH \\ S_N2}]{SH^-}$ CH_3CH_2SH

(g) CH_3Cl $\xrightarrow[DMF]{CN^-}$ CH_3CN

(h) CH_3CH_2Cl $\xrightarrow[DMF]{CN^-}$ CH_3CH_2CN

(i) CH_3OH $\xrightarrow[(-H_2)]{NaH}$ CH_3ONa $\xrightarrow[CH_3OH]{CH_3I}$ CH_3OCH_3

(j) CH_3CH_2OH $\xrightarrow[(-H_2)]{NaH}$ CH_3CH_2ONa $\xrightarrow{CH_3I}$ $CH_3CH_2OCH_3$

(k)

6.18 (a) The reaction will not take place because the leaving group would have to be a methyl anion, a very powerful base, and a very poor leaving group.

(b) The reaction will not take place because the leaving group would have to be a hydride ion, a very powerful base, and a very poor leaving group.

(c) The reaction will not take place because the leaving group would have to be a carbanion, a very powerful base, and a very poor leaving group.

$$HO^- \; + \; \substack{H_2C-CH_2 \\ |\quad\quad| \\ H_2C-CH_2} \xrightarrow{\;\;X\;\;} {}^-CH_2CH_2CH_2CH_2OH$$
$$\downarrow$$
$$CH_3CH_2CH_2CH_2OH$$

(d) The reaction will not take place by an S_N2 mechanism because the substrate is a tertiary halide, and is, therefore, not susceptible to S_N2 attack because of the steric hindrance. (A very small amount of S_N1 reaction may take place, but the main reaction will be E2 to produce an alkene.)

(e) The reaction will not take place because the leaving group would have to be a CH_3O^- ion, a strong base, and a very poor leaving group.

$$:NH_3 \; + \; CH_3-OCH_3 \xrightarrow{\;\;X\;\;} CH_3NH_3{}^+ \; + \; CH_3O^-$$
$$\downarrow$$
$$CH_3NH_2 \; + \; CH_3OH$$

(f) The reaction will not take place because the first reaction that would take place would be an acid-base reaction that would convert the ammonia to an ammonium ion. An ammonium ion, because it lacks an electron pair, is not nucleophilic.

$$NH_3 \; + \; CH_3OH_2{}^+ \; \rightleftarrows \; NH_4{}^+ \; + \; CH_3OH$$

6.19 The better yield will be obtained by using the secondary halide, 1-bromo-1-phenylethane, because the desired reaction is E2. Using the primary halide will result in substantial S_N2 reaction as well, producing the alcohol instead of the desired alkene.

6.20 Reaction (2) would give the better yield because the desired reaction is an S_N2 reaction, and the substrate is a methyl halide. Use of reaction (1) would, because the substrate is a secondary halide, result in considerable elimination by an E2 pathway.

6.21 (a) The major product would be $CH_3CH_2CH_2CH_2CH_2OCH_2CH_3$ (by an S_N2 mechanism) because the substrate is primary and the nucleophile-base is not hindered. Some $CH_3CH_2CH_2CH=CH_2$ would be produced by an E2 mechanism.

(b) The major product would be $CH_3CH_2CH_2CH=CH_2$ (by an E2 mechanism), even though the substrate is primary because the base is a hindered strong base. Some $CH_3CH_2CH_2CH_2CH_2OC(CH_3)_3$ would be produced by an S_N2 mechanism.

(c) For all practical purposes, $(CH_3)_2C=CH_2$ (by an E2 mechanism) would be the only product because the substrate is tertiary and the base is strong.

(d) Same answer as (c) above.

(e)   (formed by an S_N2 mechanism) would, for all practical purposes, be the only product. Iodide ion is a very weak base and a good nucleophile.

(f) Because the substrate is tertiary and the only nucleophile is the solvent, mechanism is S_N1. The two products that follow would be formed.

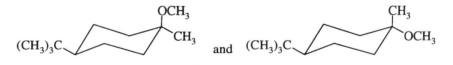

(g) $CH_3CH=CHCH_2CH_3$ (by an E2 mechanism) would be the major product because the substrate is secondary and the base/nucleophile is a strong base. Some of the ether, $CH_3CH_2CH(OCH_3)CH_2CH_3$, would be formed by an S_N2 pathway.

(h) The major product would be $CH_3CH_2CH(O_2CCH_3)CH_2CH_3$ (by an S_N2 mechanism) because the acetate ion is a weak base. Some $CH_3CH=CHCH_2CH_3$ might be formed by an E2 pathway.

(i) $CH_3CH=CHCH_3$ and $CH_2=CHCH_2CH_3$ (by E2) would be major products, and (S)-$CH_3CH(OH)CH_2CH_3$ (by S_N2) would be the minor product.

(j) (±)-$CH_3CH_2\underset{\underset{\displaystyle OCH_3}{|}}{C}(CH_3)CH_2CH_2CH_3$ (by S_N1) would be the major product.

$CH_3CH_2\underset{\underset{\displaystyle CH_3}{|}}{C}=CHCH_2CH_3$, $CH_3CH=\underset{\underset{\displaystyle CH_3}{|}}{C}CH_2CH_2CH_3$, and

$CH_3CH_2\underset{\underset{\displaystyle CH_2}{||}}{C}CH_2CH_2CH_3$ (by E1) would be minor products.

(k) (R)-$CH_3CHIC_6H_{13}$ (by S_N2) would be the only product.

6.22 (a), (b), and (c) are all S_N2 reactions and, therefore, proceed with inversion of configuration. The products are

(d) is an S_N1 reaction. The carbocation that forms can react with either nucleophile (H_2O or CH_3OH) from either the top or bottom side of the molecule. Four substitution products (below) would be obtained. (Considerable elimination by an E1 path would also occur.)

6.23 Isobutyl bromide is more sterically hindered than ethyl bromide because of the methyl groups on the β carbon atom.

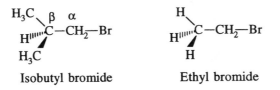

Isobutyl bromide   Ethyl bromide

This steric hindrance causes isobutyl bromide to react more slowly in S_N2 reactions and to give relatively more elimination (by an E2 path) when a strong base is used.

6.24 (a) S_N2 because the substrate is a 1° halide.

(b) Rate $= k$ [CH_3CH_2Cl][I^-]

$= 5 \times 10^{-5}$ L mol^{-1}s$^{-1} \times 0.1$ mol L$^{-1} \times 0.1$ mol L^{-1}

Rate $= 5 \times 10^{-7}$ mol L^{-1}s^{-1}

(c) 1×10^{-6} mol L^{-1}s^{-1}

(d) 1×10^{-6} mol L^{-1}s^{-1}

(e) 2×10^{-6} mol L^{-1}s^{-1}

6.25 (a) CH_3NH^- because it is the stronger base.

(b) CH_3O^- because it is the stronger base.

(c) CH_3SH because sulfur atoms are larger and more polarizable than oxygen atoms.

(d) $(C_6H_5)_3P$ because phosphorus atoms are larger and more polarizable than nitrogen atoms.

(e) H_2O because it is the stronger base.

(f) NH_3 because it is the stronger base.

(g) HS^- because it is the stronger base.

(h) OH^- because it is the stronger base.

6.26 (a)

(b)

6.27 Iodide ion is a good nucleophile and a good leaving group; it can rapidly convert an alkyl chloride or alkyl bromide into an alkyl iodide, and the alkyl iodide can then react rapidly with another nucleophile. With methyl bromide in water, for example, the following reaction can take place:

6.28 *tert*-Butyl alcohol and *tert*-butyl methyl ether are formed via an S_N1 mechanism. The rate of the reaction is independent of the concentration of methoxide ion (from sodium methoxide). This, however, is only one reaction that causes *tert*-butyl bromide to disappear. A competing reaction that also causes *tert*-butyl bromide to disappear is an E2 reaction in which methoxide ion reacts with *tert*-butyl bromide. This reaction is dependent on the concentration of methoxide ion; therefore, increasing the methoxide ion concentration causes an increase in the rate of disappearance of *tert*-butyl bromide.

6.29 (a)   You should use a strong base, such as RO^-, at a higher temperature to bring about an E2 reaction.

(b)   Here we want an S_N1 reaction. We use ethanol as the solvent *and as the nucleophile*, and we carry out the reaction at a low temperature so that elimination will be minimized.

6.30 1-Bromobicyclo[2.2.1]heptane is unreactive in an S_N2 reaction because it is a tertiary halide and its ring structure makes the backside of the carbon bearing the leaving group completely inaccessible to attack by a nucleophile.

1-Bromobicyclo[2.2.1]heptane is unreactive in an S_N1 reaction because the ring structure makes it impossible for the carbocation that must be formed to assume the required trigonal planar geometry around the positively charged carbon. Any carbocation formed from 1-bromobicyclo[2.2.1]heptane would have a trigonal pyramidal arrangement of the $-CH_2-$ groups attached to the positively charged carbon (make a model). Such a structure does not allow stabilization of the carbocation by overlap of sp^3 orbitals from the alkyl groups (see Fig. 6.9).

6.31 The cyanide ion has two nucleophilic atoms; it is what is called an ambident nucleophile.

$$^-:C{\equiv}N:$$

It can react with a substrate using either atom, although the carbon atom is more nucleophilic.

$$Br{-}CH_2CH_3 \;+\; {}^-:C{\equiv}N: \;\longrightarrow\; CH_3CH_2{-}C{\equiv}N:$$

$$^-:C{\equiv}N: \;+\; CH_3CH_2{-}Br \;\longrightarrow\; CH_3CH_2{-}\ddot{N}{=}C:$$

6.32 (a) $\underset{\overset{|}{OH}}{CH_3CH_2CHCH_3}$ $\xrightarrow[\text{Et}_2\text{O}\,(-\text{H}_2)]{\text{NaH}}$ $\underset{\overset{|}{O^-Na^+}}{CH_3CH_2CHCH_3}$

$$\underset{\overset{|}{OCH_2CH_2CH_2CH_3}}{CH_3CH_2CHCH_3} \xleftarrow[\;(-\text{NaBr})\;]{} CH_3CH_2CH_2CH_2{-}Br$$

(b) $(CH_3)_3CSH$ $\xrightarrow[\text{Et}_2\text{O}\,(-\text{H}_2)]{\text{NaH}}$ $(CH_3)_3CS^-Na^+$ $\xrightarrow[(-\text{NaBr})]{CH_3CH_2{-}Br}$

$$(CH_3)_3C{-}S{-}CH_2CH_3$$

(c)   CH₃—C(CH₃)(CH₃)—CH₂OH $\xrightarrow[\text{Et}_2\text{O} (-\text{H}_2)]{\text{NaH}}$ CH₃—C(CH₃)(CH₃)—CH₂O⁻Na⁺ $\xrightarrow[(-\text{NaI})]{\text{CH}_3\text{I}}$

CH₃—C(CH₃)(CH₃)—CH₂—OCH₂

(d)   C₆H₅OH $\xrightarrow[\text{Et}_2\text{O} (-\text{H}_2)]{\text{NaH}}$ C₆H₅O⁻Na⁺ $\xrightarrow[(-\text{NaI})]{\text{CH}_3\text{I}}$ C₆H₅OCH₃

(e)   C₆H₅CH₂—Br + Na⁺⁻:C≡N: $\xrightarrow[(-\text{NaBr})]{\text{EtOH}}$ C₆H₅CH₂—C≡N:

(f)   CH₃C(=O)—O⁻Na⁺ + C₆H₅CH₂—Br $\xrightarrow[(-\text{NaBr})]{\text{CH}_3\text{CO}_2\text{H}}$ C₆H₅CH₂OC(=O)CH₃

(g)   HO⁻ Na⁺ + (R)-2-Bromopentane $\xrightarrow[(-\text{NaBr})]{\text{acetone}}$ (S)-2-Pentanol

(h)   I⁻ Na⁺ + (S)-2-Chloro-4-methylpentane $\xrightarrow[(-\text{NaCl})]{\text{acetone}}$ (R)-2-Iodo-4-methylpentane

(i)   (CH₃)₃CCHCH₃(Br) $\xrightarrow[\substack{\text{EtOH} \\ (-\text{NaBr})}]{\text{EtO}^-\text{Na}^+}$ (CH₃)₃CCH=CH₂ + EtOH

(j)   (CH₃)₂CH—[cyclohexyl-Br] $\xrightarrow[\substack{\text{H}_2\text{O}/\text{CH}_3\text{OH} \\ (-\text{NaBr})}]{\text{Na}^+\text{OH}^-}$ (CH₃)₂CH—[cyclohexyl-OH]

(k)   Na⁺ ⁻:C≡N: + (S)-2-Bromobutane $\xrightarrow[(-\text{NaBr})]{\text{EtOH}}$:N≡C—C(CH₂CH₃)(CH₃)H

(l)   H₃C—[cyclohexyl-Cl] + Na⁺ :I:⁻ $\xrightarrow[(-\text{NaCl})]{\text{acetone}}$ H₃C—[cyclohexyl-I]

6.33 (a)

(Formation of this product depends on the fact that bromide ion is a much better leaving group than fluoride ion.)

(b) $CH_3CH_2CHCH_2CH_2CH_2$—I
$\quad\quad\quad$ |
$\quad\quad\quad$ Cl

(Formation of this product depends on the greater reactivity of 1° substrates in S_N2 reactions.)

(c)
$$S \begin{matrix} CH_2CH_2 \\ CH_2CH_2 \end{matrix} S$$

(Here two S_N2 reactions produce a cyclic molecule.)

(d) Cl—$CH_2CH_2CH_2CH_2OH$ + NaH $\xrightarrow[Et_2O]{-H_2}$ Cl—$CH_2CH_2CH_2CH_2O^-$

(e) $CH_3C{\equiv}CH$ + $NaNH_2$ $\xrightarrow[\text{liq. NH}_3]{}$ $CH_3C{\equiv}C{:}^-Na^+$ $\xrightarrow[(-NaI)]{CH_3-I}$

$$CH_3C{\equiv}CCH_3$$

6.34 The rate-determining step in the S_N1 reaction of *tert*-butyl bromide is the following:

$$(CH_3)_3C—Br \underset{\text{-x}}{\overset{\text{slow}}{\rightleftharpoons}} (CH_3)_3C^+ \quad + \quad Br^-$$
$$\quad\quad\quad\quad\quad\quad\quad \downarrow H_2O$$
$$\quad\quad\quad\quad\quad\quad\quad (CH_3)_3COH_2^+$$

$(CH_3)_3C^+$ is so unstable that it reacts almost immediately with one of the surrounding water molecules, and, for all practical purposes, no reverse reaction with Br^- takes place. Adding a common ion (Br^- from NaBr), therefore, has no effect on the rate.

$\quad$ Because the $(C_6H_5)_2CH^+$ cation is more stable, a reversible first step occurs and adding a common ion (Br^-) slows the overall reaction by increasing the rate at which $(C_6H_5)_2CH^+$ is converted back to $(C_6H_5)_2CHBr$.

$$(C_6H_5)_2CHBr \quad \rightleftharpoons \quad (C_6H_5)_2CH^+ \quad + \quad Br^-$$
$$\quad\quad\quad\quad\quad\quad\quad\quad \downarrow H_2O$$
$$\quad\quad\quad\quad\quad\quad\quad\quad (C_6H_5)_2CHOH_2^+$$

6.35 Two different mechanisms are involved. $(CH_3)_3CBr$ reacts by an S_N1 mechanism, and apparently this reaction takes place faster. The other three alkyl halids react by an S_N2 mechanism, and their reactions are slower because the nucleophile (H_2O) is weak. The reaction rates of CH_3Br, CH_3CH_2Br, and $(CH_3)_2CHBr$ are affected by the steric hindrance, and thus their order of reactivity is $CH_3Br > CH_3CH_2Br > (CH_3)_2CHBr$.

6.36   The nitrite ion is an *ambident nucleophile;* that is, it is an ion with two nucleophilic sites. The equivalent oxygen atoms and the nitrogen atom are nucleophilic.

6.37   (a)   The transition state has the form:

$$\overset{\delta^+}{Nu} ---- R ----- \overset{\delta^-}{L}$$

in which charges are developing. The more polar the solvent, the better it can solvate the transition state, thus lowering the free energy of activation and increasing the reaction rate.

(b)   The transition state has the form:

$$\overset{\delta^+}{R} ----------- \overset{\delta^+}{L}$$

in which the charge is becoming dispersed. A polar solvent is less able to solvate this transition state than it is to solvate the reactant. The free energy of activation, therefore, will become somewhat larger as the solvent polarity increases, and the rate will be slower.

6.38   (a) $Cl-CH_2-\underset{\underset{CH_3}{|}}{\overset{\overset{CH_3}{|}}{C}}-CH_2-CH_2-I$

(b) $HO-\underset{\underset{CH_3}{|}}{\overset{\overset{CH_3}{|}}{C}}-CH_2-CH_2-Cl$ +   some alkene

6.39   (a)   In an S_N1 reaction the carbocation intermediate reacts rapidly with any nucleophile it encounters in a Lewis acid–Lewis base reaction. In the case of the S_N2 reaction, the leaving group departs only when "pushed out" by the attacking nucleophile, and some nucleophiles are better than others.

(b) CN^- is a much better nucleophile than ethanol and hence the nitrile is formed in the S_N2 reaction of $CH_3CH_2CH_2CH_2Cl$. In the case of $(CH_3)_3CCl$ the *tert*-butyl cation reacts chiefly with the nucleophile present in higher concentration, here the ethanol solvent.

*6.40

	$\Delta H°$, kJ mol^{-1}	
$(CH_3)_3C—Cl \longrightarrow (CH_3)_3C\cdot + Cl\cdot$	$+328$	Homolytic bond dissoc. energy
$(CH_3)_3C\cdot \longrightarrow (CH_3)_3C^+ + e^-$	$+715$	Ionization potential
$Cl\cdot + e^- \longrightarrow Cl^-$	-330	Electron affinity
$(CH_3)_3C—Cl \longrightarrow (CH_3)_3C^+ + Cl^-$	$+713$	Heterolytic bond dissoc. energy

*6.41   (a)   The entropy term is slightly favorable. (The enthalpy term is highly unfavorable.)

(b) $\Delta G° = \Delta H° - T\Delta S°$

$$= 26.6 \text{ kJ mol}^{-1} - (298)(0.00481 \text{ kJ mol}^{-1})$$

$$= 25.2 \text{ kJ mol}^{-1}$$

(c) $\log K_{eq} = \dfrac{-\Delta G°}{2.303RT}$

$$= \dfrac{-25.2 \text{ kJ mol}^{-1}}{(2.303)(0.008314 \text{ kJ mol}^{-1} \text{ K}^{-1})(298 \text{ K})}$$

$$= -4.4165$$

$$K_{eq} = 10^{-4.4165} = 3.85 \times 10^{-5}$$

(d)   The equilibrium is very much more favorable in aqueous solution because solvation of the products (ethanol, hydronium ions, and chloride ions) takes place and thereby stabilizes them.

*6.42   The mechanism for the reaction involves the participation of the carboxylate group. In step 1 (see following reaction) an oxygen of the carboxylate group attacks the stereocenter from the backside and displaces bromide ion. (Silver ion aids in this process in much the same way that protonation assists the ionization of an alcohol.) The configuration of the stereocenter inverts in step 1, and a cyclic ester called an α-lactone forms.

The highly strained three-membered ring of the α-lactone opens when it is attacked by a water molecule in step 2. *This step also takes place with an inversion of configuration.*

The net result of two inversions (in steps 1 and 2) is an overall *retention of configuration.*

*6.43   (a) and (b)

(c)   The reaction takes place with retention of configuration.

(d)

*6.44

No change of configuration occurs, just a change in the relative priority of a group at the stereocenter.

*6.45 Comparison of the molecular formulas of starting material and product indicates a loss of HCl. The absence of IR bands in the 1620–1680 cm^{-1} region rules out the presence of the alkene function.

A nucleophilic substitution agrees with the evidence:

*6.46 The IR evidence indicates that C possesses both an alkene function and a hydroxyl group. An E2 reaction on this substrate produces enantiomeric unsaturated alcohols.

QUIZ

6.1 Which set of conditions would you use to obtain the best yield in the reaction shown?

(a) H_2O, heat   (b) CH_3CH_2ONa/CH_3CH_2OH, heat   (c) Heat alone

(d) H_2SO_4 (e) None of the above

6.2 Which of the following reactions would give the best yield?

6.3 A kinetic study yielded the following reaction rate data:

Experiment Number	Initial Concentrations		Initial Rate of Disappearance of R—Br and Formation of R—OH
	$[OH^-]$	$[R-Br]$	
1	0.50	0.50	1.00
2	0.50	0.25	0.50
3	0.25	0.25	0.25

Which of the following statements best describe this reaction?

(a) The reaction is second order.   (b) The reaction is first order.

(c) The reaction is S_N1.   (d) Increasing the concentration of OH^- has no effect on the rate.

(e) More than one of the above.

6.4 There are four compounds with the formula C_4H_9Br. List them in order of decreasing reactivity in an S_N2 reaction.

6.5 Supply the missing reactants, reagents, intermediates, or products.

6.6 Which S$_N$2 reaction will occur most rapidly. (Assume the concentrations and temperatures are all the same.)

(a) $CH_3O^- + CH_3CH_2-F \longrightarrow CH_3CH_2OCH_3 + F^-$

(b) $CH_3O^- + CH_3CH_2-I \longrightarrow CH_3CH_2OCH_3 + I^-$

(c) $CH_3O^- + CH_3CH_2-Cl \longrightarrow CH_3CH_2OCH_3 + Cl^-$

(d) $CH_3O^- + CH_3CH_2-Br \longrightarrow CH_3CH_2OCH_3 + Br^-$

6.7 Provide three-dimensional structures for the missing boxed structures and formulas for missing reagents.

7

ALKENES AND ALKYNES I: PROPERTIES AND SYNTHESIS

SOLUTIONS TO PROBLEMS

7.1   (a)   (*E*)-1-Bromo-1-chloro-1-pentene

      (b)   (*E*)-2-Bromo-1-chloro-1-iodo-1-butene

      (c)   (*Z*)-3,5-Dimethyl-2-hexene

      (d)   (*Z*)-1-Chloro-1-iodo-2-methyl-1-butene

      (e)   (*Z*,4*S*)-3,4-Dimethyl-2-hexene

      (f)   (*Z*,3*S*)-1-Bromo-2-chloro-3-methyl-1-hexene

7.2   (a), (b)

$$
\underset{\substack{\text{2-Methyl-1-butene}\\\text{(disubstituted)}}}{CH_2{=}\overset{\overset{\textstyle CH_3}{|}}{C}CH_2CH_3} \xrightarrow[\text{Pt}]{H_2} \underset{}{CH_3\overset{\overset{\textstyle CH_3}{|}}{C}HCH_2CH_3} \qquad \Delta H° = -119 \text{ kJ mol}^{-1}
$$

$$
\underset{\substack{\text{3-Methyl-1-butene}\\\text{(monosubstituted)}}}{CH_3\overset{\overset{\textstyle CH_3}{|}}{C}HCH{=}CH_2} \xrightarrow[\text{Pt}]{H_2} \underset{}{CH_3\overset{\overset{\textstyle CH_3}{|}}{C}HCH_2CH_3} \qquad \Delta H° = -127 \text{ kJ mol}^{-1}
$$

$$
\underset{\substack{\text{2-Methyl-2-butene}\\\text{(trisubstituted)}}}{CH_3\overset{\overset{\textstyle CH_3}{|}}{C}{=}CHCH_3} \xrightarrow[\text{Pt}]{H_2} \underset{}{CH_3\overset{\overset{\textstyle CH_3}{|}}{C}HCH_2CH_3} \qquad \Delta H° = -113 \text{ kJ mol}^{-1}
$$

      (c)   Yes, because hydrogenation converts each alkene into the same product.

      (d) $\underset{\text{(trisubstituted)}}{CH_3\overset{\overset{\textstyle CH_3}{|}}{C}{=}CHCH_3}$ > $\underset{\text{(disubstituted)}}{CH_2{=}\overset{\overset{\textstyle CH_3}{|}}{C}CH_2CH_3}$ > $\underset{\text{(monosubstituted)}}{CH_3\overset{\overset{\textstyle CH_3}{|}}{C}HCH{=}CH_2}$

Notice that this predicted order of stability is confirmed by the heats of hydrogenation. 2-Methyl-2-butene evolves the least heat; therefore, it is the most stable. 3-Methyl-1-butene evolves the most heat; therefore, it is the least stable.

(e) $CH_2=CHCH_2CH_2CH_3$

1-Pentene

H_3C, $C=C$,CH_2CH_3
H H

cis-2-Pentene

H_3C, $C=C$,H
H CH_2CH_3

trans-2-Pentene

(f)   Heats of combustion, because complete combustion would convert all of the alkenes to the same products. (All of the alkenes have the formula C_5H_{10}.)

$$C_5H_{10} + 7\ 1/2\ O_2 \longrightarrow 5\ CO_2 + 5\ H_2O$$

7.3   (a)   2,3-Dimethyl-2-butene would be the more stable because the double bond is tetrasubstituted. 2-Methyl-2-pentene has a trisubstituted double bond.

(b)   trans-3-Hexene would be the more stable because alkenes with trans double bonds are more stable than those with cis double bonds.

(c)   cis-3-Hexene would be more stable because its double bond is disubstituted. The double bond of 1-hexene is monosubstituted.

(d)   2-Methyl-2-pentene would be the more stable because its double bond is trisubstituted. The double bond of trans-2-hexene is disubstituted.

7.4   The relative stabilities of the pairs of alkenes in parts (b) and (c) in Problem 7.3 could be determined by measuring heats of hydrogenation, because in each instance the two alkenes would yield the same product. Heats of combustion could be used to determine the relative stabilities of the alkene pairs in parts (a) and (d) [and also those in parts (b) and (c)] because on complete combustion the alkenes produce the same number of molar equivalents of CO_2 and H_2O.

7.5   (a)

(trisubstituted,   (monosubstituted,
more stable)   less stable)
Major product   Minor product

(b)

(tetrasubstituted,   (disubstituted,
more stable)   less stable)
Major product   Minor product

7.6   An anti periplanar transition state allows the molecule to assume the more stable staggered conformation, whereas a syn periplanar transition state requires the molecule to assume the less stable eclipsed conformation:

7.7   *cis*-1-Bromo-4-*tert*-butylcyclohexane can assume an anti periplanar transition state in which the bulky *tert*-butyl group is equatorial:

The conformation (above), because it is relatively stable, is assumed by most of the molecules present, and, therefore, the reaction is rapid.

   On the other hand, for *trans*-1-bromo-4-*tert*-butylcyclohexane to assume an anti periplanar transition state, the molecule must assume a conformation in which the large *tert*-butyl group is axial:

Such a conformation is of high energy; therefore, very few molecules assume this conformation. The reaction, consequently, is very slow.

7.8   (a)   Anti periplanar elimination can occur in two ways with the cis isomer.

cis-1-Bromo-2-methylcyclohexane

(b) Anti periplanar elimination can occur in only one way with the trans isomer.

trans-1-Bromo-2-methylcyclohexane

7.9   (a) (1) $CH_3—CH—\overset{..}{\overset{..}{O}}H$ + $H—\overset{..}{\overset{..}{O}}—\overset{\overset{\overset{..}{O}..}{\parallel}}{\underset{\underset{..\overset{..}{O}..}{}}{S}}—\overset{..}{\overset{..}{O}}—H$ $\rightleftharpoons$
$\qquad\qquad\qquad\quad |$
$\qquad\qquad\qquad CH_3$

$CH_3—CH—\overset{H}{\overset{|}{\overset{+}{O}}}—H$ + $^-:\overset{..}{O}—\overset{\overset{\overset{..}{O}..}{\parallel}}{\underset{\underset{..\overset{..}{O}..}{}}{S}}—\overset{..}{O}—H$
$\qquad\quad |$
$\qquad\quad CH_3$

(2) $CH_3—CH—\overset{H}{\overset{|}{\overset{+}{O}}}—H$ $\rightleftharpoons$ $CH_3—CH^+$ + H_2O
$\qquad\qquad\quad |$
$\qquad\qquad\quad CH_3$ $\qquad\qquad\qquad |$
$\qquad\qquad\qquad\qquad\qquad\qquad CH_3$

(3) $CH_3—\overset{+}{C}H—CH_2—H$ + $^-:\overset{..}{O}SO_3H$ $\rightleftharpoons$

$\qquad\qquad\qquad\qquad\qquad\qquad CH_3—CH{=}CH_2$ + $HOSO_3H$

(b) By donating a proton to the $-OH$ group of the alcohol in step (1), the acid allows the loss of a relatively stable, weakly basic, leaving group (H_2O) in step (2). In the absence of an acid, the leaving group would have to be the strongly basic OH^- ion, and such steps almost never occur.

7.10

(1) $CH_3\overset{CH_3}{\overset{|}{C}}CH_2OH$ + HA $\rightleftharpoons$ $CH_3\overset{CH_3}{\overset{|}{C}}CH_2\overset{+}{O}H_2$ + $:A^-$
$\qquad\quad |$
$\qquad CH_3$ $\qquad\qquad\qquad\qquad\qquad |$
$\qquad\qquad\qquad\qquad\qquad\qquad CH_3$

(2) $CH_3\overset{CH_3}{\overset{|}{C}}CH_2\overset{+}{O}H_2$ $\rightleftharpoons$ $CH_3\overset{CH_3}{\overset{|}{C}}\overset{+}{C}H_2$ + H_2O
$\qquad\quad |$
$\qquad CH_3$ $\qquad\qquad\qquad\quad |$
$\qquad\qquad\qquad\qquad\qquad CH_3$
$\qquad\qquad\qquad\qquad\qquad$ **1° Carbocation**

(3) $CH_3\overset{CH_3}{\overset{|}{\overset{+}{C}}}CH_2$ $\longrightarrow$ $\left[CH_3\overset{CH_3}{\overset{|}{C}}—CH_2\right]^{‡}$ $\longrightarrow$ $CH_3\overset{CH_3}{\overset{|}{C}}—CH_2—CH_3$
$\qquad\quad |$
$\qquad CH_3$ $\qquad\qquad\qquad\overset{\cdot}{\underset{+}{\overset{+}{C}H_3}}$ $\qquad\qquad\qquad\quad +$

1° Carbocation   **Transition state**   **3° Carbocation**

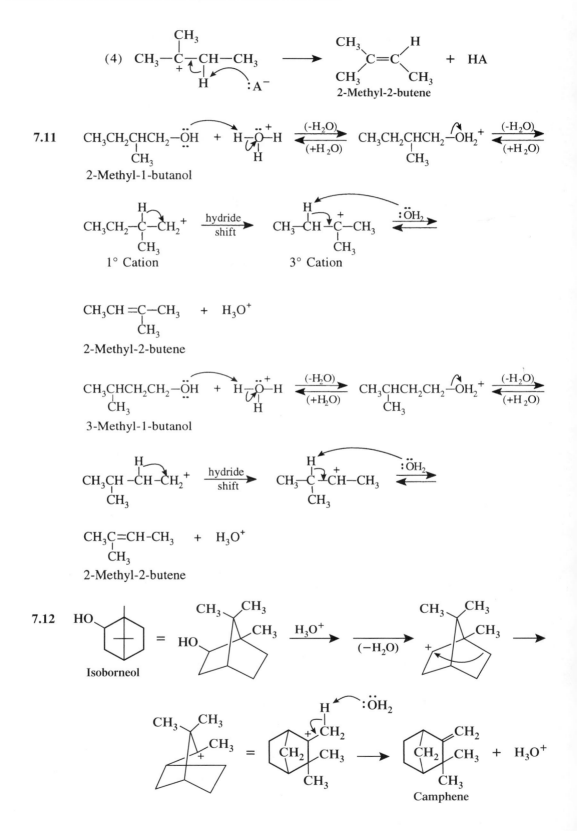

(4) 2-Methyl-2-butene + HA

7.11
2-Methyl-1-butanol

1° Cation   3° Cation

2-Methyl-2-butene

3-Methyl-1-butanol

2-Methyl-2-butene

7.12 Isoborneol

Camphene

7.13 (a) $CH_3\overset{O}{\overset{\|}{C}}CH_3 \xrightarrow[0°C]{PCl_5} CH_3CCl_2CH_3 \xrightarrow[\substack{\text{mineral oil, heat} \\ (2)\ NH_4^+}]{(1)\ 3\ NaNH_2} CH_3C\equiv CH$

(b) $CH_3CH_2CHBr_2 \xrightarrow[\substack{\text{mineral oil, heat} \\ (2)\ NH_4^+}]{(1)\ 3\ NaNH_2} CH_3C\equiv CH$

(c) $CH_3CHBrCH_2Br \xrightarrow{\text{[same as (b)]}} CH_3C\equiv CH$

(d) $CH_3CH{=}CH_2 \xrightarrow[CCl_4]{Br_2} CH_3\underset{\underset{Br}{|}}{C}HCH_2Br \xrightarrow{\text{[same as (b)]}} CH_3C\equiv CH$

7.14 (a) $CH_3CH{=}CH_2\ +\ NaNH_2 \longrightarrow$ No reaction

(b) $\underset{\substack{\text{Stronger} \\ \text{acid}}}{CH_3C\equiv C{-}H}\ +\ \underset{\substack{\text{Stronger} \\ \text{base}}}{Na^+\ :\!\ddot{N}H_2} \longrightarrow \underset{\substack{\text{Weaker} \\ \text{base}}}{CH_3C\equiv C\!:^-\ Na^+}\ +\ \underset{\substack{\text{Weaker} \\ \text{acid}}}{:NH_3}$

(c) $CH_3CH_2CH_3\ +\ NaNH_2 \longrightarrow$ No reaction

(d) $\underset{\substack{\text{Stronger} \\ \text{base}}}{CH_3C\equiv C\!:^-}\ +\ \underset{\substack{\text{Stronger} \\ \text{acid}}}{H{-}\ddot{O}CH_2CH_3} \longrightarrow \underset{\substack{\text{Weaker} \\ \text{acid}}}{CH_3C\equiv CH}\ +\ \underset{\substack{\text{Weaker} \\ \text{base}}}{:\ddot{O}CH_2CH_3}$

(e) $\underset{\substack{\text{Stronger} \\ \text{base}}}{CH_3C\equiv C\!:^-}\ +\ \underset{\text{Stronger acid}}{H{-}\overset{+}{N}H_3} \longrightarrow \underset{\substack{\text{Weaker} \\ \text{acid}}}{CH_3C\equiv CH}\ +\ \underset{\substack{\text{Weaker} \\ \text{base}}}{:NH_3}$

7.15 $CH_3\underset{\underset{CH_3}{|}}{\overset{\overset{CH_3}{|}}{C}}{-}C\equiv C{-}H\ +\ Na^+\ :\ddot{N}H_2 \xrightarrow{(-NH_3)} CH_3\underset{\underset{CH_3}{|}}{\overset{\overset{CH_3}{|}}{C}}{-}C\equiv C\ :^-\ Na^+$

$\downarrow CH_3{-}I$

$CH_3\underset{\underset{CH_3}{|}}{\overset{\overset{CH_3}{|}}{C}}{-}C\equiv C{-}CH_3$

7.16 (a) C_6H_{14} = formula of alkane
$\underline{C_6H_{12}}$ = formula of 2-hexene
H_2 = difference = 1 pair of hydrogen atoms

Index of hydrogen deficiency = 1

(b) C_6H_{14} = formula of alkane
$\underline{C_6H_{12}}$ = formula of methylcyclopentane
H_2 = difference = 1 pair of hydrogen atoms

Index of hydrogen deficiency = 1

(c) No, all isomers of C_6H_{12}, for example, have the same index of hydrogen deficiency.

(d) No

(e) C_6H_{14} = formula of alkane
$\underline{C_6H_{10}}$ = formula of 2-hexyne
    H_4 = difference = 2 pairs of hydrogen atoms

Index of hydrogen deficiency = 2

(f) $C_{10}H_{22}$ (alkane)
$\underline{C_{10}H_{16}}$ (compound)
    H_6 = difference = 3 pairs of hydrogen atoms

Index of hydrogen deficiency = 3

The structural possibilities are thus

> 3 double bonds
> 1 double bond and one triple bond
> 2 double bonds and 1 ring
> 1 double bond and 2 rings
> 3 rings
> 1 triple bond and one ring

7.17 (a) $C_{15}H_{32}$ = formula of alkane
$\underline{C_{15}H_{24}}$ = formula of zingiberene
    H_8 = difference = 4 pairs of hydrogen atoms

Index of hydrogen deficiency = 4

(b) Since 1 mol of zingiberene absorbs 3 mol of hydrogen, one molecule of zingiberene must contain three double bonds. (We are told that molecules of zingiberene do not contain any triple bonds.)

(c) If a molecule of zingiberene has three double bonds and an index of hydrogen deficiency equal to 4, it must have one ring. (The structural formula for zingiberene can be found in Problem 23.2.)

7.18 (a) We designate the position of the double bond by using the *lower* number of the two numbers of the doubly bonded carbon atoms, and the chain is numbered from the end nearer the double bond. The correct name is *trans*-2-pentene

(b) We must choose the longest chain for the base name. The correct name is 2-methyl-propene.

(c) We use the lower number of the two doubly bonded carbon atoms to designate the position of the double bond. The correct name is 1-methylcyclohexene.

(d) We must number the ring starting with the double bond in the direction that gives the substituent the lower number. The correct name is 3-methylcyclobutene.

(e) We number in the way that gives the double bond *and the substituent* the lower number. The correct name is (Z)-2-chloro-2-butene.

(f) We number the ring starting with the double bond so as to give the substituents the lower numbers. The correct name is 3,4-dichlorocyclohexene.

7.20   (a)

(Z,4R)-4-Bromo-2-hexene

(Z,4S)-4-Bromo-2-hexene

(E,4R)-4-Bromo-2-hexene

(E,4S)-4-Bromo-2-hexene

(b)

(3R,4Z)-3-Chloro-1,4-hexadiene

(3S,4Z)-3-Chloro-1,4-hexadiene

(3R,4E)-3-Chloro-1,4-hexadiene

(3S,4E)-3-Chloro-1,4-hexadiene

(c)

(2E,4R)-2,4-Dichloro-2-pentene

(2Z,4R)-2,4-Dichloro-2-pentene

(2E,4S)-2,4-Dichloro-2-pentene

(2Z,4S)-2,4-Dichloro-2-pentene

(d)

(3R,4Z)-5-Bromo-3-chloro-4-
hexen-1-yne

(3S,4Z)-5-Bromo-3-chloro-4-
hexen-1-yne

(3R,4E)-5-Bromo-3-chloro-4-
hexen-1-yne

(3S,4E)-5-Bromo-3-chloro-4-
hexen-1-yne

7.21   (a)   (*E*)-3,5-Dimethyl-2-hexene

(b)   4-Chloro-3-methylcyclopentene

(c)   6-Methyl-3-heptyne

(d)   1-*sec*-Butyl-2-methylcyclohexene or 1-methyl-2-(1-methylpropyl)cyclohexene

(e)   (*Z*,3*R*)-3-Chloro-4-hepten-1-yne

(f)   2-Pentyl-1-heptene

7.22   (a) $CH_3CH_2CH_2Cl$ $\xrightarrow[\text{(CH}_3)_3\text{COH}]{\text{(CH}_3)_3\text{COK}}$ $CH_3CH=CH_2$

(b) $CH_3\underset{\underset{Cl}{|}}{C}HCH_3$ $\xrightarrow[\text{CH}_3\text{CH}_2\text{OH}]{\text{CH}_3\text{CH}_2\text{ONa}}$ $CH_3CH=CH_2$

(c) $CH_3CH_2CH_2OH$ $\xrightarrow{\text{H}^+,\text{ heat}}$ $CH_3CH=CH_2$

(d) $CH_3\underset{\underset{OH}{|}}{C}HCH_3$ $\xrightarrow{\text{H}^+,\text{ heat}}$ $CH_3CH=CH_2$

(e) $CH_3\underset{\underset{Br}{|}}{C}HCH_2Br$ $\xrightarrow[\text{or NaI/acetone}]{\text{Zn, CH}_3\text{CO}_2\text{H}}$ $CH_3CH=CH_2$

(f) $CH_3C\equiv CH$ $\xrightarrow[\text{Ni}_2\text{B (P-2)}]{\text{H}_2}$ $CH_3CH=CH_2$

(e) $CH_3-C\equiv C-CH_3$ [from(c)] $\xrightarrow{\text{(1) Li, } CH_3CH_2NH_2}{\text{(2) } NH_4Cl}$

$$\underset{H}{\overset{CH_3}{>}}C=C\underset{CH_3}{\overset{H}{<}}$$

(f) $HC\equiv C:^- Na^+$ [from(a)] $\xrightarrow[(-NaBr)]{CH_3CH_2CH_2-Br}$ $HC\equiv C-CH_2CH_2CH_3$

(g) $CH_3CH_2CH_2C\equiv CH$ [from(f)] $\xrightarrow[\text{liq. } NH_3]{NaNH_2}$ $CH_3CH_2CH_2C\equiv C:^- Na^+$ $\xrightarrow[(-NaI)]{CH_3-I}$

$$CH_3CH_2CH_2C\equiv CCH_3$$

(h) $CH_3CH_2CH_2C\equiv CCH_3$ [from(g)] $\xrightarrow[Ni_2B \ (P-2)]{H_2}$

$$\underset{H}{\overset{CH_3CH_2CH_2}{>}}C=C\underset{H}{\overset{CH_3}{<}}$$

(i) $CH_3CH_2CH_2C\equiv CCH_3$ [from(g)] $\xrightarrow{\text{(1) Li, } CH_3CH_2NH_2}{\text{(2) } NH_4Cl}$

$$\underset{H}{\overset{CH_3CH_2CH_2}{>}}C=C\underset{CH_3}{\overset{H}{<}}$$

(j) $HC\equiv C:^- Na^+$ [from(a)] $\xrightarrow[(-NaBr)]{CH_3CH_2-Br}$ $HC\equiv CCH_2CH_3$ $\xrightarrow[\text{liq. } NH_3]{NaNH_2}$ $CH_3CH_2C\equiv C:^- Na^+$

$$\xrightarrow{CH_3CH_2-Br} \quad CH_3CH_2C\equiv CCH_2CH_3$$

(k) $CH_3CH_2C\equiv C:^- Na^+$ [from(j)] $\xrightarrow{D_2O}$ $CH_3CH_2C\equiv CD$

(l) $CH_3C\equiv CCH_3$ [from(c)] $\xrightarrow[Ni_2B \ (P-2)]{D_2}$

$$\underset{D}{\overset{CH_3}{>}}C=C\underset{D}{\overset{CH_3}{<}}$$

7.25 We notice that the deuterium atoms are cis to each other, and we conclude, therefore, that we need to choose a method that will cause a syn addition of deuterium. One way would be to use D_2 and a metal catalyst (Section 7.13)

7.26 Dehydration of *trans*-2-methylcyclohexanol proceeds through the formation of a carbocation (through an E1 reaction of the protonated alcohol) and leads preferentially to the more stable alkene. 1-Methylcyclohexene (below) is more stable than 3-methylcyclohexene (the minor product of the dehydration) because its double bond is more highly substituted.

(major)   (minor)
Trisubstituted   Disubstituted
double bond   double bond

Dehydrohalogenation of *trans*-1-bromo-2-methylcyclohexane is an E2 reaction and must proceed through an anti periplanar transition state. Such a transition state is possible only for the elimination leading to 3-methylcyclohexene (cf. Problem 7.8).

3-Methylcyclohexene

7.27 (a) $C_6H_5-\underset{\underset{Br}{|}}{\overset{\overset{Br}{|}}{C}}-CH_3$ $\xrightarrow[\text{mineral oil, heat}]{\text{3 NaNH}_2}$ $C_6H_5-C\equiv C\overset{..}{:}Na^+$ $\xrightarrow{\text{NH}_4\text{Cl}}$ $C_6H_5-C\equiv CH$

(b) $C_6H_5CH_2-CHBr_2$ $\xrightarrow[\text{mineral oil, heat}]{\text{3 NaNH}_2}$ $C_6H_5C\equiv C\overset{..}{:}Na^+$ $\xrightarrow{\text{NH}_4\text{Cl}}$ $C_6H_5-C\equiv CH$

(c) $C_6H_5CH=CH_2$ $\xrightarrow[\text{CCl}_4]{\text{Br}_2}$ $C_6H_5-\underset{\underset{Br}{|}}{CH}-\underset{\underset{Br}{|}}{CH_2}$ $\xrightarrow[\text{mineral oil, heat}]{\text{3 NaNH}_2}$ $C_6H_5C\equiv C\overset{..}{:}Na^+$

$\xrightarrow{\text{NH}_4\text{Cl}}$ $C_6H_5-C\equiv CH$

(d) $C_6H_5-\overset{\overset{O}{\|}}{C}-CH_3$ $\xrightarrow{\text{PCl}_5}$ $C_6H_5-\underset{\underset{Cl}{|}}{\overset{\overset{Cl}{|}}{C}}-CH_3$ $\xrightarrow[\text{mineral oil, heat}]{\text{3 NaNH}_2}$ $C_6H_5C\equiv C\overset{..}{:}Na^+$

$\xrightarrow{\text{NH}_4\text{Cl}}$ $C_6H_5-C\equiv CH$

7.28 Cyclobutane is less stable than any of the butene isomers.

7.29 1-Pentene, 3375 kJ mol^{-1}

cis-2-Pentene, 3369 kJ mol^{-1}

trans-2-Pentene, 3365 kJ mol^{-1}

2-Methyl-1-butene, 3361 kJ mol^{-1}

2-Methyl-2-butene, 3355 kJ mol^{-1}

7.30   1-Pentanol > 1-pentyne > 1-pentene > pentane
(See Section 3.7 for the explanation.)

7.33 (a) $CH_3CH_2CH_2CH_2CH_2Br$ $\xrightarrow[\text{(CH}_3)_3\text{COH}]{\text{(CH}_3)_3\text{COK}}$ $CH_3CH_2CH_2CH=CH_2$

(b) $CH_3CHCH_2CH_2Br$ $\xrightarrow[\text{(CH}_3)_3\text{COH}]{\text{(CH}_3)_3\text{COK}}$ $CH_3CHCH=CH_2$
      $|$ $|$
      CH_3 CH_3

(c) $CH_3\overset{\overset{\displaystyle CH_3}{|}}{CH}CHCH_2Br$ $\xrightarrow[\text{(CH}_3)_3\text{COH}]{\text{(CH}_3)_3\text{COK}}$ $CH_3\overset{\overset{\displaystyle CH_3}{|}}{CH}C=CH_2$
            $|$ $|$
           CH_3 CH_3

(d) H_3C—⬡—Br $\xrightarrow[\text{(CH}_3)_3\text{COH}]{\text{(CH}_3)_3\text{COK}}$ H_3C—⬡

(e) $\underset{H_3C}{\overset{Br}{\diagdown}}$⬠ $\xrightarrow[\text{CH}_3\text{CH}_2\text{OH}]{\text{CH}_3\text{CH}_2\text{ONa}}$ H_3C—⬠

7.34

$\underset{\underset{3°}{\overset{\displaystyle |}{OH}}}{CH_3\overset{\overset{\displaystyle CH_3}{|}}{C}CH_2CH_3}$ > $\underset{\underset{2°}{\overset{\displaystyle |}{OH}}}{CH_3\overset{\overset{\displaystyle CH_3}{|}}{CH}CHCH_3}$ > $\underset{1°}{CH_3CH_2CH_2CH_2CH_2OH}$

7.35 (a) $\underset{\overset{\displaystyle |}{OH}}{CH_3\overset{\overset{\displaystyle CH_3}{|}}{C}CH_3}$ $\xrightarrow[\text{heat,}\ (-\text{H}_2\text{O})]{\text{H}^+}$ $\overset{\overset{\displaystyle CH_3}{|}}{CH_2=CCH_3}$

(b) $\underset{\overset{\displaystyle |}{OH}}{CH_3\overset{\overset{\displaystyle CH_3}{|}}{CH}CHCH_3}$ $\xrightarrow[\text{heat,}\ (-\text{H}_2\text{O})]{\text{H}^+}$ $\underset{\text{(minor)}}{\overset{\overset{\displaystyle CH_3}{|}}{CH_2=CHCCH_3}}$ + $\underset{\text{(major)}}{\overset{\overset{\displaystyle CH_3}{|}}{CH_3CH=CCH_3}}$

(c) $\underset{\overset{\displaystyle |}{OH}}{CH_3CH_2\overset{\overset{\displaystyle CH_3}{|}}{C}CH_2CH_3}$ $\xrightarrow[\text{heat,}\ (-\text{H}_2\text{O})]{\text{H}^+}$ $\underset{\text{(major)}}{\overset{\overset{\displaystyle CH_3}{|}}{CH_3CH=CCH_2CH_3}}$ + $\underset{\text{(minor)}}{\overset{\overset{\displaystyle CH_2}{||}}{CH_3CH_2CCH_2CH_3}}$

(d) $\underset{\overset{\displaystyle |}{CH_3}}{CH_3\overset{\overset{\displaystyle CH_3}{|}}{C}CH_2OH}$ $\xrightarrow[\substack{\text{heat,}\ (-\text{H}_2\text{O})\\ \text{rearrangement}}]{\text{H}^+}$ $\underset{\text{(major)}}{\underset{\overset{\displaystyle |}{CH_3}}{CH_3C=CHCH_3}}$ + $\underset{\text{(minor)}}{\underset{\overset{\displaystyle |}{CH_3}}{CH_2=CCH_2CH_3}}$

(e) $\underset{H_3C}{\overset{HO}{\diagdown}}$⬡—$CH_3$ $\xrightarrow[\text{heat,}\ (-\text{H}_2\text{O})]{\text{H}^+}$ H_3C—⬡—CH_3 $\underset{\text{(major)}}{}$ + $H_2C=$⬡—CH_3 $\underset{\text{(minor)}}{}$

7.36 The alkene cannot be formed because the double bond in the product is too highly strained. Recall that the atoms at each carbon of a double bond prefer to be in the same plane.

7.37 Only the deuterium atom can assume the anti periplanar orientation necessary for an E2 reaction to occur.

7.38 (a) A hydride shift occurs.

$$CH_3CH_2CH_2CH_2OH \xrightarrow{H^+} CH_3CH_2CH_2CH_2\overset{+}{\underset{OH_2}{\frown}} \xrightarrow{(-H_2O)}$$

$$CH_3CH_2\underset{\underset{H}{|}}{CH}-\overset{+}{CH_2} \xrightarrow[\text{shift}]{\text{hydride}} CH_3\underset{\underset{H}{|}}{CH}-\overset{+}{CH}CH_3 \longrightarrow \underset{H}{\overset{CH_3}{\diagdown}}C=C\underset{CH_3}{\overset{H}{\diagup}} + H_3O^+$$

(1° cation) $:\overset{..}{O}H_2$ (major product)

(b) A methanide shift occurs.

(c) A methanide shift occurs.

(d) The required anti periplanar transtition state leads only to (E) alkene:

7.39   In the first step, cholesterol reacts with bromine to form the *vic*-dibromide. This product is then purified by crystallization, and then treatment with zinc in ethanol converts the pure *vic*-dibromide back to cholesterol. (Recrystallization of the *vic*-dibromide is especially easy because it has a higher melting point than cholesterol.)

Crude cholesterol   Careful recrystallization to purify

Pure cholesterol

7.40   (a)   Caryophyllene has the same molecular formula as zingiberene (Problem 7.17); thus it, too, has an index of hydrogen deficiency equal to 4. That 1 mol of caryophyllene absorbs 2 mol of hydrogen on catalytic hydrogenation indicates the presence of two double bonds per molecule.

(b)   Caryophyllene molecules must also have two rings. (See Problem 23.2 for the structure of caryophyllene.)

7.41   (a) $C_{30}H_{62}$ = formula of alkane
$\underline{C_{30}H_{50}}$ = formula of squalene
H_{12} = difference = 6 pairs of hydrogen atoms

Index of hydrogen deficiency = 6

(b)   Molecules of squalene contain six double bonds.

(c)   Squalene molecules contain no rings. (See Problem 23.2 for the structural formula of squalene.)

7.42   (a)   We are given (Section 7.3A) the following heats of hydrogenation:

$$cis\text{-2-Butene} + H_2 \xrightarrow{\text{Pt}} \text{butane} \qquad \Delta H° = -120 \text{ kJ mol}^{-1}$$

$$trans\text{-2-Butene} + H_2 \xrightarrow{\text{Pt}} \text{butane} \qquad \Delta H° = -115 \text{ kJ mol}^{-1}$$

Thus, for

$$cis\text{-2-Butene} \longrightarrow trans\text{-2-butene} \qquad \Delta H° = -5.0 \text{ kJ mol}^{-1}$$

(b)   Converting *cis*-2-butene into *trans*-2-butene involves breaking the π bond. Therefore, we would expect the energy of activation to be at least as large as the π-bond strength, that is, at least 264 kJ mol^{-1}.

(c)

7.43 (a)

E
Optically active (the
enantiomeric form is an
equally valid answer)

F
Optically inactive and
nonresolvable

(b)

CH_3CH_2 C=C=C CH_3 / H $\xrightarrow[Pt]{H_2}$ $CH_3CH_2CH_2CH_2CH_2CH_3$

G
Optically active (the
enantiomeric form is an
equally valid answer)

H
Optically inactive and
nonresolvable

7.44 That **I** and **J** rotate plane-polarized light in the same direction tells us that **I** and **J** are not
enantiomers of each other. Thus, the following are possible structures for **I, J,** and **K.** (The
enantiomers of **I, J,** and **K** would form another set of structures, and other answers are
possible as well.)

I
Optically active

J
Optically active

K
Optically
active

7.45   The following are possible structures:

L

M

(other answers are possible as well)

7.46   (a)   With either the (1*R*,2*R*)- or the (1*S*,2*S*)-1,2-dibromo-1,2-diphenylethane, only one conformation will allow an anti periplanar arrangement of the H- and Br-. In either case, the elimination leads only to (*Z*)-1-bromo-1,2-diphenylethene:

(1*R*,2*R*)-1,2-Dibromo-1,2-diphenylethane
(anti periplanar orientation of H- and -Br)

(*Z*)-1-Bromo-1,2-diphenylethene

(1*S*,2*S*)-1,2-Dibromo-1,2-diphenylethane
(anti periplanar orientation of H- and -Br)

(*Z*)-1-Bromo-1,2-diphenylethene

(b)   With (1*R*,2*S*)-1,2-dibromo-1,2-diphenylethane, only one conformation will allow an anti periplanar arrangement of the H- and Br-. In this case, the elimination leads only to (*E*)-1-bromo-1,2-diphenylethene:

(1*R*,2*S*)-1,2-Dibromo-1,2-diphenylethane
(anti periplanar orientation of H- and -Br)

(*E*)-1-Bromo-1,2-diphenylethene

(c)   With (1*R*,2*S*)-1,2-dibromo-1,2-diphenylethane, only one conformation will allow an anti periplanar arrangement of both bromine atoms. In this case, the elimination leads only to (*E*)-1,2-diphenylethene:

(1*R*,2*S*)-1,2-Dibromo-1,2- diphenylethane   (*E*)-1,2- Diphenylethene
(anti periplanar orientation of both -Br atoms)

*7.47   (a)

(b)   No, tetrasubstituted double bonds usually show no C=C stretching absorption in their infrared spectra.

*7.48

*7.49   (a)   Three
        (b)   Six

QUIZ

7.1 Which conditions/reagents would you employ to obtain the best yields in the following reaction?

$$CH_3CH_2\underset{\underset{Br}{|}}{C}HCH_3 \xrightarrow{\;?\;} CH_3CH_2CH{=}CH_2$$

(a) H_2O/heat

(c) $(CH_3)_3COK/(CH_3)_3COH$, heat

(b) CH_3CH_2ONa/CH_3CH_2OH, heat

(d) Reaction cannot occur as shown

7.2 Which of the following names is incorrect?

(a) 1-Butene   (b) *trans*-2-Butene   (c) (Z)-2-Chloro-2-pentene

(d) 1,1-Dimethylcyclopentene   (e) Cyclohexene

7.3 Select the major product of the reaction

(a) $CH_3CH_2\underset{\underset{CH_3}{|}}{C}{=}C(CH_3)_2$

(b) $CH_3CH_2\underset{\underset{CH_2}{||}}{C}{-}CH(CH_3)_2$

(c) $CH_3CH{=}\underset{\underset{CH_3}{|}}{C}{-}CH(CH_3)_2$

(d) $CH_2{=}CH{-}\underset{\underset{CH_3}{|}}{C}H{-}CH(CH_3)_2$

(e) $CH_3CH_2\underset{\underset{OC_2H_5}{|}}{\overset{\overset{CH_3}{|}}{C}}{-}CH(CH_3)_2$

7.4 Supply the missing reagents.

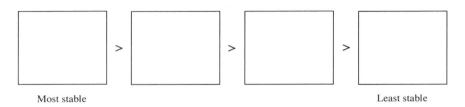

7.5 Arrange the following alkenes in order of decreasing stability.

1-Pentene, *cis*-2-pentene, *trans*-2-pentene, 2-methyl-2-butene

	>		>		>	

Most stable   Least stable

7.6 Complete the following synthesis.

8

ALKENES AND ALKYNES II: ADDITION REACTIONS

SOLUTIONS TO PROBLEMS

8.1 CH_3CHCH_2I
       |
       Br 2-Bromo-1-iodopropane

(b)

3-Chloro-2,2-dimethylbutane
(from unrearranged carbocation)

2° Carbocation

1,2-methanide
shift

2-Chloro-2,3-dimethylbutane
(from rearranged carbocation)

8.4 $CH_2=CH_2 + H_2SO_4 \longrightarrow CH_3CH_2OSO_3H \xrightarrow[\text{heat}]{H_2O} CH_3CH_2OH + H_2SO_4$

8.5   (a)   Use a high concentration of water because we want the carbocation produced to react with water. And use a strong acid whose conjugate base is a very weak nucleophile. (For this reason we would not use HI, HBr, or HCl.) An excellent method, therefore, is to use dilute sulfuric acid.

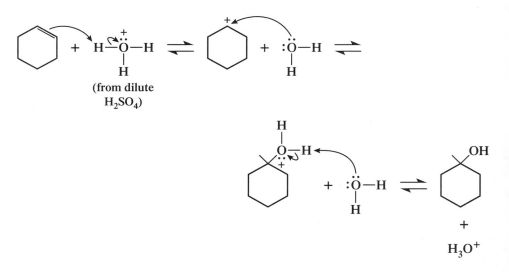

(b)   Use a low concentration of water (i.e., use concentrated H_2SO_4) and use a higher temperature to encourage elimination.

(c) 1-Methylcyclohexanol would be the product because a 3° carbocation would be formed as the intermediate.

8.6

8.7 The order reflects the relative ease with which these alkenes accept a proton and form a carbocation. $(CH_3)_2C=CH_2$ reacts faster because it leads to a tertiary cation,

$$(CH_3)_2C=CH_2 \xrightarrow{H_3O^+} CH_3-\underset{+}{\overset{\overset{\displaystyle CH_3}{|}}{C}}-CH_3 \quad \text{3° Carbocation}$$

$CH_3CH=CH_2$ leads to a secondary cation,

$$CH_3CH=CH_2 \xrightarrow{H_3O^+} CH_3-\underset{+}{CH}-CH_3 \quad \text{2° Carbocation}$$

and $CH_2=CH_2$ reacts most slowly because it leads to a primary carbocation.

$$CH_2=CH_2 \xrightarrow{H_3O^+} CH_3-CH_2^+ \quad \text{1° Carbocation}$$

Recall that formation of the carbocation is the rate-determining step in acid-catalyzed hydration and that the order of stabilities of carbocations is the following:

$$3° > 2° > 1° > {}^+CH_3$$

8.8

8.9

8.10

Bromonium
ion

Because paths (a) and (b) occur
at equal rates, these enantiomers
are formed at equal rates.

Racemic form

8.11

8.12 (a)   + enantiomer   (b)

(c)

8.13 $\xrightarrow[\text{KOC(CH}_3)_3]{\text{CHBr}_3}$

8.14 $\xrightarrow[\text{Zn(Cu)}]{\text{CH}_3\text{CHI}_2}$ +

8.15 (a) $\xrightarrow[\text{(2) NaHSO}_3]{\text{(1) OsO}_4\text{,pyridine, 25°C}}$

(b) $\xrightarrow[\text{(2) NaHSO}_3]{\text{(1) OsO}_4\text{,pyridine, 25°C}}$ + enantiomer

(c) $\xrightarrow[\text{(2) NaHSO}_3]{\text{(1) OsO}_4\text{,pyridine, 25°C}}$ + enantiomer

8.16 (a)   Syn-hydroxylation at either face of (Z)- or cis-alkene leads to the meso compound (2R,3S)-2,3-butanediol.

(b)   Syn-hydroxylation at one face of the (E)- or trans-alkene leads to the (2R,3R)-enantiomer; at the other face, which is equally likely, it leads to the (2S,3S)-enantiomer.

8.17 (a)

(b) $CH_3CH_2CH{=}CHCH_2CH_3$ $\xrightarrow[\text{(2) Zn, HOAc}]{\text{(1) O}_3}$ $CH_3CH_2CH{=}O$

(c)

8.18 Ordinary alkenes *are* more reactive toward electrophilic reagents. But the alkenes obtained from the addition of an electrophilic reagent to an alkyne have at least one electronegative atom (Cl, Br, etc.) attached to a carbon atom of the double bond.

These alkenes àre less reactive than alkynes toward electrophilic addition because the electronegative group makes the double bond "electron poor."

8.19   The molecular formula and the formation of octane from **A** and **B** indicate that both compounds are unbranched octynes. Since **A** yields only $CH_3CH_2CH_2CO_2H$ on ozonolysis, **A** must be the symmetrical octyne $CH_3CH_2CH_2C{\equiv}CCH_2CH_2CH_3$. The IR absorption for **B** shows the presence of a terminal triple bond. Hence **B** is $CH_3(CH_2)_5C{\equiv}CH$.

Since **C** (C_8H_{12}) gives $HO_2C(CH_2)_6CO_2H$ on ozonolysis, **C** must be cyclooctyne. This is supported by the molecular formula of **C** and the fact that it is converted to C_8H_{16} (cyclooctane) on catalytic hydrogenation.

Answer to the Study Problem related to cholesterol biosynthesis, page 355.

In cholesterol biosynthesis an anti-Markovnikov addition occurs between C10 (once it becomes like a tertiary carbocation) and C15, forming the six-membered "C" ring of cholesterol (see the reactions in the text, page 355). This process results in a developing *secondary* carbocation at C14, which adds to C18 to form the five-membered "D" ring of cholesterol. If, on the other hand, Markovnikov addition occurred between C10 and C14 instead of anti-Markovnikov addition between C10 and C15, as shown in the following scheme, a developing *tertiary* carbocation would result at C15, along with formation of a five-membered "C" ring. Then, Markovnikov addition of C15 (as it becomes like a tertiary carbocation) to C18 would lead to a four-membered "D" ring and a tertiary carbocation at C19.

8.20 By converting the 3-hexyne to *cis*-3-hexene using H$_2$/Ni$_2$B (P-2).

Then, addition of bromine to *cis*-3-hexene will yield (3R,4R), and (3S,4S)-3,4-dibromo-hexane as a racemic form.

8.21 (a) CH$_3$CH$_2$CHCH$_3$
           |
           I

(b) CH$_3$CH$_2$CH$_2$CH$_3$

(c) CH$_3$CH$_2$CHCH$_3$
           |
           OH

(d) CH$_3$CH$_2$CHCH$_3$
           |
           OSO$_3$H

(e) CH$_3$CH$_2$CHCH$_3$
           |
           OH

(f) CH$_3$CH$_2$CHCH$_3$
           |
           Br

(g) CH$_3$CH$_2$CHCH$_2$Br
           |
           Br

(h) CH$_3$CH$_2$CH=CH$_2$

(i) CH$_3$CH$_2$CHCH$_2$Br
           |
           OH

(j) CH$_3$CH$_2$CHCH$_3$
           |
           Cl

(k) CH$_3$CH$_2$CHCH$_2$OH
           |
           OH

(l) CH$_3$CH$_2$C̈H + HC̈H
        O   O

(m) CH$_3$CH$_2$CHCH$_2$OH
           |
           OH

(n) CH$_3$CH$_2$CO$_2$H + CO$_2$

8.22 (a) [cyclohexane with I]
(b) [cyclohexane]
(c) [cyclohexane with OH]

(d) [cyclohexane with OSO$_3$H]
(e) [cyclohexane with OH]
(f) [cyclohexane with Br]

(g) [cyclohexane with Br, Br] + enantiomer
(h) [cyclohexane with double substituent]
(i) [cyclohexane with Br, OH] + enantiomer

(j) [cyclohexane with Cl]
(k) [cyclohexane with OH, OH]
(l) HC̈(CH$_2$)$_4$C̈H

(m) [cyclohexane ring with OH and OH substituents]

(n) $HO\overset{O}{\overset{\|}{C}}(CH_2)_4\overset{O}{\overset{\|}{C}}OH$

8.23 (a) $CH_3CH_2\underset{\underset{Br}{|}}{C}=CHBr$ (b) $CH_3CH_2\underset{\underset{Br}{|}}{C}=CH_2$ (c) $CH_3CH_2CBr_2CH_3$

(d) $CH_3CH_2CH_2CH_3$ (e) $CH_3CH_2CH=CH_2$ (f) $CH_3CH_2C\equiv CCH_3$

(g) $CH_3CH_2C\equiv CH$ and $CH_2\!=\!\overset{\overset{\displaystyle CH_3}{|}}{C}CH_3$ [An E2 reaction would take place when $CH_3CH_2C\equiv CNa$ is treated with $(CH_3)_3CBr$.]

8.24 (a) $CH_3\underset{\underset{Br}{|}}{C}=CHCH_3$ (b) $CH_3CBr_2CH_2CH_3$ (c) $\underset{Br}{\overset{H_3C}{\diagdown}}C=C\underset{\diagup CH_3}{\overset{Br}{\diagup}}$

(d) $CH_3CBr_2CBr_2CH_3$ (e) $\overset{H_3C}{\diagdown}\underset{H}{\diagup}C=C\overset{CH_3}{\diagdown}\underset{H}{\diagup}$ (f) $\overset{H_3C}{\diagdown}\underset{H}{\diagup}C=C\overset{Cl}{\diagdown}\underset{CH_3}{\diagup}$

(g) $\overset{H_3C}{\diagdown}\underset{H}{\diagup}C=C\overset{H}{\diagdown}\underset{CH_3}{\diagup}$ (h) $CH_3CH_2CH_2CH_3$ (i) $CH_3CH_2CH_2CH_3$

(j) CH_3CO_2H (2 molar equivalents)   (k) CH_3CO_2H (2 molar equivalents)

(l) no reaction

8.25 (a) $CH_3CH_2CH=CH_2 \xrightarrow[CCl_4]{Br_2} CH_3CH_2\underset{\underset{Br}{|}}{CH}-\underset{\underset{Br}{|}}{CH_2} \xrightarrow[\substack{mineral\ oil,\\heat}]{3\,NaNH_2}$

$CH_3CH_2C\equiv CNa \xrightarrow{NH_4Cl} CH_3CH_2C\equiv CH$

(b) $CH_3CH_2CH_2CH_2Cl \xrightarrow[\text{t-BuOH, heat}]{\text{t-BuOK}} CH_3CH_2CH=CH_2$ Then as in (a)

(c) $CH_3CH_2CH=CHCl \xrightarrow[\substack{mineral\ oil,\\heat}]{2NaNH_2} CH_3CH_2C\equiv CNa \xrightarrow{NH_4Cl}$

$CH_3CH_2C\equiv CH$

(d) $CH_3CH_2CH_2\underset{\underset{Cl}{|}}{CH}-Cl \xrightarrow[\substack{mineral\ oil,\\heat}]{3NaNH_2} CH_3CH_2C\equiv CNa \xrightarrow{NH_4Cl}$

$CH_3CH_2C\equiv CH$

(e) $H-C\equiv C-H \xrightarrow[\text{liq.}\,NH_3]{NaNH_2} H-C\equiv CNa \xrightarrow{CH_3CH_2-Br} CH_3CH_2C\equiv CH$

8.26   (a) $\underset{\displaystyle CH_3\overset{\displaystyle \overset{CH_3}{|}}{C}=CH_2}{}$ $\xrightarrow{H_3O^+,\ H_2O}$ $\underset{\displaystyle CH_3\overset{\overset{CH_3}{|}}{\underset{\underset{OH}{|}}{C}}CH_3}{}$

(b) $CH_3\overset{\overset{CH_3}{|}}{C}=CH_2$ $\xrightarrow{HCl}$ $CH_3\overset{\overset{CH_3}{|}}{\underset{\underset{Cl}{|}}{C}}CH_3$

(c) $CH_3\overset{\overset{CH_3}{|}}{C}=CH_2$ $\xrightarrow[\text{(no peroxides)}]{HBr}$ $CH_3\overset{\overset{CH_3}{|}}{\underset{\underset{Br}{|}}{C}}CH_3$

(d) $CH_3\overset{\overset{CH_3}{|}}{C}=CH_2$ $\xrightarrow{HF}$ $CH_3\overset{\overset{CH_3}{|}}{\underset{\underset{F}{|}}{C}}CH_3$

(e) $CH_3\overset{\overset{CH_3}{|}}{C}=CH_2$ $\xrightarrow{Cl_2,\ H_2O}$ $CH_3\overset{\overset{CH_3}{|}}{\underset{\underset{OH}{|}}{C}}CH_2Cl$

8.27   (a) $C_{10}H_{22}$ (saturated alkane)

$\underline{C_{10}H_{16}}$ (formula of myrcene)

H_6 = 3 pairs of hydrogen atoms

Index of hydrogen deficiency (IHD) = 3

(b)   Myrcene contains no rings because complete hydrogenation gives $C_{10}H_{22}$, which corresponds to an alkane.

(c)   That myrcene absorbs three molar equivalents of H_2 on hydrogenation indicates that it contains three double bonds.

(d)   Three structures are possible; however, only one gives 2,6-dimethyloctane on complete hydrogenation. Myrcene is therefore

$$CH_3\overset{\overset{CH_3}{|}}{C}=CHCH_2CH_2\overset{\overset{CH_2}{||}}{C}CH=CH_2$$

(e) $O=CHCH_2CH_2\overset{\overset{O}{||}}{C}CH=O$

8.28

8.29 The rate-determining step in each reaction is the formation of a carbocation when the alkene accepts a proton from HI. When 2-methylpropene reacts, it forms a 3° carbocation (the most stable); therefore, it reacts fastest. When ethene reacts, it forms a 1° carbocation (the least stable); therefore, it reacts the slowest.

8.30 CH₃CHCH₂CH₂CH₂CHCH₂CH₂CH₂CHCH₂CH₃
             | | |
            CH₃   CH₃   CH₃

2,6,10-Trimethyldodecane

8.31

8.32

Limonene

8.33

8.34

2° Carbocation

3° Carbocation

8.35   (a)   The hydrogenation experiment discloses the carbon skeleton of the pheromone.

$C_{13}H_{24}O$ $\xrightarrow[\text{Pt}]{\text{2 H}_2}$

Codling moth
pheromone

3-Ethyl-7-methyl-1-decanol

The ozonolysis experiment allows us to locate the position of the double bonds.

(1) O₃
(2) Zn, HOAc

(b)

8.36   *Retrosynthetic   analysis*

$CH_3(CH_2)_{12}\overset{(CH_2)_7CH_3}{\underset{H}{C=C}}H$ $\Longrightarrow$ $CH_3(CH_2)_{12}C\equiv C(CH_2)_7CH_3$ **Syn addition of H₂**

$CH_3(CH_2)_{12} \!\!\vdash\!\! C\equiv C \!\!\vdash\!\! (CH_2)_7CH_3$ $\Longrightarrow$ $CH_3(CH_2)_{11}CH_2X$ $+$ $HC\equiv CH$ $+$

$X-CH_2(CH_2)_6CH_3$

Synthesis

*8.37   *Retrosynthetic   analysis*

$$CH_3CBr_2CH_2CH_2CH_2CH_2CH_3 \xrightarrow[\text{nikov}]{\text{Markov-}} CH_2{=}CBrCH_2CH_2CH_2CH_2CH_3 + HBr$$
                                           addition

$$\xrightarrow[\text{nikov}]{\text{Markov-}} HC{\equiv}CCH_2CH_2CH_2CH_2CH_3 \implies HC{\equiv}CH + BrCH_2CH_2CH_2CH_2CH_3$$
addition $+$ HBr

Synthesis

8.38   Syn hydrogenation of the triple bond is required. So use H_2 and $Ni_2B(P-2)$ or H_2 and Lindlar's catalyst.

8.39   (a)

CH₃
H — C — OH   and
     | enantiomer
H — C — OH   through syn
     | addition
  CH₂CH₃

(b)

CH₃
H — C — OH   and
     | enantiomer
HO — C — H   through syn
     | addition
  CH₂CH₃

(c)

CH₃
H — C — Br   and
     | enantiomer
H — C — Br   through anti
     | addition
  CH₂CH₃

(d)

CH₃
H — C — Br   and
     | enantiomer
Br — C — H   through anti
     | addition
  CH₂CH₃

8.40 (a) (2S,3R)- [the enantiomer is (2R,3S)-]

(b) (2S,3S)- [the enantiomer is (2R,3R)-]

(c) (2S,3R)- [the enantiomer is (2R,3S)-]

(d) (2S,3S)- [the enantiomer is (2R,3R)-]

8.41

Bromonium ion

The bromonium ion reacts with a chloride ion to produce the *trans*-1-bromo-2-chloro-cyclohexane enantiomers.

8.42 (a) 1-Pentyne has IR absorption at about 3300 cm^{-1} due to its terminal triple bond. Pentane does not absorb in that region.

(b) 1-Pentene absorbs in the 1620–1680 cm^{-1} region due to the alkene function. Pentane does not exhibit absorption in that region.

(c) See parts (a) and (b).

(d) 1-Bromopentane shows C-Br absorption in the 690-515 cm^{-1} region while pentane does not.

(e) For 1-pentyne, see (a). The interior triple bond of 2-pentyne gives relatively weak absorption in the 2100–2260 cm^{-1} region.

(f) For 1-pentene, see (b). 1-Pentanol has a broad absorption band in the 3200–3550 cm^{-1} region.

(g) See (f)

(h) 1-Bromo-2-pentene has double bond absorption in the 1620–1680 cm^{-1} region which 1-bromopentane lacks.

(i) 2-Penten-1-ol has double bond absorption in the 1620–1680 cm^{-1} region not found in 1-pentanol.

8.43 Because of the electron-withdrawing nature of chlorine, the electron density at the double bond is greatly reduced and attack by the electrophilic bromine does not occur.

*8.44 The index of hydrogen deficiency of **A, B,** and **C** is two.

$$C_6H_{14}$$
$$\underline{C_6H_{10}}$$
$$H_4 = 2 \text{ pairs of hydrogen atoms}$$

This result suggests the presence of a triple bond, two double bonds, a double bond and a ring, or two rings. The fact that **A, B,** and **C** all decolorize Br$_2$/CCl$_4$ and dissolve in concd. H$_2$SO$_4$ suggests they all have a carbon-carbon multiple bond.

A must be a terminal alkyne, because of IR absorption at about 3300 cm^{-1}.

Since **A** gives hexane on catalytic hydrogenation, **A** must be 1-hexyne.

$$CH_3(CH_2)_3C{\equiv}CH \xrightarrow[\text{Pt}]{2\,H_2} CH_3(CH_2)_4CH_3$$
$$\textbf{A}$$

This is confirmed by the oxidation experiment

$$CH_3(CH_2)_3C{\equiv}CH \xrightarrow[\text{(2) H}_3\text{O}^+]{\text{(1) KMnO}_4, \text{ OH}^-, \text{ heat}} CH_3(CH_2)_3CO_2H \;+\; CO_2{\uparrow}$$

Hydrogenation of **B** to hexane shows that its chain is unbranched, and the oxidation experiment shows that **B** is 3-hexyne.

$$\underset{\textbf{B}}{CH_3CH_2C{\equiv}CCH_2CH_3} \xrightarrow[\text{(2) H}_3\text{O}^+]{\text{(1) KMnO}_4, \text{ OH}^-, \text{ heat}} 2\;CH_3CH_2CO_2H$$

Oxidation of **C** shows that it is cyclohexene.

$$\xrightarrow[\text{(2) H}_3\text{O}^+]{\text{(1) KMnO}_4, \text{ OH}^-, \text{ heat}} HO_2C(CH_2)_4CO_2H$$

C

8.45 (a) Four

(b)

8.46 Hydroxylations by $KMnO_4$ are syn hydroxylations (cf. Section 8.9). Thus, maleic acid must be the *cis*-dicarboxylic acid:

Maleic acid   *meso*-Tartaric acid

Fumaric acid must be the *trans*-dicarboxylic acid:

Fumaric acid   (±)-Tartaric acid

8.47 (a) The addition of bromine is an anti addition. Thus, fumaric acid yields a meso compound.

A meso compound

(b) Maleic acid adds bromine to yield a racemic modification.

*8.48

8.49

8.50

D

Optically active
(the other enantiomer is
an equally valid answer)

E

Optically inactive
(nonresolvable)

***8.51** (a,b)

(c)   **C**, in contrast to its *cis* isomer, would exhibit no intramolecular hydrogen-bonding. This would be proven by the absence of infrared absorption in the 3500- to 3600- cm^{-1} region when studied as a very dilute solution in CCl_4. **C** would only show free O—H stretch at about 3625 cm^{-1}

***8.52**

QUIZ

8.1   A hydrocarbon whose molecular formula is C_7H_{12}, on catalytic hydrogenation (excess H_2/Pt), yields C_7H_{16}. The original hydrocarbon adds bromine and also exhibits an IR absorption band at 3300 cm^{-1}. Which of the following is a plausible choice of structure for the original hydrocarbon?

(a)   (b)

(c) $CH_3CH{=}CHCH{=}CHCH_2CH_3$

(d) $CH_3CH_2CH_2C{\equiv}CCH_2CH_3$

(e) $CH_3CH_2CH_2CH_2CH_2C{\equiv}CH$

8.2   Select the major product of the dehydration of the alcohol, $CH_3\overset{\underset{\displaystyle CH_3}{|}}{\underset{}{C}}{\overset{\displaystyle CH_3\quad OH}{}}{-}CHCH_3$

(a) $CH_3\overset{\underset{\displaystyle CH_3}{|}}{\underset{\displaystyle CH_3}{C}}{-}CH{=}CH_2$

(b) $CH_3\overset{\underset{\displaystyle CH_3}{|}}{\underset{\displaystyle CH_3}{C}}{=}CHCH_3$

(c) $CH_3\overset{\underset{\displaystyle CH_3}{|}}{C}{=}CCH_3$

(d) $CH_3CH{-}\overset{\underset{\displaystyle CH_3}{|}}{\overset{\displaystyle CH_3}{C}}{=}CH_2$

(e) $CH_2{=}\overset{\underset{\displaystyle CH_3}{|}}{C}{-}CHCH_3$

8.3   Give the major product of the reaction of *cis*-2-pentene with bromine.

(a)   (b)   (c)   (d)   (e) A racemic mixture of (c) and (d)

8.4   The compound shown here is best prepared by which sequence of reactions?

(a)   [cyclopentane with CH_3]${-}C{\equiv}CH$ + $NaNH_2$ $\longrightarrow$ then CH_3CH_2Br $\longrightarrow$ product

(b) $CH_3CH_2C{\equiv}CH$ + $NaNH_2$ $\longrightarrow$ then [cyclopentane with CH_3 and Br] $\longrightarrow$ product

(c)   [cyclopentane with CH_3 and $CH{=}CHCH_2CH_3$] + H_2 $\xrightarrow{Pt}$ product

(d)   [cyclopentane with CH_3 and $CH_2{-}CHCH_2CH_3$ / Br] $\xrightarrow[C_2H_5OH]{NaOC_2H_5}$ product

8.5   A compound whose formula is C_6H_{10} (Compound **A**) reacts with H_2/Pt in excess to give a product C_6H_{12}, which does not decolorize Br_2/CCl_4. Compound **A** does not show IR absorption in the 3200–3400 cm^{-1} region.

Ozonolysis of **A** gives 1 mol of HCH (with O double bonded) and 1 mol of . Give the structure of **A**.

(a)

(b) $CH_3CH_2CH_2C{\equiv}CCH_3$

(d)

(c) $CH_3CH_2CH_2CH_2C{\equiv}CH$

(e) $CH_2{=}CHCH_2CH_2CH{=}CH_2$

8.6   Compound **B** (C_5H_{10}) does not dissolve in cold, concentrated H_2SO_4. What is **B**?

(a) $CH_2{=}CHCH_2CH_2CH_3$

(b) $CH_3CH{=}CHCH_2CH_3$

(c)

(d)

8.7   Which reaction sequence converts cyclohexene to *cis*-1,2-cyclohexanediol? That is,

(a) Cold, dilute, aqueous $KMnO_4$, OH^- (b) (1) O_3 (2) Zn/HOAc

(c) (1) OsO_4 (2) $NaHSO_3$

(d) (1) $R\overset{O}{\underset{||}{C}}{-}OOH$ (2) H_3O^+/H_2O

(e) More than one of these

8.8   Which of the following sequences leads to the best synthesis of the compound $CH_3CH_2C{\equiv}CH$? (Assume that the quantities of reagents are sufficient to carry out the desired reaction.)

(a) $CH_3CH_2CH{=}CH_2$ $\xrightarrow{Br_2}$ $\xrightarrow[H_2O]{NaOH}$

(b) $CH_3CH_2CH{=}CH_2$ $\xrightarrow{Br_2}$ $\xrightarrow{NaNH_2}$

(c) $CH_3CH_2CH_2CHBr_2$ $\xrightarrow{H_2SO_4}$

(d) $CH_3CH_2CH_2CH_3$ $\xrightarrow[light]{Br_2}$ $\xrightarrow{NaNH_2}$

(e) $CH_3CH_2CH{=}CH_2$ $\xrightarrow{O_3}$ $\xrightarrow{Zn,HOAc}$

9

NUCLEAR MAGNETIC RESONANCE AND MASS SPECTROMETRY: TOOLS FOR STRUCTURE DETERMINATION

SOLUTIONS TO PROBLEMS

9.1   We see below that if we replace a methyl hydrogen by some group Z, then rotation of the methyl group or turning the whole molecule end-for-end gives structures that represent the same compound. This means that all of the methyl hydrogens are equivalent.

All of these are the same compound

Replacing a ring hydrogen by Z gives another compound, but replacing each ring hydrogen in turn gives the same compound. This shows that all four ring hydrogens are equivalent and that 1,4-dimethylbenzene has only two sets of chemical shift equivalent protons.

9.2   (a)   One   (d)   One

      (b)   Two   (e)   Two

      (c)   Two   (f)   Two

9.3   (a)

(b)   Six   (c)   Six signals

9.4   (a)   Two, CH_3—CH_2—CH_2—CH_3 (b)   Three, CH_3—CH_2—O—H

      (c)   Four,   (d)   Two,

      (e)   Four, CH_3—$CHBr$—C—Br (f)   Two,

(g)   Three,

(h)   Four,

(i)   Six,

(j)   Five,

9.5   The ^{1}H NMR spectrum of $CHBr_2CHCl_2$ consists of two doublets. The doublet from the proton of the $-CHCl_2$ group should occur at lower magnetic field strength because the greater electronegativity of chlorine reduces the electron density in the vicinity of the $-CHCl_2$ proton, and consequently, reduces its shielding relative to $-CHBr_2$.

9.6   The determining factors here are the number of chlorine atoms attached to the carbon atoms bearing protons and the deshielding that results from chlorine's electronegativity. In 1,1,2-trichloroethane the proton that gives rise to the triplet is on a carbon atom that bears two chlorines, and the signal from this proton is downfield. In 1,1,2,3,3-penta-chloropropane, the proton that gives rise to the triplet is on a carbon atom that bears only one chlorine; the signal from this proton is upfield.

9.7   The signal from the three equivalent protons designated *(a)* should be split into a doublet by the proton *(b)*. This doublet, because of the electronegativity of the attached chlorines, should occur downfield.

$$
\begin{array}{cc}
(a) & (b)\\
(Cl_2CH)\overline{_3}-CH
\end{array}
$$

The signal for the proton designated *(b)* should be split into a quartet by the three equivalent protons *(a)*. The quartet should occur upfield.

9.8

$$
\begin{array}{ccc}
& (a) & (b) & (a)\\
A & C_3H_7I \text{ is } CH_3-CH-CH_3\\
& & |\\
& & I
\end{array}
$$

(a)   doublet δ 1.9
(b)   septet δ 4.35

$$
\begin{array}{cc}
& (a) & (b)\\
B & C_2H_4Cl_2 \text{ is } CH_3-CH-Cl\\
& & |\\
& & Cl
\end{array}
$$

(a)   doublet δ 2.08
(b)   quartet δ 5.9

$$
\begin{array}{ccc}
& (a) & (b) & (a)\\
C & C_3H_6Cl_2 \text{ is } Cl-CH_2-CH_2-CH_2-Cl
\end{array}
$$
(a)   triplet δ 3.7
(b)   quintet δ 2.2

9.9 (a) $J_{ab} = 2 J_{bc}$

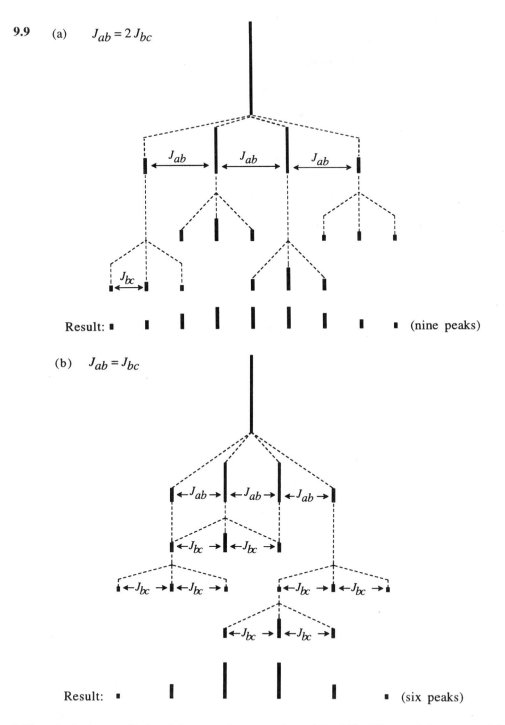

Result: (nine peaks)

(b) $J_{ab} = J_{bc}$

Result: (six peaks)

9.10 A single unsplit signal, because the proton is rapidly shifted from axial to equatorial positions.

9.11   **A** is 1-bromo-3-methylbutane. The following are the signal assignments:

$$(d) \ (c) \ (b) \ (a)$$
$$BrCH_2CH_2CH(CH_3)_2$$

(a) $\delta\,23$
(b) $\delta\,27$
(c) $\delta\,32$
(d) $\delta\,42$

B is 2-bromo-2-methylbutane. The following are the signal assignments:

$$(a) \ (c) \ \overset{Br}{|} \ (b)$$
$$CH_3CH_2C(CH_3)_2$$
$$(d)$$

(a) $\delta\,11$
(b) $\delta\,33$
(c) $\delta\,40$
(d) $\delta\,68$

C is 1-bromopentane. The following are the signal assignments.

$$(e) \ (d) \ (c) \ (b) \ (a)$$
$$BrCH_2CH_2CH_2CH_2CH_3$$

(a) $\delta\,14$
(b) $\delta\,23$
(c) $\delta\,30$
(d) $\delta\,33$
(e) $\delta\,34$

9.12   The compound is methane, CH_4. The molecular ion is at m/z 16. (This peak also happens to be the base peak.)

The peaks at m/z 15, 14, 13, and 12 are caused by successive losses of hydrogen atoms.

The small peak at m/z 17 ($M^{\ddagger} + 1$) comes mainly from methane molecules that contain ^{13}C.

9.13 The compound is water.

$$H-\ddot{O}-H + e^{-} \longrightarrow H-\overset{\bullet+}{\underset{\cdot\cdot}{O}}-H + 2\,e^{-}$$

m/z 18

($M\ddagger$)

$$H-\overset{\bullet+}{\underset{\cdot\cdot}{O}}-H \longrightarrow H-\ddot{O}^{+} + H\bullet$$

m/z 17

$$H-\ddot{O}^{+} \longrightarrow \bullet\ddot{O}^{+} + H\bullet$$

m/z 16

The peaks at m/z 19 and m/z 20 are due (primarily) to naturally occurring oxygen isotopes.

$$H-\overset{17\,\cdot\cdot}{\underset{\cdot\cdot}{O}}-H + e^{-} \longrightarrow H-\overset{17\,\bullet+}{\underset{\cdot\cdot}{O}}-H + 2\,e^{-}$$

m/z 19

($M\ddagger + 1$)

$$H-\overset{18\,\cdot\cdot}{\underset{\cdot\cdot}{O}}-H + e^{-} \longrightarrow H-\overset{18\,\bullet+}{\underset{\cdot\cdot}{O}}-H + 2\,e^{-}$$

m/z 20

($M\ddagger + 2$)

9.14   The compound is methyl fluoride, CH_3F.

$$CH_3\text{–}F \; + \; e^- \; \longrightarrow \; [CH_3\text{–}F]^{\ddot +} \; + \; 2\,e^-$$

m/z 34

$(M^{\ddot +})$

$$[CH_3F]^{\ddot +} \; \longrightarrow \; [CH_2F]^+ \; + \; H\bullet$$

m/z 33

$$[CH_2F]^+ \; \longrightarrow \; [CHF]^{\ddot +} \; + \; H\bullet$$

m/z 32

$$[CHF]^{\ddot +} \; \longrightarrow \; [CF]^+ \; + \; H\bullet$$

m/z 31

$$[CH_3F]^{\ddot +} \; \longrightarrow \; [F]^+ \; + \; CH_3\bullet$$

m/z 19

$$[CH_3F]^{\ddot +} \; \longrightarrow \; [CH_3]^+ \; + \; F\bullet$$

m/z 15

$$[CH_3]^+ \; \longrightarrow \; [CH_2]^{\ddot +} \; + \; H\bullet$$

m/z 14

9.15 (a)

$$\underset{CH_3\overset{\textstyle O}{\overset{\|}{C}}CH_2CH_3}{} \qquad CH_3CH_2CH_2\overset{\textstyle O}{\overset{\|}{C}}H \qquad CH_3\underset{CH_3}{\overset{\textstyle O}{\overset{\|}{C}H}}CH \qquad CH_2{=}CHCH_2OCH_3$$

$$CH_3CH{=}CHCH_2OH \qquad CH_2{=}\underset{}{\overset{CH_3}{C}}{-}CH_2OH$$

(b)   Only the first three. (The peak at 1730 cm⁻¹ is due to a C=O group.)

9.16   First, we recalculate the intensities of the peaks so as to base them on the M^+ peak:

m/z		INTENSITY
		% of M^+
86 M^+	$10.0/10.0 \times 100 =$	100
87	$0.56/10.0 \times 100 =$	5.6
88	$0.04/10.0 \times 100 =$	0.4

1.   Since M^+ is even, the compound must contain an even number of nitrogen atoms (i.e., 0, 2, 4, etc.).

2.   The value of the $M^+ + 1$ peak gives the number of carbon atoms.

Number of carbon atoms $= 5.6/1.1 \simeq 5$

The compound must contain no nitrogen atoms because $C_5N_2 = (5 \times 12) + (2 \times 14) = 88$, and the molecular weight of the compound (from the M^+ peak) is only 86.

3.   The very low value of the $M^+ + 2$ peak (0.4%) tells us that the compound does not contain S, Cl, or Br.

4.   If the compound were composed only of C and H, it would have to be C_5H_{26}:

$$H = 86 - (5 \times 12) = 26$$

But C_5H_{26} is impossible.

However, a formula with one oxygen gives a reasonable number of hydrogen atoms,

$$H = 86 - (5 \times 12) - 16 = 10$$

and thus our compound has the formula $C_5H_{10}O$.

9.17   (a)   The $M^+ + 2$ peak due to $CH_3-^{37}Cl$ (at m/z 52) should be almost one-third (32.5%) as large as the M^+ peak at m/z 50.

(b)   The peaks due to $CH_3-^{79}Br$ and $CH_3-^{81}Br$ (at m/z 94 and m/z 96, respectively) should be of nearly equal intensity.

(c)   That the M^+ and $M^+ + 2$ peaks are of nearly equal intensity tells us that the compound contains bromine. C_3H_7Br is therefore a likely molecular formula.

$C_3 = 36$	$C_3 = 36$
$H_7 = 7$	$H_7 = 7$
$^{79}Br = \underline{79}$	$^{81}Br = \underline{81}$
$m/z = 122$	$m/z = 124$

9.18   Recalculating the intensities to base on M^+

PEAK	m/z	% of BASE PEAK	% of M^+
M^+	73	86.1	100
$M^+ + 1$	74	3.2	3.72
$M^+ + 2$	75	0.2	0.23

These data best fit the formula C_3H_7NO.

9.19   (a)   First recalculating the intensities so as to base them on the M^+ peak:

m/z INTENSITY
                                               % of M^+

78 M^+ 24/24 $\times$ 100 =   100
79   0.8/24 $\times$ 100 =   3.3
80   8/24 $\times$ 100 =   33

1.   Since M^+ is even, the compound contains an even number of nitrogen atoms.

2.   Number of carbon atoms = $(M^+ + 1)/1.1 = 3.3/1.1 = 3$.

3.   The intensity of the $M^+ + 2$ peak (33%) tells us that the compound contains one chlorine atom.

4.   We use the molecular weight (from the M^+ peak) to calculate the number of hydrogen atoms.

$$H = 78 - (3 \times 12) - 35 = 7$$

Thus, the formula for the compound is C_3H_7Cl.

(b) CH_3CHCH_3
              $|$
              Cl

9.20   (a)   A *tert*-butyl cation, $(CH_3)_3C^+$.

(b)
$$\left[CH_3-\underset{\underset{CH_3}{|}}{\overset{\overset{CH_3}{|}}{C}}-CH_2CH_3 \right]^{\ddagger} \longrightarrow CH_3-\underset{\underset{CH_3}{|}}{\overset{\overset{CH_3}{|}}{C^+}} + \cdot CH_2CH_3$$

m/z 57

9.21   A peak at M^+ - 15 involves the loss of a methyl radical and the formation of a 1° or 2° carbocation.

$$[CH_3CH_2\underset{\overset{|}{CH_3}}{\overset{\overset{CH_3}{|}}{C}}HCH_2CH_3]^{\ddagger} \longrightarrow CH_3CH_2\overset{+}{C}HCH_2CH_3 + CH_3^{\bullet}$$

(**M**$\ddagger$)   (**M**$\ddagger$ – 15)

or

$$[CH_3CH_2\underset{\overset{|}{CH_3}}{\overset{\overset{CH_3}{|}}{C}}HCH_2CH_3]^{\ddagger} \longrightarrow CH_3CH_2\underset{\overset{|}{CH_3}}{\overset{\overset{CH_3}{|}}{C}}HCH_2^+ + CH_3^{\bullet}$$

(**M**$\ddagger$)   (**M**$\ddagger$ – 15)

A peak at M^+ - 29 arises from the loss of an ethyl radical and the formation of a 2° carbocation.

$$[CH_3CH_2\underset{\overset{|}{CH_3}}{\overset{\overset{CH_3}{|}}{C}}HCH_2CH_3]^{\ddagger} \longrightarrow CH_3CH_2\overset{+}{\underset{\overset{|}{CH_3}}{\overset{\overset{CH_3}{|}}{C}}}H + CH_3CH_2^{\bullet}$$

(**M**$\ddagger$)   (**M**$\ddagger$ – 29)

Since a 2° carbocation is more stable, the peak at M^+ - 29 is more intense.

9.22 Both peaks arise from allylic fragmentations:

$^+CH_2-\overset{\bullet}{C}H-CH_2-CHCH_2CH_3 \longrightarrow \overset{\bullet}{C}H_2-CH=CH_2 + {}^+CHCH_2CH_3$

$\overset{|}{C}H_3$ Allyl radical $\overset{|}{C}H_3$

                                                        m/z 57

$CH_2^+-\overset{\bullet}{C}H-CH_2 : CHCH_2CH_3 \longrightarrow {}^+CH_2-CH=CH_2 + \bullet CHCH_2CH_3$

$\overset{|}{C}H_3$ *m/z* 41 $\overset{|}{C}H_3$

                                     Allyl cation

9.23 (a) Alcohols undergo rapid cleavage of a carbon-carbon bond next to oxygen because this leads to a resonance-stabilized cation.

1° alcohol $R : CH_2 \overset{\bullet}{\to} \overset{..}{O}H^+ \xrightarrow{-R\bullet} CH_2=\overset{+}{O}H \longleftrightarrow \overset{+}{C}H_2-\overset{..}{O}H$

2° alcohol $R-CH \overset{..}{\to} \overset{..}{O}H^+ \xrightarrow{-R\bullet} RCH=\overset{+}{O}H \longleftrightarrow R\overset{+}{C}H-\overset{..}{O}H$

3° alcohol $R-\overset{|}{\underset{R}{C}} \overset{..}{O}H^+ \xrightarrow{-R\bullet} R\overset{|}{\underset{R}{C}}=\overset{+}{O}H \longleftrightarrow R\overset{+}{\underset{R}{C}}-\overset{..}{O}H$

The cation obtained from a tertiary alcohol is the most stable (because of the electron-releasing R groups).

(b) Primary alcohols give a peak at *m/z* 31 due to $CH_2=\overset{+}{O}H$.

(c) Secondary alcohols give peaks at *m/z* 45, 59, 73, and so forth, because ions like the following are produced.

    $CH_3CH=\overset{+}{O}H$ $CH_3CH_2CH=\overset{+}{O}H$ $CH_3CH_2CH_2CH=\overset{+}{O}H$

     m/z 45   *m/z* 59   *m/z* 73

(d) Tertiary alcohols give peaks at *m/z* 59, 73, 87, and so forth, because ions like the following are produced.

    $CH_3\overset{+}{C}=\overset{+}{O}H$ $CH_3CH_2\overset{+}{C}=\overset{+}{O}H$ $CH_3CH_2CH_2\overset{+}{C}=\overset{+}{O}H$

       $\overset{|}{C}H_3$ $\overset{|}{C}H_3$ $\overset{|}{C}H_3$

     m/z 59   *m/z* 73   *m/z* 87

9.24   The spectrum given in Fig. 9.41 is that of butyl isopropyl ether. The main clues are the peaks at m/z 101 and m/z 73 due to the following fragmentations.

$$\left[CH_3\!-\!\underset{\underset{CH_3}{|}}{CH}\!-\!OCH_2CH_2CH_2CH_3 \right]^{\ddagger} \xrightarrow{-CH_3^{\bullet}} CH_3CH\!=\!\overset{+}{O}CH_2CH_2CH_2CH_3$$
$$m/z\ 101$$

$$\left[CH_3\!-\!\underset{\underset{CH_3}{|}}{CH}\!-\!OCH_2CH_2CH_2CH_3 \right]^{\ddagger} \xrightarrow{-CH_3CH_2CH_2^{\bullet}} \underset{\underset{CH_3}{|}}{CH_3}\overset{+}{CH}\overset{}{O}\!=\!CH_2$$
$$m/z\ 73$$

Butyl propyl ether (Fig. 9.42) has no peak at m/z 101 but has a peak at m/z 87 instead.

$$[CH_3CH_2CH_2\!-\!O\!-\!CH_2CH_2CH_2CH_3]^{\ddagger} \xrightarrow{-CH_3CH_2^{\bullet}} CH_2\!=\!\overset{+}{O}CH_2CH_2CH_2CH_3$$
$$m/z\ 87$$

Butyl propyl ether also has a peak at m/z 73.

$$[CH_3CH_2CH_2\!-\!O\!-\!CH_2CH_2CH_2CH_3]^{\ddagger} \xrightarrow{-CH_3CH_2CH_2^{\bullet}} CH_3CH_2CH_2\overset{+}{O}\!=\!CH_2$$
$$m/z\ 73$$

[Although the observation does not help us decide, it is interesting to notice that both spectra have intense peaks at m/z 43 and m/z 57 corresponding to propyl (or isopropyl) and butyl cations formed by carbon-oxygen bond cleavage.]

9.25   The compound is butanal. The peak at m/z 44 arises from a McLafferty rearrangement.

$$\underset{(M\ddagger)}{m/z\ 72} \qquad\qquad \underset{(M\ddagger-28)}{m/z\ 44}$$

The peak at m/z 29 arises from a fragmentation producing an acylium ion.

$$H\!-\!C\!\equiv\!\overset{+}{O} \quad+\quad CH_3CH_2CH_2^{\bullet}$$
$$m/z\ 29$$

9.26   The ion, $CH_2\!=\!\overset{+}{N}H_2$, produced by the following fragmentation.

$$R\!:\!CH_2\!\overset{+}{\overset{\bullet}{N}}H_2 \xrightarrow{-R^{\bullet}} CH_2\!=\!\overset{+}{N}H_2 \longleftrightarrow \overset{+}{C}H_2\!-\!\overset{\cdot\cdot}{N}H_2$$
$$m/z\ 30$$

9.27 Compound **A** is *tert*-butylamine. Our first clue is the molecular ion at *m/z* 73 (an odd-numbered mass unit) indicating the presence of an odd number of nitrogen atoms. The base peak at *m/z* 58 is our second important clue. It arises from the following fragmentation.

The ^{1}H NMR spectrum confirms the structure

$$
\begin{array}{c} (a) \qquad (b) \\ (CH_3)_3C{-}NH_2 \end{array}
$$

(a) Singlet δ 1.2 (9H)
(b) Singlet δ 1.3 (2H)

9.28 The compound is 2-methyl-2-butanol. Although the molecular ion is not discernible, we are given that it is at *m/z* 88. This information gives us the molecular weight of **B** and rules out the possibility of a structure with an odd number of nitrogen atoms.

The IR absorption (3200–3550 cm^{-1}) suggests the presence of an —OH group.

Two important peaks in the mass spectrum are the intense peaks at *m/z* 59 and *m/z* 73. These peaks correspond to fragmentation reactions that produce resonance-stabilized oxonium ions and strongly suggest that we have a tertiary alcohol [see Problem 9.23, part (d)].

The peak at *m/z* 70 corresponds to the loss of a molecule of water from the molecular ion, and the peak at *m/z* 55 probably arises from a subsequent allylic cleavage.

The ^{1}H NMR spectrum of **B** confirms that it is 2-methyl-2-butanol.

$$\underset{(d)}{\overset{(c)}{\underset{OH}{\overset{CH_3\,(c)}{\underset{(a)\quad(b)}{CH_3-CH_2-\underset{|}{\overset{|}{C}}-CH_3}}}}}$$

(a) Triplet, δ 0.9 (3H)
(b) Quartet, δ 1.6 (2H)
(c) and (d) Overlapping singlets, δ 1.1 (7H)

9.29 (a)

$$\underset{CH_3\,(a)}{\overset{CH_3\,(a)}{CH_3-\underset{|}{\overset{|}{C}}-OH\,(b)}}$$

(a) Singlet, δ 1.28 (9H)
(b) Singlet, δ 1.35 (1H)

(b)

$$\underset{Br}{\overset{(a)\;\;(b)\;\;(a)}{CH_3-\underset{|}{CH}-CH_3}}$$

(a) Doublet, δ 1.71 (6H)
(b) Septet, δ 4.32 (1H)

(c)

$$\overset{(b)\;\overset{O}{\overset{\|}{}}\;(c)\;\;(a)}{CH_3-C-CH_2-CH_3}$$

(a) Triplet, δ 1.05 (3H)   C=O, 1720 cm^{-1}
(b) Singlet, δ 2.13 (3H)
(c) Quartet, δ 2.47 (2H)

(d)

$$\overset{(b)\;\;(a)}{-CH_2-OH}$$
(c)

(a) Singlet, δ 2.43 (1H)   O–H, 3200-3550 cm^{-1}
(b) Singlet, δ 4.58 (2H)
(c) Multiplet, δ 7.28 (5H)

(e)

$$\underset{(a)}{\overset{(b)\;\;\;(c)}{\underset{CH_3}{CH_3-CH-CH_2Cl}}}$$

(a) Doublet, δ 1.04 (6H)
(b) Multiplet, δ 1.95 (1H)
(c) Doublet, δ 3.35 (2H)

(f)

$$\underset{(c)}{\overset{(b)\;\overset{O}{\overset{\|}{}}\;(a)}{\underset{C_6H_5}{C_6H_5-CH-C-CH_3}}}$$

(a) Singlet, δ 2.20 (3H)   C=O, near 1720 cm^{-1}
(b) Singlet, δ 5.08 (1H)
(c) Multiplet, δ 7.25 (10H)

(g)

$$\underset{Br}{\overset{(a)\;\;(b)\;\;(c)\;\;(d)}{CH_3-CH_2-\underset{|}{CH}CO_2H}}$$

(a) Triplet, δ 1.08 (3H)   C=O (acid) 1715 cm^{-1}
(b) Multiplet, δ 2.07 (2H)   O–H, 2500-3000 cm^{-1}
(c) Triplet, δ 4.23 (1H)
(d) Singlet, δ 10.97 (1H)

(h)

$$\overset{(b)\;\;(a)}{-CH_2-CH_3}$$
(c)

(a) Triplet, δ 1.25 (3H)
(b) Quartet, δ 2.68 (2H)
(c) Multiplet, δ 7.23 (5H)

(i)

$$\overset{(a)\;\;(b)\;\;\;(c)\;\;\;(d)}{CH_3-CH_2-O-CH_2CO_2H}$$

(a) Triplet, δ 1.27 (3H)   C=O (acid) 1715 cm^{-1}
(b) Quartet, δ 3.66 (2H)   O–H, 2500-3000 cm^{-1}
(c) Singlet, δ 4.13 (2H)
(d) Singlet, δ 10.95 (1H)

(j)
$$\overset{(a)\ \ (b)\ \ (a)}{CH_3-CH-CH_3}$$
$$|$$
$$NO_2$$

(a) Doublet, δ 1.55 (6H)
(b) Septet, δ 4.67 (1H)

(k)
$$\overset{(a)\ \ \ \ (b)\ (b)\ \ \ \ (a)}{CH_3O-CH_2CH_2-OCH_3}$$

(a) Singlet, δ 3.25 (6H)
(b) Singlet, δ 3.45 (4H)

(l)

$$\overset{(b)\ \ \ \ O\ \ \ \ (c)}{CH_3-\overset{||}{C}-CH-CH_3}$$
$$\underset{(a)}{\overset{|}{CH_3}}$$

(a) Doublet, δ 1.10 (6H)   C=O, near 1720 cm^{-1}
(b) Singlet, δ 2.10 (3H)
(c) Septet, δ 2.50 (1H)

(m)

(a) Doublet, δ 2.00 (3H)
(b) Quartet, δ 5.15 (1H)
(c) Multiplet, δ 7.35 (5H)

9.30 Compound **E** is phenylacetylene, C$_6$H$_5$C≡CH. We can make the following assignments in the IR spectrum:

The IR spectrum of compound **E** (Problem 9.30). (Spectrum courtesy of Sadtler Research Laboratories Inc., Philadelphia.)

9.31 Compound **G** is 2-bromobutane. Assignments are as follows:

$$\overset{(a)\ \ (b)\ \ (d)\ \ (c)}{CH_3-CH_2-CH-CH_3}$$
$$|$$
$$Br$$

(a) triplet δ 1.05
(b) multiplet δ 1.82
(c) doublet δ 1.7
(d) multiplet δ 4.1

Compound **H** is 2,3-dibromopropene. Assignments are as follows:

(a) δ 4.2
(b) δ 6.05
(c) δ 5.64

9.32 Compound **J** is *cis*-1,2-dichloroethene.

We can make the following IR assignments:

3125 cm^{-1}, alkene C—H stretching
1625 cm^{-1}, C=C stretching
695 cm^{-1}, out-of-plane bending of cis double bond

9.33 (a) Compound **K** is,

(a) (b) (c)   (a) Singlet, δ 2.15   C=O, 1720 cm^{-1}
CH$_3$–C–CH–CH$_3$ (b) Quartet, δ 4.25
         | (c) Doublet, δ 1.35
        OH (d)   (d) Singlet, δ 3.75

(b) When the compound is dissolved in D$_2$O, the —OH proton (*d*) is replaced by a deuteron, and thus the ^{1}H NMR absorption peak disappears.

*9.34 Run the spectrum with the spectrometer operating at a different magnetic field strength (i.e., at 30 or at 100 MHz). If the peaks are two singlets the distance between them—*when measured in hertz*—will change because chemical shifts *expressed in hertz* are proportional to the strength of the applied field (Section 9.6). If, however, the two peaks represent a doublet, then the distance that separates them, expressed in hertz, will not change because this distance represents the magnitude of the coupling constant and coupling constants are independent of the applied magnetic field (Section 9.8).

9.35   Compound **O** is 1,4-cyclohexadiene and **P** is cyclohexane.

(a) δ 26.0

(b) δ 124.5

O   **P**

*9.36   The molecular formula of **Q** (C_7H_8) indicates an index of hydrogen deficiency (Section 7.8) of four. The hydrogenation experiment suggests that **Q** contains two double bonds (or one triple bond). Compound **Q**, therefore, must contain two rings.

Bicyclo[2.2.1]hepta-2,5-diene. The following reasoning shows one way to arrive at this conclusion: There is only one signal (δ 143) in the region for a doubly bonded carbon. This fact indicates that the doubly bonded carbon atoms are all equivalent. That the signal at δ 143 is a CH group in the DEPT spectra indicates that each of the doubly bonded carbon atoms bears one hydrogen atom. Information from the DEPT spectra tells us that the signal at δ 75 is a —CH_2— group and the signal at δ 50 is a $\overset{\diagdown}{\underset{\diagup}{C}}$—H group.

The molecular formula tells us that the compound must contain two $\overset{\diagdown}{\underset{\diagup}{C}}$—H groups, and since only one signal occurs in the ^{13}C spectrum, these $\overset{\diagdown}{\underset{\diagup}{C}}$—H groups must be equivalent. Putting this all together, we get the following:

(a) δ 50

(b) δ 75

(c) δ 143

Q   **R**

*9.37   That **S** decolorizes bromine indicates that it is unsaturated. The molecular formula of **S** allows us to calculate an index of hydrogen deficiency equal to 1. Therefore, we can conclude that **S** has one double bond.

The ^{13}C spectrum shows the doubly bonded carbon atoms at δ 130 and δ 135. In the DEPT spectra, one of these signals (δ 130) is a carbon that bears no hydrogen atoms; the other (δ 135) is a carbon that bears one hydrogen atom. We can now arrive at the following partial structure.

$$\overset{\textstyle C\diagdown\qquad\diagup C}{\underset{\textstyle C\diagup\qquad\diagdown H}{C=C}}$$

The three most upfield signals (δ 19, δ 28, and δ 31) all arise from methyl groups. The signal at δ 32 is a carbon atom with no hydrogen atoms. Putting these facts together allows us to arrive at the following structure.

(a) δ 19   (d) δ 32

(b) δ 28   (e) δ 130

(c) δ 31   (f) δ 135

S

Although the structure just given is the actual compound, other reasonable structures that one might be led to are

and

9.38   The IR absorption band at 1745 cm^{-1} indicates the presence of a $>$C=O group in a five-member ring, and the signal at δ 218.2 can be assigned to the carbon of the carbonyl group.

There are only two other signals in the ^{13}C spectrum; the DEPT spectra suggest two equivalent sets of two –CH$_2$– groups each. Putting these facts together, we arrive at cyclopentanone as the structure for **T**.

(a) δ 23.5
(b) δ 38.0
(c) δ 218.2

T

***9.39**

***9.40**

QUIZ

9.1   Propose a structure that is consistent with each set of following data.

    (a) C_4H_9Br ^1H NMR spectrum
                               Singlet δ 1.7

    (b) $C_4H_7Br_3$ ^1H NMR spectrum
                               Singlet δ 1.95 (3H)
                               Singlet δ 3.9 (4H)

    (c) C_8H_{16} ^1H NMR spectrum   IR spectrum
                               Singlet δ 1.0 (9H)   3040, 2950, 1640 cm^{-1}
                               Singlet δ 1.75 (3H)   and other peaks.
                               Singlet δ 1.9 (2H)
                               Singlet δ 4.6 (1H)
                               Singlet δ 4.8 (1H)

    (d) $C_9H_{10}O$ ^1H NMR spectrum   IR spectrum
                               Singlet δ 2.0 (3H)   3100, 3000, 1720,
                               Singlet δ 3.75 (2H)   740, 700 cm^{-1}
                               Singlet δ 7.2 (5H)   and other peaks.

    (e) $C_5H_7NO_2$ ^1H NMR spectrum   IR spectrum
                               Triplet δ 1.2 (3H)   2980, 2260, 1750 cm^{-1}
                               Singlet δ 3.5 (2H)   and other peaks.
                               Quartet δ 4.2 (2H)   This compound has a nitro group.

9.2   How many ^1H NMR signals would the following compound give?

$$CH_3CHCH_2Cl$$
$$|$$
$$CH_3$$

    (a)   One   (b)   Two   (c)   Three   (d)   Four   (e)   Five

9.3   How many ^1H NMR signals would 1,1-dichlorocyclopropane give?

    (a)   One   (b)   Two   (c)   Three   (d)   Four   (e)   Five

10 RADICAL REACTIONS

SOLUTIONS TO PROBLEMS

10.1 **(a)** $\quad$ H—H $\quad$ + $\quad$ F—F $\quad\longrightarrow\quad$ 2 H—F

$\quad\quad\quad$ ($DH° = 435$) $\quad$ ($DH° = 159$) $\quad\quad\quad\quad$ 2($DH° = 569$)

$\quad\quad\quad$ +594 kJ mol^{-1} $\quad\quad\quad\quad\quad\quad\quad$ −1138 kJ mol^{-1}
$\quad\quad\quad$ is required for $\quad\quad\quad\quad\quad\quad\quad\quad$ is evolved in
$\quad\quad\quad$ bond cleavage $\quad\quad\quad\quad\quad\quad\quad\quad$ bond formation

$$\Delta H° = +594 - 1138$$
$$= -544 \text{ kJ mol}^{-1}$$

(b) $\quad$ CH$_3$—H $\quad$ + $\quad$ F—F $\quad\longrightarrow\quad$ CH$_3$—F $\quad$ + $\quad$ H—F

$\quad\quad$ ($DH° = 435$) $\quad$ ($DH° = 159$) $\quad\quad\quad$ ($DH° = 452$) $\quad\quad$ ($DH° = 569$)

$\quad\quad\quad$ +594 kJ mol^{-1} $\quad\quad\quad\quad\quad\quad\quad$ −1021 kJ mol^{-1}
$\quad\quad\quad$ is required for $\quad\quad\quad\quad\quad\quad\quad\quad$ is evolved in
$\quad\quad\quad$ bond cleavage $\quad\quad\quad\quad\quad\quad\quad\quad$ bond formation

$$\Delta H° = +594 - 1021$$
$$= -427 \text{ kJ mol}^{-1}$$

(c) $\quad$ CH$_3$—H $\quad$ + $\quad$ Cl—Cl $\quad\longrightarrow\quad$ CH$_3$—Cl $\quad$ + $\quad$ H—Cl

$\quad\quad$ ($DH° = 435$) $\quad$ ($DH° = 243$) $\quad\quad\quad$ ($DH° = 349$) $\quad\quad$ ($DH° = 431$)

$\quad\quad\quad$ +678 kJ mol^{-1} $\quad\quad\quad\quad\quad\quad\quad$ −780 kJ mol^{-1}
$\quad\quad\quad$ is required for $\quad\quad\quad\quad\quad\quad\quad\quad$ is evolved in
$\quad\quad\quad$ bond cleavage $\quad\quad\quad\quad\quad\quad\quad\quad$ bond formation

$$\Delta H° = +678 - 780$$
$$= -102 \text{ kJ mol}^{-1}$$

(d) $\quad$ CH$_3$—H $\quad$ + $\quad$ Br—Br $\quad\longrightarrow\quad$ CH$_3$—Br $\quad$ + $\quad$ H—Br

$\quad\quad$ ($DH° = 435$) $\quad$ ($DH° = 192$) $\quad\quad\quad$ ($DH° = 293$) $\quad\quad$ ($DH° = 366$)

$\quad\quad\quad$ +627 kJ mol^{-1} $\quad\quad\quad\quad\quad\quad\quad$ −659 kJ mol^{-1}
$\quad\quad\quad$ is required for $\quad\quad\quad\quad\quad\quad\quad\quad$ is evolved in
$\quad\quad\quad$ bond cleavage $\quad\quad\quad\quad\quad\quad\quad\quad$ bond formation

$$\Delta H° = +627 - 659$$
$$= -32 \text{ kJ mol}^{-1}$$

(e) $\underset{(DH° = 435)}{CH_3—H}$ + $\underset{(DH° = 151)}{I—I}$ $\longrightarrow$ $\underset{(DH° = 234)}{CH_3—I}$ + $\underset{(DH° = 297)}{H—I}$

$+586$ kJ mol^{-1} -531 kJ mol^{-1}
is required for   is evolved in
bond cleavage   bond formation

$$\Delta H° = +586 - 531$$
$$= +55 \text{ kJ mol}^{-1}$$

(f) $\underset{(DH° = 410)}{CH_3CH_2—H}$ + $\underset{(DH° = 243)}{Cl—Cl}$ $\longrightarrow$ $\underset{(DH° = 341)}{CH_3CH_2—Cl}$ + $\underset{(DH° = 431)}{H—Cl}$

$+653$ kJ mol^{-1} -772 kJ mol^{-1}
is required for   is evolved in
bond cleavage   bond formation

$$\Delta H° = +653 - 772$$
$$= -119 \text{ kJ mol}^{-1}$$

(g) $\underset{(DH° = 395)}{(CH_3)_2CH—H}$ + $\underset{(DH° = 243)}{Cl—Cl}$ $\longrightarrow$ $\underset{(DH° = 339)}{(CH_3)_2CH—Cl}$ + $\underset{(DH° = 431)}{H—Cl}$

$+638$ kJ mol^{-1} -770 kJ mol^{-1}
is required for   is evolved in
bond cleavage   bond formation

$$\Delta H° = +638 - 770$$
$$= -132 \text{ kJ mol}^{-1}$$

(h) $\underset{(DH° = 381)}{(CH_3)_3C—H}$ + $\underset{(DH° = 243)}{Cl—Cl}$ $\longrightarrow$ $\underset{(DH° = 328)}{(CH_3)_3C—Cl}$ + $\underset{(DH° = 431)}{H—Cl}$

$+624$ kJ mol^{-1} -759 kJ mol^{-1}
is required for   is evolved in
bond cleavage   bond formation

$$\Delta H° = +624 - 759$$
$$= -135 \text{ kJ mol}^{-1}$$

10.2 $\underset{CH_3}{\overset{CH_3}{CH_3CH_2\overset{|}{\underset{|}{C}}•}}$ $>$ $\underset{CH_3}{CH_3CH_2\overset{|}{CH}•}$ $>$ $(CH_3)_2CHCH_2•$ $>$ $CH_3•$

     $3°$ $>$ $2°$ $>$ $1°$ $>$ **Methyl**

10.3   The compounds all have different boiling points. They could, therefore, be separated by careful fractional distillation. Or, because the compounds have different vapor pressures, they could easily be separated by gas chromatography. GC/MS (gas chromatography/mass spectrometry) could be used to separate the compounds as well as provide structural information from their mass spectra. Their mass spectra would show contributions from the naturally occurring ^{35}Cl and ^{37}Cl isotopes. The natural abundance of

^{35}Cl is approximately 75% and that of ^{37}Cl is approximately 25%. Thus, for CH_3Cl, containing only one chlorine atom, there will be an $M\overset{+}{\cdot}$ peak and an $M\overset{+}{\cdot}$ +2 peak in roughly a 3:1 (0.75:0.25) ratio of intensities. For CH_2Cl_2 there will be $M\overset{+}{\cdot}$, $M\overset{+}{\cdot}$ +2, and $M\overset{+}{\cdot}$ +4 peaks in roughly a 9:6:1 ratio, respectively. [The probability of a molecular ion $M\overset{+}{\cdot}$ with both chlorine atoms as ^{35}Cl is (.75)(.75) = .56, the probability of an $M\overset{+}{\cdot}$ +2 ion from one ^{35}Cl and one ^{37}Cl is 2(.75)(.25) = .38, and the probability of an $M\overset{+}{\cdot}$ +4 ion peak from both chlorine atoms as ^{37}Cl is (.25)(.25) = 0.06; thus, their ratio is 9:6:1.] For $CHCl_3$ there will be $M\overset{+}{\cdot}$, $M\overset{+}{\cdot}$ +2, and $M\overset{+}{\cdot}$ +4, and $M\overset{+}{\cdot}$ +6 peaks in approximately a 27:27:9:1 ratio, respectively (based on a calculation by the same method). (This calculation does not take into account the contribution of ^{13}C, ^2H, and other isotopes, but these are much less abundant.)

10.4 A small amount of ethane is formed by the combination of two methyl radicals:

$$2CH_3\cdot \longrightarrow CH_3 : CH_3$$

This ethane then reacts with chlorine in a substitution reaction (see Section 10.6) to form chloroethane.

The significance of this observation is that it is evidence for the proposal that the combination of methyl radicals is one of the chain-terminating steps in the chlorination of methane.

10.5 The use of a large excess of chlorine allows all of the chlorinated methanes (CH_3Cl, CH_2Cl_2, and $CHCl_3$) to react with chlorine.

10.6 *Chain Initiation*

Step 1 $\quad$ F—F $\quad\longrightarrow\quad$ 2 F· $\quad$ $\Delta H° = +159\ \text{kJ mol}^{-1}$
$\quad\quad$ ($DH° = 159$)

Chain Propagation

Step 2 $\quad$ CH_3—H + F· $\quad\longrightarrow\quad$ CH_3· + H—F $\quad$ $\Delta H° = -134\ \text{kJ mol}^{-1}$
$\quad\quad$ ($DH° = 435$) $\quad\quad\quad\quad\quad\quad\quad\quad$ ($DH° = 569$)

Step 3 $\quad$ CH_3· + F—F $\quad\longrightarrow\quad$ CH_3—F + F· $\quad$ $\Delta H° = -293\ \text{kJ mol}^{-1}$
$\quad\quad\quad\quad$ ($DH° = 159$) $\quad\quad$ ($DH° = 452$)

Chain Termination

$\quad\quad$ CH_3· + F· $\quad\longrightarrow\quad$ CH_3—F $\quad$ $\Delta H° = -452\ \text{kJ mol}^{-1}$
$\quad\quad\quad\quad\quad\quad\quad\quad$ ($DH° = 452$)

$\quad\quad$ CH_3· + CH_3· $\quad\longrightarrow\quad$ CH_3—CH_3 $\quad$ $\Delta H° = -368\ \text{kJ mol}^{-1}$
$\quad\quad\quad\quad\quad\quad\quad\quad$ ($DH° = 368$)

$\quad\quad$ F· + F· $\quad\longrightarrow\quad$ F—F $\quad$ $\Delta H° = -159\ \text{kJ mol}^{-1}$
$\quad\quad\quad\quad\quad\quad\quad\quad$ ($DH° = 159$)

10.7 CH_3—H + F̶· $\quad\longrightarrow\quad$ C̶H̶$_3$· $\quad$ + H—F $\quad$ $\Delta H° = -134\ \text{kJ mol}^{-1}$

$\quad$ C̶H̶$_3$· $\quad$ + F—F $\quad\longrightarrow\quad$ CH_3—F + F̶· $\quad$ $\Delta H° = -293\ \text{kJ mol}^{-1}$

$\quad$ ——

$\quad$ CH_3—H + F—F $\quad\longrightarrow\quad$ CH_3—F + H—F $\quad$ $\Delta H° = -427\ \text{kJ mol}^{-1}$

10.8 (a) Reactions (3), (5), and (6) should have $E_{act} = 0$ because these are gas-phase reactions in which small radicals combine to form molecules.

(b) Reactions (1), (2), and (4) should have $E_{act} > 0$ because in them covalent bonds are broken.

(c) Reactions (1) and (2) should have $E_{act} = \Delta H°$ because in them bonds are broken homolytically but no bonds are formed.

10.9 (a)

(c)

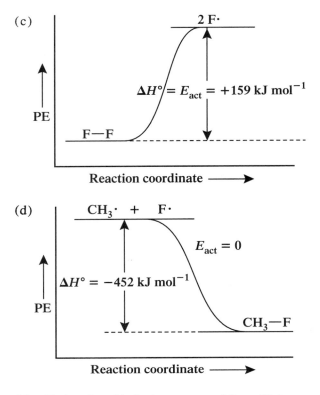

(d)

(e) Notice that this is the reverse of Step (2) in part (a)

10.10 (a) $CH_3CH_2-H + Cl\cdot \longrightarrow CH_3CH_2\cdot + H-Cl$
$(DH° = 410)$ $\qquad\qquad\qquad (DH° = 431)$

$\Delta H° = -431 + 410 = -21$ kJ mol^{-1}

(b)

$$\text{Transition state}$$

$$\overset{\delta\cdot}{Cl}\text{---}H\text{---}\overset{\delta\cdot}{CH_2CH_3}$$

$CH_3CH_3 + Cl\cdot$

$E_{act} = +4.2 \text{ kJ mol}^{-1}$

PE

$\Delta H° = -21 \text{ kJ mol}^{-1}$

$CH_3CH_2\cdot + H\text{---}Cl$

Reaction coordinate ⟶

(c) The hydrogen abstraction step for ethane,

$$CH_3CH_2\text{--}H + Cl\cdot \longrightarrow CH_3CH_2\cdot + HCl \quad (E_{act} = 4.2 \text{ kJ mol}^{-1})$$

has a much lower energy of activation than the corresponding step for methane:

$$CH_3\text{--}H + Cl\cdot \longrightarrow CH_3\cdot + HCl \quad (E_{act} = 15.9 \text{ kJ mol}^{-1})$$

Therefore, ethyl radicals form much more rapidly in the mixture than methyl radicals, and this leads to the more rapid formation of ethyl chloride.

10.11

$$Cl_2 \xrightarrow[\text{or heat}]{h\nu} 2\,Cl\cdot$$

Step 2a $\quad Cl\cdot + H:\overset{\displaystyle Cl}{\underset{\displaystyle H}{C}}\text{-}CH_3 \longrightarrow H:Cl + \cdot\overset{\displaystyle Cl}{\underset{\displaystyle H}{C}}\text{-}CH_3$

Step 3a $\quad CH_3\overset{\displaystyle Cl}{\underset{\displaystyle H}{C}}\cdot + Cl:Cl \longrightarrow CH_3\text{--}\overset{\displaystyle Cl}{\underset{\displaystyle H}{C}}\text{-}Cl + Cl\cdot$

1,1-dichloroethane

Step 2b $\quad Cl\cdot + H\text{--}CH_2CH_2Cl \longrightarrow H:Cl + \cdot CH_2CH_2Cl$

Step 3b $\quad \cdot CH_2CH_2Cl + Cl:Cl \longrightarrow ClCH_2CH_2Cl + Cl\cdot$

1,2-dichloroethane

10.12 (a) There is a total of eight hydrogen atoms in propane. There are six equivalent 1° hydrogen atoms, replacement of any one of which leads to propyl chloride, and there are two equivalent 2° hydrogen atoms, replacement of any one of which leads to isopropyl chloride.

$$CH_3CH_2CH_3 + Cl_2 \longrightarrow CH_3CH_2CH_2Cl + CH_3\underset{\displaystyle Cl}{CH}CH_3$$

If all the hydrogen atoms were equally reactive, we would expect to obtain 75% propyl chloride and 25% isopropyl chloride:

$$\% \text{ Propyl chloride} \; = \; \tfrac{6}{8} \; \times \; 100 \; = \; 75\%$$

$$\% \text{ Isopropyl chloride} \; = \; \tfrac{2}{8} \; \times \; 100 \; = \; 25\%$$

(b) Reasoning in the same way as in part (a), we would expect 90% isobutyl chloride and 10% *tert*-butyl chloride, if the hydrogen atoms were equally reactive.

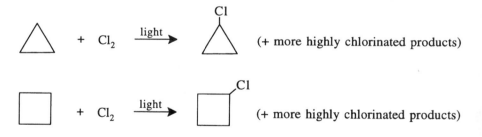

$$\% \text{ Isobutyl chloride} \; = \; \tfrac{9}{10} \; \times \; 100 \; = \; 90\%$$

$$\% \text{ tert-Butyl chloride} \; = \; \tfrac{1}{10} \; \times \; 100 \; = \; 10\%$$

(c) In the case of propane (see Section 10.6), we actually get more than twice as much isopropyl chloride (55%) than we would expect if the 1° and 2° hydrogen atoms were equally reactive (25%). Clearly, then, 2° hydrogen atoms are more than twice as reactive as 1° hydrogen atoms.

In the case of isobutane, we get almost four times as much *tert*-butyl chloride (37%) as we would get (10%) if the 1° and 3° hydrogen atoms were equally reactive. The order of reactivity of the hydrogens then must be

$$3° \; > \; 2° \; > \; 1°$$

10.13 The hydrogen atoms of these molecules are all equivalent. Replacing any one of them yields the same product.

We can minimize the amounts of more highly chlorinated products formed by using a large excess of the cyclopropane or cyclobutane. (And we can recover the unreacted cyclopropane or cyclobutane after the reaction is over.)

10.14 (a) ⬠   (b) $CH_3{-}\underset{\underset{H_3C}{|}}{\overset{\overset{H_3C}{|}}{C}}{-}\underset{\underset{CH_3}{|}}{\overset{\overset{CH_3}{|}}{C}}{-}CH_3$ (c) $CH_3{-}\underset{\underset{CH_3}{|}}{\overset{\overset{CH_3}{|}}{C}}{-}CH_3$

10.15 At lower temperatures, isomer distribution accurately reflects the inherent reactivities of the hydrogens of the alkanes. As the temperature is raised, chlorine atoms become more reactive, and hence less discriminating. If the temperature is high enough, hydrogens are replaced by chlorine on a purely statistical basis.

10.16 (a)

(2S,4S)-2,4-Dichloro-
pentane   (2R,4S)-2,4-Dichloro-
pentane

(b) They are diastereomers. (They are stereoisomers, but they are not mirror images of each other.)

(c) No, (2R,4S)-2,4-dichloropentane is achiral because it is a meso compound. (It has a plane of symmetry passing through C3.)

(d) No, the achiral meso compound would not be optically active.

(e) Yes, by fractional distillation or by gas chromatography. (Diastereomers have different physical properties. Therefore, the two isomers would have different vapor pressures.)

(f and g) In addition to the (2S,4S)-2,4-dichloropentane and (2R,4S)-2,4-dichloropentane isomers described previously, we would also get (2S,3S)-2,3-dichloropentane, (2S,3R)-2,3-dichloropentane and the following:

(optically active)   (optically inactive)   (optically active)

10.17 (a) Seven (see below)

(b)

(a)   (b)   (c)

(d)   (e)   (f)   (g)

(c) None of the fractions would show optical activity. Fractions (b), (d), and (f) are racemic forms; all others contain achiral molecules.

(d) Ions likely to be formed are the molecular ion (M^+), M^+-Br, and others. The M^+ peak would be accompanied by $M^+ + 2$ and $M^+ + 4$ peaks due to the isotopes of bromine. The natural abundance of ^{79}Br is 50.5% and the natural abundance of ^{81}Br is 49.5%. Thus, the M^+ peak (where both bromines are ^{79}Br) has probability $(.505)(.505) = .255$, the $M^+ + 2$ peak (one ^{79}Br and one ^{81}Br) has probability $2(.495)(.505) = .500$, and the probability of the $M^+ + 4$ peak (where both bromine atoms are ^{81}Br) is $(.495)(.495) = .245$. Thus, the M^+, $M^+ + 2$, and $M^+ + 4$ peaks will have an intensity ratio of approximately $1:2:1$, respectively. (This calculation does not take into account the contribution of ^{13}C, ^2H, and other isotopes.) The M^+-Br peak would also be accompanied by an M^+-Br+2 peak.

10.18 (a) No, the only fractions that would contain chiral molecules (as enantiomers) would be those containing 1-chloro-2-methylbutane and the one containing 2-chloro-3-methylbutane. These fractions would not show optical activity, however, because they would contain racemic forms of the enantiomers.

(b) Yes, the fractions containing 1-chloro-2-methylbutane and the one containing 2-chloro-3-methylbutane.

(c) Yes, each fraction from the distillation could be identified on the basis of ^1H NMR spectroscopy. The signals related to the carbons where the chlorine atom is bonded would be sufficient to distinguish them. The protons at C1 of 1-chloro-2-methyl butane would be a doublet due to splitting from the single hydrogen at C2. There would be no proton signal for C2 of 2-chloro-2-methylbutane since there are no hydrogens bonded at C2 in this compound; however, there would be a strong sin-glet for the six hydrogens of the geminal methyl groups. The protons at C2 of 2-chloro-3-methylbutane would approximately be a pentet, due to combined splitting from the three hydrogens at C1 and the single hydrogen at C3. The protons at C1 of 1-chloro-3-methylbutane would be a triplet due to splitting by the two hydro-gens at C2.

10.19 *Chain-Initiating Step*

$$Cl_2 \xrightarrow[\text{light}]{\text{heat,} h\nu} 2\ Cl\cdot$$

Chain-Propagating Steps

10.20 (a) Three

$$CH_3CH_2CH_2CH_3 \xrightarrow{Cl_2} CH_3CH_2CH_2CH_2Cl + II + III$$

Enantiomers as a racemic form

(b) Only two: one fraction containing I, and another fraction containing the enantiomers II and III as a racemic form. (The enantiomers, having the same boiling points, would distill in the same fraction.)

(c) Both of them.

(d) The fraction containing the enantiomers.

(e) The ^1H spectrum for 1-chlorobutane would show a triplet at approximately $\delta 3.6$-3.8, whereas the hydrogens at C2 of 2-chlorobutane (either enantiomer) would approximately be a pentet at $\delta 3.6$-3.8. The DEPT spectra for 1-chlorobutane would have three positive signals corresponding to the carbons of the three CH_2 groups and one negative signal corresponding to the CH_3. The DEPT spectra for the 2-chlorobutane enantiomers would have only one positive signal corresponding to the single CH_2 carbon, and three negative signals corresponding to the two CH_3 groups and the CH carbon.

(f) Molecular ions from both 1-chlorobutane and the 2-chlorobutane enantiomers would be present (but probably of different intensities). $M^{\ddot{+}}$ +2 peaks would also be present. Both compounds would likely undergo C-Cl bond cleavage to yield $C_4H_9^+$ cations. The mass spectrum of 1-chlorobutane would probably show loss of a propyl radical by C-C bond cleavage adjacent to the chlorine, resulting in an m/z 49 peak for CH_2Cl^+ (and m/z 51 from ^{37}Cl). Similar fragmentation in 2-chlorobutane would produce an m/z 63 peak for CH_3CHCl^+ (and m/z 65).

10.21 (a) Five

A   B   C   D   E

(b) Five. None of the fractions would be a racemic form.

(c) The fractions containing **A, D,** and **E.** The fraction containing **B** and **C** would be optically inactive. (**B** contains no stereocenter and **C** is a meso compound.)

10.22 (a) Oxygen-oxygen bonds are especially weak, that is,

$$HO-OH \qquad DH° = 213 \text{ kJ mol}^{-1}$$
$$CH_3CH_2O-OCH_3 \qquad DH° = 184 \text{ kJ mol}^{-1}$$

This means that a peroxide will dissociate into radicals at a relatively low temperature.

$$RO-OR \xrightarrow{100\text{-}200°C} 2\ RO\bullet$$

Oxygen-hydrogen single bonds, on the other hand, are very strong. (For HO-H, $DH° = 498 \text{ kJ mol}^{-1}$.) This means that reactions like the following will be highly exothermic.

$$RO\bullet\ +\ R-H\ \longrightarrow\ RO-H\ +\ R\bullet$$

(b) *Step 1* $(CH_3)_3CO-OC(CH_3)_3 \xrightarrow{\text{heat}} 2\ (CH_3)_3CO\bullet$ $\left.\begin{array}{c} \\ \\ \end{array}\right\}$ Chain Initiation

Step 2 $(CH_3)_3CO\bullet\ +\ R-H\ \longrightarrow\ (CH_3)_3COH\ +\ R\bullet$

Step 3 $R\bullet\ +\ Cl-Cl\ \longrightarrow\ R-Cl\ +\ Cl\bullet$ $\left.\begin{array}{c} \\ \\ \end{array}\right\}$ Chain Propagation

Step 4 $Cl\bullet\ +\ R-H\ \longrightarrow\ H-Cl\ +\ R\bullet$

10.23
$$\underset{(3°)}{CH_3\overset{\overset{\displaystyle CH_3}{|}}{\underset{\bullet}{C}}CH_2CH_3} > \underset{(2°)}{CH_3\overset{\overset{\displaystyle CH_3}{|}}{\underset{\bullet}{C}}HCHCH_3} > \underset{(1°)}{CH_3\overset{\overset{\displaystyle CH_3}{|}}{C}HCH_2CH_2\bullet} \sim \underset{(1°)}{CH_3\overset{\overset{\displaystyle CH_2\bullet}{|}}{C}HCH_2CH_3}$$

10.24 (1) $Cl_2 \longrightarrow 2\ Cl\bullet$ $\Delta H° = +243 \text{ kJ mol}^{-1}$

(2) $Cl\bullet\ +\ CH_4 \longrightarrow CH_3Cl\ +\ H\bullet$ $\Delta H° = +85.8 \text{ kJ mol}^{-1}$

(3) $H\bullet\ +\ Cl_2 \longrightarrow HCl\ +\ Cl\bullet$ $\Delta H° = -188 \text{ kJ mol}^{-1}$

This mechanism is highly unlikely to compete with one given in Section 10.4 because
step (2) of this mechanism is highly endothermic ($\Delta H° = +85.8 \text{ kJ mol}^{-1}$). This means
that the energy of activation for this step, a chain-propagating step, will have to be larger
than $+85.8 \text{ kJ mol}^{-1}$. Notice in Section 10.5B that neither of the chain-propagating steps
for the mechanism given in Section 10.4 has an energy of activation greater than $+15.9$

kJ mol^{-1}. The alternative mechanism given in this problem, consequently, will proceed at a rate that is very much slower than the one given in Section 10.4 and will, for all practical purposes, not compete with it at all.

10.25   (a) CH$_3$CH$_3$ $\xrightarrow[\substack{\text{heat,}\\\text{light}}]{\text{Br}_2}$ CH$_3$CH$_2$Br $\xrightarrow[\text{(S}_\text{N}\text{2)}]{\text{NaI}}$ CH$_3$CH$_2$I

(b) CH$_3$CH$_2$Br $\xrightarrow[\text{(S}_\text{N}\text{2)}]{\text{NaOH}}$ CH$_3$CH$_2$OH $\xrightarrow[\text{(--H}_2\text{)}]{\text{Na}}$ CH$_3$CH$_2$O$^-$ Na$^+$
[from part (a)]

$\downarrow$ CH$_3$CH$_2$Br

CH$_3$CH$_2$OCH$_2$CH$_3$

(c) $\xrightarrow[\substack{\text{heat,}\\\text{light}}]{\text{Br}_2}$ ⬠—Br $\xrightarrow[\text{heat (E2)}]{\text{(CH}_3)_3\text{COK}}$

(d) CH$_3$CHCH$_2$CH$_3$ with CH$_3$ above $\xrightarrow[\substack{\text{heat,}\\\text{light}}]{\text{Br}_2}$ CH$_3$CCH$_2$CH$_3$ (CH$_3$ above, Br below) $\xrightarrow[\substack{\text{EtOH,}\\\text{heat}\\\text{(E2)}}]{\text{NaOEt}}$ CH$_3$C=CHCH$_3$ (CH$_3$ above)

$\downarrow$ HBr, ROOR, heat, light

CH$_3$C—CHCH$_3$ (CH$_3$ above, H and Br below)
(by anti-Markovnikov addition)

(e) CH$_4$ $\xrightarrow[\substack{\text{heat,}\\\text{light}}]{\text{Br}_2}$ CH$_3$Br

HC≡CH $\xrightarrow[\text{2) CH}_3\text{Br}]{\text{1) NaNH}_2}$ HC≡C—CH$_3$ $\xrightarrow[\text{2) CH}_3\text{Br}]{\text{1) NaNH}_2}$ CH$_3$—C≡C—CH$_3$

(f) HC≡CH $\xrightarrow[\text{2) CH}_3\text{CH}_2\text{Br}]{\text{1) NaNH}_2}$ CH$_3$CH$_2$C≡CH $\xrightarrow[\substack{\text{or} \\ \text{H}_2, \text{Lindlar's} \\ \text{catalyst}}]{\text{H}_2, \text{Ni}_2\text{B}}$
[from part (a)]

CH$_3$CH$_2$CH=CH$_2$

$\downarrow$ HA, H$_2$O

CH$_3$CH$_2$CHCH$_3$
|
OH

(g) CH$_3$CH$_2$Br + Na$^+$ $^-$N=$\overset{+}{\text{N}}$=N$^-$ $\xrightarrow{\text{S}_\text{N}2}$ CH$_3$CH$_2$—N=$\overset{+}{\text{N}}$=N$^-$

10.26 Chlorine atoms are electronegative and abstract hydrogens according to their "electron richness." The electronegative fluorine atom reduces electron density in proportion to its proximity to the different CH$_2$ groups. The CH$_3$ group is the least reactive site because the bond dissociation energy for a CH$_3$ group is greater than that for a CH$_2$ group.

*10.27 Besides direct H· abstraction from C5 there would be many H· abstractions from the three methyl groups, leading to:

Any of these radicals could then, besides directly attacking chlorine, intramolecularly abstract H· from C5 (analogous to the "back biting" that explains branching during alkene radical polymerization).

*10.28

X

Use the single-bond dissociation energies of Table 10.1:

Table 10.1 Single-bond homolytic dissociation energies $DH°$ at 25°C

Compound	A : B $\longrightarrow$ A· + B· kJ mol^{-1}	Compound	kJ mol^{-1}
H–H	435	$(CH_3)_2CH–Br$	285
D–D	444	$(CH_3)_2CH–I$	222
F–F	159	$(CH_3)_2CH–OH$	385
Cl–Cl	243	$(CH_3)_2CH–OCH_3$	337
Br–Br	192	$(CH_3)_2CHCH_2–H$	410
I–I	151	$(CH_3)_3C–H$	381
H–F	569	$(CH_3)_3C–Cl$	328
H–Cl	431	$(CH_3)_3C–Br$	264
H–Br	366	$(CH_3)_3C–I$	207
H–I	297	$(CH_3)_3C–OH$	379
$CH_3–H$	435	$(CH_3)_3C–OCH_3$	326
$CH_3–F$	452	$C_6H_5CH_2–H$	356
$CH_3–Cl$	349	$CH_2=CHCH_2–H$	356
$CH_3–Br$	293	$CH_2=CH–H$	452
$CH_3–I$	234	$C_6H_5–H$	460
$CH_3–OH$	383	$HC≡C–H$	523
$CH_3–OCH_3$	335	$CH_3–CH_3$	368
$CH_3CH_2–H$	410	$CH_3CH_2–CH_3$	356
$CH_3CH_2–F$	444	$CH_3CH_2CH_2–CH_3$	356
$CH_3CH_2–Cl$	341	$CH_3CH_2–CH_2CH_3$	343
$CH_3CH_2–Br$	289	$(CH_3)_2CH–CH_3$	351
$CH_3CH_2–I$	224	$(CH_3)_3C–CH_3$	335
$CH_3CH_2–OH$	383	HO–H	498
$CH_3CH_2–OCH_3$	335	HOO–H	377
$CH_3CH_2CH_2–H$	410	HO–OH	213
$CH_3CH_2CH_2–F$	444	$(CH_3)_3CO—OC(CH_3)_3$	157
$CH_3CH_2CH_2–Cl$	341		
$CH_3CH_2CH_2–Br$	289	$C_6H_5\overset{O}{\overset{\|}{C}}O—O\overset{O}{\overset{\|}{C}}C_6H_5$	139
$CH_3CH_2CH_2–I$	224	$CH_3CH_2O–OCH_3$	184
$CH_3CH_2CH_2–OH$	383	$CH_3CH_2O–H$	431
$CH_3CH_2CH_2–OCH_3$	335		
$(CH_3)_2CH–H$	395	$CH_3\overset{O}{\overset{\|}{C}}–H$	364
$(CH_3)_2CH–F$	439		
$(CH_3)_2CH–Cl$	339		

QUIZ

10.1   On the basis of Table 10.1, what is the order of decreasing stability of the radicals,

HC≡C•   CH$_2$=CH•   CH$_2$=CHCH$_2$• ?

(a) HC≡C• > CH$_2$=CH• > CH$_2$=CHCH$_2$•

(b) CH$_2$=CH• > HC≡C• > CH$_2$=CHCH$_2$•

(c) CH$_2$=CHCH$_2$• > HC≡C• > CH$_2$=CH•

(d) CH$_2$=CHCH$_2$• > CH$_2$=CH• > HC≡C•

(e) CH$_2$=CH• > CH$_2$=CHCH$_2$• > HC≡C•

10.2   In the radical chlorination of methane, one propagation step is shown as

Cl• + CH$_4$ $\longrightarrow$ HCl + •CH$_3$

Why do we eliminate the possibility that this step goes as shown below?

Cl• + CH$_4$ $\longrightarrow$ CH$_3$Cl + H•

(a) Because in the next propagation step, H• would have to react with Cl$_2$ to form Cl• and HCl; this reaction is not feasible.

(b) Because this alternative step has a more endothermic $\Delta H°$ than the first.

(c) Because free hydrogen atoms cannot exist.

(d) Because this alternative step is not consistent with the high photochemical efficiency of this reaction.

10.3   Pure (S)-CH$_3$CH$_2$CHBrCH$_3$ is subjected to monobromination to form several isomers of C$_4$H$_8$Br$_2$. Which of the following is not produced?

(a)

CH$_3$
H──Br
H──Br
CH$_3$

(b)

CH$_3$
H──Br
Br──H
CH$_3$

(c)

CH$_3$
Br──H
H──Br
CH$_3$

(d) CH$_3$CH$_2$CBr$_2$CH$_3$

(e) (R)-CH$_3$CH$_2$CHBrCH$_2$Br

10.4   Using the data of Table 10.1, calculate the heat of reaction, $\Delta H°$, of the reaction,

CH$_3$CH$_3$ + Br$_2$ $\longrightarrow$ CH$_3$CH$_2$Br + HBr

(a) 52.3 kJ mol^{-1} (b) −52.3 kJ mol^{-1} (c) 1257 kJ mol^{-1}

(d) −1257 kJ mol^{-1} (e) −244.8 kJ mol^{-1}

10.5 Which gas-phase reaction would have $E_{act} = 0$?

(a) $CH_3\bullet + (CH_3)_3C-H \longrightarrow CH_4 + (CH_3)_3C\bullet$

(b) $CH_3\bullet + CH_3CH_3 \longrightarrow CH_4 + CH_3CH_2\bullet$

(c) $CH_3CH_2\bullet + CH_3CH_2\bullet \longrightarrow CH_3CH_2CH_2CH_3$

(d) $Br\bullet + H-Cl \longrightarrow H-Br + Cl\bullet$

(e) $Br\bullet + H-I \longrightarrow H-Br + I\bullet$

10.6 What is the most stable radical that would be formed in the following reaction?

$$Cl\bullet + CH_3CH_2\overset{\overset{\textstyle CH_3}{|}}{C}HCH_3 \longrightarrow \boxed{} + HCl$$

10.7 The reaction of 2-methylbutane with chlorine would yield a total of _____ different monochloro products (including stereoisomers).

10.8 For which reaction would the transition state most resemble the products?

(a) $CH_4 + F\bullet \longrightarrow CH_3\bullet + HF$

(b) $CH_4 + Cl\bullet \longrightarrow CH_3\bullet + HCl$

(c) $CH_4 + Br\bullet \longrightarrow CH_3\bullet + HBr$

(d) $CH_4 + I\bullet \longrightarrow CH_3\bullet + HI$

A.1   Head-to-tail polymerization leads to a more stable radical on the growing polymer chain. In head-to-tail coupling, the radical is 2° (actually 2° benzylic, and as we shall see in Section 15.12A this makes it even more stable). In head-to-head coupling, the radical is 1°.

A.2   (a)

(b)   R• + CH₂=CCl₂ ⟶ R—CH₂—CCl₂• ...

A.3   In the cationic polymerization of isobutylene (see text), the growing polymer chain has a stable 3° carbocation at the end. In the cationic polymerization of ethene, for example, the intermediates would be much less stable 1° cations.

With vinyl chloride and acrylonitrile, the cations at the end of the growing chain would be destabilized by electron-withdrawing groups.

A.4

A.5   (a)

Atactic polystyrene
$(Ph = C_6H_5)$

Syndiotactic polystyrene
$(Ph = C_6H_5)$

Isotactic polystyrene
$(Ph = C_6H_5)$

(b) The solution of isotactic polystyrene.

11 ALCOHOLS AND ETHERS

SOLUTIONS TO PROBLEMS

11.1 These names mix two systems of nomenclature (radicofunctional and substitutive; see Section 4.3F). The proper names are: isopropyl alcohol (radicofunctional) or 2-propanol (substitutive), and *tert*-butyl alcohol (radicofunctional) and 2-methyl-2-propanol (substitutive). Names with mixed systems of nomenclature should not be used.

11.2 (a)

1-Propanol   2-Propanol   Methoxyethane
(Propyl alcohol)   (Isopropyl alcohol)   (Ethyl methyl ether)

(b)

1-Butanol   2-Methyl-1-propanol   2-Butanol   2-Methyl-2-propanol
(Butyl alcohol)   (Isobutyl alcohol)   (*sec*-Butyl alcohol)   (*tert*-Butyl alcohol)

1-Methoxypropane   Ethoxyethane   2-Methoxypropane
(Methyl propyl ether)   (Diethyl ether)   (Isopropyl methyl ether)

11.3 The presence of two —OH groups in each molecule of 1,2-propanediol and 1,3-propanediol allows their molecules to form more hydrogen bonds. Greater hydrogen-bond formation means that the molecules of 1,2-propanediol and 1,3-propanediol are more highly associated, and, consequently, their boiling points are higher.

11.4 (a) CH_3CH_2OH (b) CH_3CHCH_3 (c) $CH_3\overset{\underset{\textstyle CH_3}{|}}{\overset{\textstyle CH_3}{C}}{-}OH$ (d) $CH_3\overset{\underset{\textstyle OH}{|}}{\overset{\textstyle CH_3}{C}}CH_2CH_3$

                                       $\underset{\textstyle OH}{|}$

11.5 A rearrangement takes place.

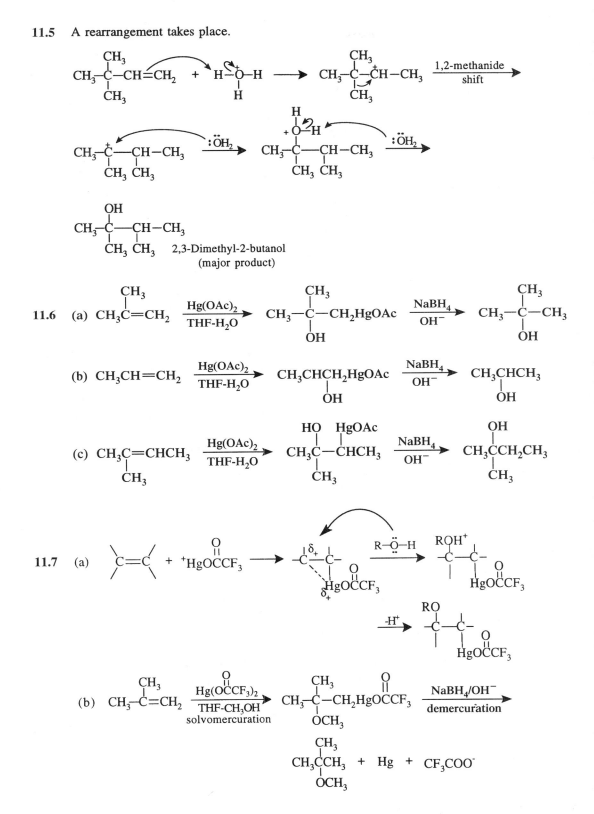

11.6 (a), (b), (c) and **11.7** (a), (b) reaction schemes as shown.

11.8 (a) $3 \; CH_3CH_2CH{=}CH_2 \xrightarrow{\text{THF: BH}_3} (CH_3CH_2CH_2CH_2)_3B$

(b) $3 \; CH_3{-}\overset{\overset{\displaystyle CH_3}{|}}{C}{=}CH_2 \xrightarrow{\text{THF: BH}_3} (CH_3\overset{\overset{\displaystyle CH_3}{|}}{CH}CH_2)_3B$

(c) $3 \; CH_3CH{=}CHCH_3 \xrightarrow{\text{THF:BH}_3} (CH_3CH_2\overset{\overset{\displaystyle CH_3}{|}}{CH})_3B$

(d) 3 $\xrightarrow[\substack{\text{syn addition}\\ \text{anti Markovnikov}}]{\text{THF: BH}_3}$ + enantiomer

11.9 $CH_3\overset{\overset{\displaystyle CH_3}{|}}{C}{=}CHCH_3 \xrightarrow{\text{THF: BH}_3} \left(CH_3\overset{\overset{\displaystyle CH_3}{|}}{CH}{-}\underset{\underset{\displaystyle CH_3}{|}}{CH}{-}\right)_2 BH$

Disiamylborane

11.10 (a) $CH_3CH_2CH_2CH{=}CH_2 \xrightarrow[\text{(2) } H_2O_2, \text{ OH}]{\text{(1) THF:BH}_3} CH_3CH_2CH_2CH_2CH_2OH$
1-Pentene

(b) $CH_3CH_2CH_2\underset{\underset{\displaystyle CH_3}{|}}{C}{=}CH_2 \xrightarrow[\text{(2) } H_2O_2, \text{ OH}]{\text{(1) THF:BH}_3} CH_3CH_2CH_2\underset{\underset{\displaystyle CH_3}{|}}{CH}CH_2OH$
2-Methyl-1-pentene

(c) $CH_3CH_2\underset{\underset{\displaystyle CH_3}{|}}{C}{=}CHCH_3 \xrightarrow[\text{(2) } H_2O_2, \text{ OH}]{\text{(1) THF:BH}_3} CH_3CH_2\underset{\underset{\displaystyle CH_3}{|}}{CH}\overset{\overset{\displaystyle OH}{|}}{CH}CH_3$
3-Methyl-2-pentene

(d) $CH_3\underset{\underset{\displaystyle CH_3}{|}}{C}{=}CHCH_2CH_3 \xrightarrow[\text{(2) } H_2O_2, \text{ OH}]{\text{(1) THF:BH}_3} CH_3\underset{\underset{\displaystyle CH_3}{|}}{CH}\overset{\overset{\displaystyle OH}{|}}{CH}CH_2CH_3$
2-Methyl-2-pentene

(e) $-CH_3 \xrightarrow[\text{(2) } H_2O_2, \text{ OH}]{\text{(1) THF:BH}_3}$ + enantiomer
1-Methylcyclobutene

11.11 (a) $3 \; CH_3\underset{\underset{\displaystyle CH_3}{|}}{CH}CH{=}CH_2 \xrightarrow{\text{THF:BH}_3} \left(CH_3\underset{\underset{\displaystyle CH_3}{|}}{CH}CH_2CH_2{-}\right)_3 B \xrightarrow{CH_3CO_2D}$

$3 \; CH_3\underset{\underset{\displaystyle CH_3}{|}}{CH}CH_2CH_2D$

11.12 (a)

$$CH_3CH_2OH \quad + \quad NaNH_2 \quad \longrightarrow \quad CH_3CH_2ONa \quad + \quad NH_3$$

Stronger acid   Stronger base   Weaker base   Weaker acid

(b) $CH_3CH_2OH \quad + \quad HC{\equiv}CNa \quad \longrightarrow \quad CH_3CH_2ONa \quad + \quad HC{\equiv}CH$

(c) $CH_3CH_2OH \quad + \quad NaOAc \quad \xrightarrow{\quad\times\quad} \quad CH_3CH_2ONa \quad + \quad HOAc$

Weaker acid   Weaker base   Stronger base   Stronger acid

11.13 Use an alcohol containing labeled oxygen. If all of the label appears in the sulfonate ester, then one can conclude that the alcohol C–O bond does not break during the reaction:

$$R{-}^{18}O{-}H \quad + \quad R'{-}SO_2Cl \quad \xrightarrow[\text{(-HCl)}]{\text{base}} \quad R{-}^{18}O{-}SO_2{-}R'$$

11.14 (a)

$$H_3C{-}\underset{}{\bigcirc}{-}SO_2OH \quad \xrightarrow[(-POCl_3, -HCl)]{PCl_5} \quad H_3C{-}\underset{}{\bigcirc}{-}SO_2Cl$$

$$\xrightarrow[\text{base} (-HCl)]{CH_3OH} \quad H_3C{-}\underset{}{\bigcirc}{-}SO_2OCH_3$$

(b) CH_3SO_2OH $\xrightarrow[(-POCl_3, -HCl)]{PCl_5}$ CH_3SO_2Cl $\xrightarrow[base\ (-HCl)]{(CH_3)_2CHCH_2OH}$

$CH_3SO_2OCH_2CH(CH_3)_2$

(c) CH_3SO_2Cl $\xrightarrow[base\ (-HCl)]{(CH_3)_3COH}$ $CH_3SO_2OC(CH_3)_3$

11.15   (a)

(a)
(R)-2-Butanol

(b)

(c)

cis-4-Methyl-
cyclohexanol

trans-1-Chloro-4-
methylcyclohexane

11.16   (a) Tertiary alcohols react faster than secondary alcohols because they form more stable carbocations; that is, 3° rather than 2°:

(b) CH_3OH reacts faster than 1° alcohols because it offers less hindrance to S_N2 attack. (Recall that CH_3OH and 1° alcohols must react through an S_N2 mechanism.)

11.17

This reaction succeeds because a 3° carbocation is much more stable than a 1° carbocation. Consequently, mixing the 1° alcohol and H_2SO_4 does not lead to formation of appreciable amounts of a 1° carbocation. However, when the 3° alcohol is added, it is rapidly converted to a 3° carbocation, which then reacts with the 1° alcohol that is present in the mixture.

11.19 (a) (1) CH_3CHO^- Na^+ + CH_3—L $\longrightarrow$ CH_3CHO—CH_3 + L$^-$ + Na^+
(L = X, OSO$_2$R, or OSO$_2$OR)

(2) CH_3O^- + CH_3—CH—L $\longrightarrow$ CH_3O-CHCH$_3$ + L$^-$
(L = X, OSO$_2$R, or OSO$_2$OR)

(b) Both methods involve S_N2 reactions. Therefore, method (1) is better because substitution takes place at an unhindered methyl carbon atom. In method (2) where substitution must take place at a relatively hindered secondary carbon atom, the reaction would be accompanied by considerable elimination.

11.20 Reaction of the alcohol with K and then of the resulting salt with C_2H_5Br does not break bonds to the stereocenter, and these reactions therefore occur with retention of configuration at the stereocenter.

Reaction of the tosylate, $C_6H_5CH_2CHCH_3$, with C_2H_5OH in K_2CO_3 solution, however,
|
OTs

is an S_N2 reaction that takes place at the stereocenter and thus it occurs with inversion at the stereocenter.

11.21 Cl—CH$_2$ CH$_2$—CH$_2$ | CH$_2$ | OH $\xrightarrow{OH^-}$ Cl—CH$_2$ CH$_2$—CH$_2$ | CH$_2$ | :O:$^-$ + H$_2$O $\longrightarrow$ CH$_2$—CH$_2$ | CH$_2$ O + Cl$^-$

11.22   (a)   HO⁻ + HOCH₂-CH₂-Cl $\rightleftharpoons$ H₂O + ⁻O-CH₂-CH₂-Cl $\longrightarrow$

(b)   The —Ö⁻: group must displace the Cl⁻ from the backside,

trans-2-Chlorocyclohexanol

Backside attack is not possible with the cis isomer (below); therefore, it does not form an epoxide.

cis-2-Chlorocyclohexanol

11.23   (a)

$$CH_2{=}\underset{\underset{CH_3}{|}}{\overset{\overset{CH_3}{|}}{C}} \xrightarrow{H^+} CH_3{-}\underset{\underset{CH_3}{|}}{\overset{\overset{CH_3}{|}}{C^+}} \xrightarrow{R\text{-}OH} CH_3{-}\underset{\underset{CH_3}{|}}{\overset{\overset{CH_3}{|}}{\underset{}{C}}}{-}\overset{+}{O}R \xrightarrow{-H^+} CH_3{-}\underset{\underset{CH_3}{|}}{\overset{\overset{CH_3}{|}}{C}}{-}OR$$

(b)   The *tert*-butyl group is easily removed because, in acid, it is easily converted to a relatively stable, tertiary carbocation.

(c)

11.24   (a)

CH₃O—C‴CH₃ + HI $\longrightarrow$ I⁻ + CH₃—Ö—C‴CH₃ $\longrightarrow$
with CH₂CH₃ substituent

ICH₃ + HO—C‴CH₃
with CH₂CH₃ substituent

S_N2 attack of I^- occurs at the methyl carbon atom because it is less hindered; therefore, the bond between the *sec*-butyl group and the oxygen is not broken.

(b) $CH_3-O-C(CH_3)_3$ + HI $\longrightarrow$ $CH_3-\overset{+}{O}-C(CH_3)_3$ + $\overset{-}{I}$

In this reaction the much more stable *tert*-butyl cation is produced. It then combines with I^- to form *tert*-butyl iodide.

11.25 (a)

$\xrightarrow{A^-}$ $HOCH_2CH_2OCH_3$
Methyl Cellosolve

(b) An analogous reaction yields ethyl cellosolve, $HOCH_2CH_2OCH_2CH_3$.

(c) H_2C-CH_2 (epoxide) $\xrightarrow{I^-}$ $\underset{O^-}{CH_2CH_2}$ (with I) $\xrightarrow{H_2O}$ $HOCH_2CH_2I$ + OH^-

(d) H_2C-CH_2 (epoxide) $\xrightarrow{:NH_3}$ $\underset{O^-}{CH_2CH_2}$ (with NH_3^+) $\longrightarrow$ $HOCH_2CH_2NH_2$

(e) H_2C-CH_2 (epoxide) $\xrightarrow{CH_3O^-}$ $\underset{O^-}{CH_2CH_2}$ (with OCH_3) $\xrightarrow{CH_3OH}$ $HOCH_2CH_2OCH_3$ + CH_3O^-

11.26 The reaction is an S_N2 reaction, and thus nucleophilic attack takes place much more rapidly at the primary carbon atom than at the more hindered secondary carbon atom.

11.27 Ethoxide ion attacks the epoxide ring at the primary carbon because it is less hindered, and the following reactions take place.

11.28

11.29

11.30 (a)

15-Crown-5

(b)

12-Crown-4

11.31 (a) 3,3-Dimethyl-1-butanol

(b) 4-Penten-2-ol

(c) 2-Methyl-1,4-butanediol

(d) 2-Phenylethanol

(e) 1-Methyl-2-cyclopenten-1-ol

(f) *cis*-3-Methylcyclohexanol

11.32   (a)

(b)

(c)

(d)

(e) CH₃CH₂C≡CCHCH₂OH
              |
              Cl

(f) H₂C——CH₂
     H₂C   CH₂
         O

(g) CH₃CHCH₂CH₂CH₃
        |
        OCH₂CH₃

(h) CH₃CH₂—O—⬡

(i) CH₃CH—O—CHCH₃
        | |
        CH₃   CH₃

(j) CH₃CH₂—O—CH₂CH₂OH

11.33   (a) CH₃CH₂CH=CH₂ $\xrightarrow[\text{(hydroboration)}]{\text{THF}:\text{BH}_3}$ (CH₃CH₂CH₂CH₂)₃B

$\xrightarrow[\text{(oxidation)}]{\text{H}_2\text{O}_2/\text{OH}^-}$ CH₃CH₂CH₂CH₂OH

(b) CH₃CH₂CH₂CH₂Cl $\xrightarrow{\text{OH}^-}$ CH₃CH₂CH₂CH₂OH

(c) CH₃CH₂CHCH₃ $\xrightarrow{(\text{CH}_3)_3\text{COK}/(\text{CH}_3)_3\text{COH}}$ CH₃CH₂CH=CH₂
        |
        Cl

$\xrightarrow[\text{ROOR}]{\text{HBr}}$ CH₃CH₂CH₂CH₂Br $\xrightarrow{\text{OH}^-}$ CH₃CH₂CH₂CH₂OH

(d) CH₃CH₂C≡CH $\xrightarrow[\text{Ni}_2\text{B (P-2)}]{\text{H}_2}$ CH₃CH₂CH=CH₂

$\xrightarrow{\text{[as in (a)]}}$ CH₃CH₂CH₂CH₂OH

11.34   (a) 3 CH₃CH₂CHCH₃ + PBr₃ ⟶ 3 CH₃CH₂CHCH₃ + H₃PO₃
                | |
                OH   Br

(b) CH₃CH₂CH₂CH₂OH $\xrightarrow{\text{PBr}_3}$ CH₃CH₂CH₂CH₂Br $\xrightarrow[(\text{CH}_3)_3\text{COH}]{(\text{CH}_3)_3\text{COK}}$

CH₃CH₂CH=CH₂ $\xrightarrow[\text{(no peroxides)}]{\text{HBr}}$ CH₃CH₂CHCH₃
                                                            |
                                                            Br

(c)   See (b) above.

(d)

$CH_3CH_2C\equiv CH$ $\xrightarrow[\text{Ni}_2\text{B (P-2)}]{\text{H}_2}$ $CH_3CH_2CH=CH_2$ $\xrightarrow[\text{(no peroxides)}]{\text{HBr}}$

$\underset{\underset{\text{Br}}{|}}{CH_3CH_2CHCH_3}$

11.35   (a)   [cyclohexanol] $+$ $SOCl_2$ $\longrightarrow$ [chlorocyclohexane] $+$ SO_2 $+$ HCl

(b)   [cyclohexene] $+$ HCl $\longrightarrow$ [chlorocyclohexane]

(c)   [1-methylcyclohexene with CH$_3$] $\xrightarrow[\text{(no peroxides)}]{\text{HBr}}$ [1-bromo-1-methylcyclohexane]

(d)   [1-methylcyclohexene with CH$_3$] $\xrightarrow[\text{(2) H}_2\text{O}_2,\text{OH}^-]{\text{(1) THF : BH}_3}$ [2-methylcyclohexanol] $+$ enantiomer

(e)   [1-bromo-1-methylcyclohexane] $\xrightarrow[\text{t-BuOH}]{\text{t-BuOK}}$ [methylenecyclohexane] $\xrightarrow[\text{(2) H}_2\text{O}_2,\text{ OH}^-]{\text{(1) THF : BH}_3}$ [cyclohexylmethanol CH$_2$OH]

11.36   (a) $CH_3CH_2CH_2CH_2ONa$ Sodium butoxide

(b) $CH_3CH_2CH_2CH_2OCH_2CH_2CH_3$ Butyl propyl ether

(c) $CH_3SO_2OCH_2CH_2CH_2CH_3$ Butyl mesylate

(d) H_3C-⟨benzene ring⟩$-SO_2OCH_2CH_2CH_2CH_3$ Butyl tosylate

(e) $CH_3OCH_2CH_2CH_2CH_3$ 1-Methoxybutane

(f) $CH_3CH_2CH_2CH_2I$ Butyl iodide

(g) $CH_3CH_2CH_2CH_2Cl$ Butyl chloride

(h)   same as (g)

(i) $CH_3CH_2CH_2CH_2OCH_2CH_2CH_2CH_3$ Dibutyl ether

(j) $CH_3CH_2CH_2CH_2Br$ Butyl bromide

$$\text{(k)} \quad CH_3CH_2CH_2CH_2O-\underset{\underset{CH_3}{|}}{\overset{\overset{CH_3}{|}}{Si}}-C(CH_3)_3 \qquad \text{Butyl } \textit{tert}\text{-butyldimethylsilyl ether}$$

(l) $CH_3CH_2CH_2CH_2OH$ 1-Butanol

11.37 (a) $CH_3CH_2\overset{\overset{CH_3}{|}}{C}HONa$ Sodium *sec*-butoxide

(b) $CH_3CH_2\overset{\overset{CH_3}{|}}{C}HOCH_2CH_2CH_3$ *sec*-Butyl propyl ether

(c) $CH_3SO_2O\overset{\overset{CH_3}{|}}{C}HCH_2CH_3$ *sec*-Butyl mesylate

(d) $H_3C-\langle\bigcirc\rangle-SO_2O\overset{\overset{CH_3}{|}}{C}HCH_2CH_3$ *sec*-Butyl tosylate

(e) $CH_3O\overset{\overset{CH_3}{|}}{C}HCH_2CH_3$ 2-Methoxybutane

(f) $CH_3\overset{\underset{I}{|}}{C}HCH_2CH_3$ *sec*-Butyl iodide

(g) $CH_3\overset{\underset{Cl}{|}}{C}HCH_2CH_3$ *sec*-Butyl chloride

(h) same as (g)

$$\text{(i)} \quad \underset{H}{\overset{H_3C}{}}C=C\underset{CH_3}{\overset{H}{}} \qquad \text{mainly, } \textit{trans}\text{-2-Butene}$$

(j) $CH_3\overset{\underset{Br}{|}}{C}HCH_2CH_3$ *sec*-Butyl bromide

$$\text{(k)} \quad CH_3CH_2CHO-\underset{\underset{CH_3}{|}}{\overset{\overset{CH_3}{|}}{Si}}-C(CH_3)_3 \qquad \textit{sec}\text{-Butyl } \textit{tert}\text{-butyldimethylsilyl ether}$$
$$\qquad\qquad\quad \underset{CH_3}{|}$$

(l) $CH_3CH_2\overset{\underset{OH}{|}}{C}HCH_3$ 2-Butanol

11.38 (a) $CH_3Br + CH_3CH_2Br$ (c) $Br-CH_2CH_2CH_2CH_2-Br$

(b) $CH_3\overset{\overset{CH_3}{|}}{\underset{\underset{CH_3}{|}}{C}}-Br + CH_3CH_2Br$ (d) $Br-CH_2CH_2-Br$ (2 molar equivalents)

(d) $CH_3CHCH_2Br \xrightarrow[\text{acetone}]{KI} CH_3CHCH_2I$ (with CH_3 substituents)

(e) $CH_3CHCH_2Br \xrightarrow[H_2O]{OH^-} CH_3CHCH_2OH$

or

$CH_3C=CH_2 \xrightarrow[\text{(2) } H_2O_2,OH^-]{\text{(1) } (BH_3)_2} CH_3CHCH_2OH$

(f) $CH_3C=CH_2 \xrightarrow[\text{heat}]{H_3O^+, H_2O} CH_3CCH_3$ (OH)

(g) $CH_3CHCH_2Br \xrightarrow[CH_3OH]{CH_3ONa} CH_3CHCH_2OCH_3$

(h) $CH_3CHCH_2Br \xrightarrow[CH_3COH]{CH_3CONa} CH_3CHCH_2OCCH_3$

(i) $CH_3CHCH_2Br \xrightarrow{NaCN} CH_3CHCH_2CN$

(j) $CH_3CHCH_2Br \xrightarrow{CH_3SNa} CH_3CHCH_2SCH_3$

or

$CH_3C=CH_2 \xrightarrow[\text{peroxides}]{CH_3SH} CH_3CHCH_2SCH_3$

(k) $CH_3C=CH_2 \xrightarrow[\text{peroxides}]{CBr_4} CH_3CCH_2CBr_3$ (Br)

11.43 (a)

(b) The trans product because the Cl$^-$ attacks anti to the epoxide and an inversion of configuration occurs.

11.44 $HC\equiv CH$ $\xrightarrow[\text{liq. NH}_3]{\text{NaNH}_2}$ $HC\equiv CNa$ $\xrightarrow{\underset{}{\overset{CH_3}{CH_3\overset{|}{C}H(CH_2)_4Br}}}$ $CH_3\overset{\overset{CH_3}{|}}{C}H(CH_2)_4C\equiv CH$

A (C_9H_{16})

$\xrightarrow[\text{liq. NH}_3]{\text{NaNH}_2}$ $CH_3\overset{\overset{CH_3}{|}}{C}H(CH_2)_4C\equiv CNa$ $\xrightarrow{CH_3(CH_2)_9Br}$ $CH_3\overset{\overset{CH_3}{|}}{C}H(CH_2)_4C\equiv C(CH_2)_9CH_3$

B ($C_9H_{15}Na$) **C** ($C_{19}H_{36}$)

$\xrightarrow[\text{Ni}_2\text{B (P-2)}]{\text{H}_2}$ $\underset{H \qquad\qquad H}{\overset{\overset{CH_3}{|}}{CH_3\overset{}{C}H(CH_2)_4}}C=C\,(CH_2)_9CH_3$ **D** ($C_{19}H_{38}$) $\xrightarrow{C_6H_5CO_3H}$

$\underset{H \quad O \quad H}{\overset{\overset{CH_3}{|}}{CH_3\overset{}{C}H(CH_2)_4}}\underset{}{C-C}\,(CH_2)_9CH_3$ **E** ($C_{19}H_{38}O$)

11.45 (a) $CH_3\overset{\overset{CH_3}{|}}{C}=CH_2$ $\xrightarrow[\text{(2) H}_2\text{O}_2,\ \text{OH}^-]{\text{(1) THF:BH}_3}$ $CH_3\overset{\overset{CH_3}{|}}{C}HCH_2OH$

(b) $CH_3\overset{\overset{CH_3}{|}}{C}=CH_2$ $\xrightarrow[\text{(2) CH}_3\text{CO}_2\text{T}]{\text{(1) THF: BH}_3}$ $CH_3\overset{\overset{CH_3}{|}}{C}HCH_2T$

(c) $CH_3\overset{\overset{CH_3}{|}}{C}=CH_2$ $\xrightarrow{\text{THF: BD}_3}$ $CH_3\underset{D}{\overset{\overset{CH_3}{|}}{C}}-CH_2-B\diagdown$ $\xrightarrow{CH_3CO_2T}$ $CH_3\underset{D}{\overset{\overset{CH_3}{|}}{C}}CH_2T$

(d) $CH_3\overset{\overset{CH_3}{|}}{C}HCH_2OH$ $\xrightarrow{\text{Na}}$ $CH_3\overset{\overset{CH_3}{|}}{C}HCH_2ONa$ $\xrightarrow{CH_3CH_2Br}$

$CH_3\overset{\overset{CH_3}{|}}{C}HCH_2-O-CH_2CH_3$

11.46 (a) $CH_3CH_2CH_2CH=CH_2$ $\xrightarrow{\text{Hg(OAc)}_2}$ $CH_3CH_2CH_2\underset{OH}{\overset{}{C}}HCH_2HgOAc$

$\xrightarrow[\text{OH}^-]{\text{NaBH}_4}$ $CH_3CH_2CH_2\underset{OH}{\overset{}{C}}HCH_3$

(b) ⬠$-CH=CH_2$ $\xrightarrow[\text{(2) NaBH}_4/\text{OH}^-]{\text{(1) Hg(OAc)}_2}$ ⬠$-\underset{OH}{\overset{}{C}}HCH_3$

(c) CH₃CH=CCH₂CH₃ (with CH₃ substituent) →(1) Hg(OAc)₂ (2) NaBH₄/OH⁻→ CH₃CH₂CCH₂CH₃ (with CH₃ and OH)

(d) cyclopentane=CHCH₃ →(1) Hg(OAc)₂ (2) NaBH₄/OH⁻→ cyclopentane with CH₂CH₃ and OH

*11.47 (a) **A** = cyclobutane with ⁗CH₃ and OH + enantiomer **B** = cyclobutane with ⁗CH₃ and OTs + enantiomer

C = cyclobutane with ⁗CH₃ and ⁗OH + enantiomer

(b) Diastereomers

(c) **D** = cyclobutane with ⁗CH₃ and ⁗I + enantiomer

(d) **E** = H₃C–cyclohexane with H and OMs

F = H₃C–cyclohexane with C≡CH and H ⇌ H–cyclohexane with CH₃ and C≡CH

(e) **H** = CH₃CH₂–C(CH₃)(⁗H)–ONa   **J** = CH₃CH₂–C(CH₃)(⁗H)–OCH₃

(f) **K** = CH₃CH₂–C(CH₃)(⁗H)–OMs   **L** = CH₃CH₂–C(CH₃)(⁗OCH₃)–H

(g) Enantiomers

*11.48   The reactions proceed through the formation of bromonium ions identical to those formed in the bromination of *trans-* and *cis*-2-butene (see Section 8.7A).

meso-2,3-Dibromobutane

(Attack at the other carbon atom of the bromonium ion gives the same product.)

(±)-2,3-Dibromobutane

*11.49

*11.50

A and B are enantiomers
A, C, and D are all diastereomers
B, C, and D are all diastereomers
C is meso
D is meso

QUIZ

11.1 Which set of reagents would effect the conversion,

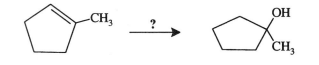

(a) BH$_3$:THF, then H$_2$O$_2$/OH$^-$ (b) H$_2$O/Hg(OAc)$_2$, THF, then NaBH$_4$/OH$^-$

(c) H$_3$O$^+$, H$_2$O, heat   (d) More than one of these   (e) None of these

11.2 Which of the reagents in item 11.1 would effect the conversion,

11.3   The following compounds have identical molecular weights. Which would have the lowest boiling point?

(a) 1-Butanol

(b) 2-Butanol

(c) 2-Methyl-1-propanol

(d) 1,1-Dimethylethanol

(e) 1-Methoxypropane

11.4   Complete the following synthesis:

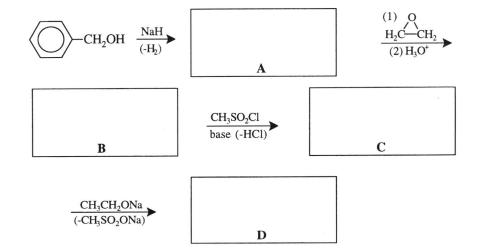

12 ALCOHOLS FROM CARBONYL COMPOUNDS: OXIDATION–REDUCTION AND ORGANOMETALLIC COMPOUNDS

SOLUTIONS TO PROBLEMS

12.1 (a)

$$\begin{array}{ccc} & H & \\ & | & \\ H-&C&-O-H \\ & | & \\ & H & \end{array}$$

3 **H** = -3
1 **O** = +1
Total = -2   = oxidation state of C

$$\begin{array}{c} O \\ \parallel \\ H-C-O-H \end{array}$$

1 **H** = -1
3 **O** = +3
Total = +2   = oxidation state of C

$$\begin{array}{c} O \\ \parallel \\ H-C-H \end{array}$$

2 **H** = -2
2 **O** = +2
Total = 0   = oxidation state of C

(b)

CH_4 CH_3OH $\overset{O}{\overset{\parallel}{H-C-H}}$ $\overset{O}{\overset{\parallel}{H-C-O-H}}$ CO_2

-4   -2   0   +2   +4

(c) A change from − 2 to 0

(d) An oxidation, since the oxidation state increases

(e) A reduction from +6 to +3

12.2 (a)

$$\begin{array}{ccc} H & H & \\ | & | & \\ H-C-&C&-OH \\ | & | & \\ H & H & \end{array}$$

3 **H** = -3
1 **C** = 0
Total = -3

2 **H** = -2
1 **C** = 0
1 **O** = +1
Total = -1

$$\begin{array}{ccc} H & O & \\ | & \parallel & \\ H-C-&C&-H \\ | & & \\ H & & \end{array}$$

3 **H** = -3
1 **C** = 0
Total = -3

1 **H** = -1
1 **C** = 0
2 **O** = +2
Total = +1

(b) Only the carbon atom of the $-CH_2OH$ group of ethanol undergoes a change in oxidation state. The oxidation state of the carbon atom in the CH_3- group remains unchanged.

(c)

The oxygen-bearing carbon atom increases its oxidation state from +1 (in acetaldehyde) to +3 (in acetic acid).

$$3 \ H = -3 \qquad 1 \ C = 0$$
$$1 \ C = 0 \qquad 3 \ O = +3$$
$$\overline{Total = -3} \qquad \overline{Total = +3}$$

12.3   (a) If we consider the hydrogenation of ethene as an example, we find that the oxidation state of carbon decreases. Thus, because the reaction involves the *addition* of *hydrogen*, it is both an *addition reaction* and a *reduction*.

$$2 \ H = -2 \qquad\qquad 3 \ H = -3$$
$$2 \ C = 0 \qquad\qquad 1 \ C = 0$$
$$\overline{Total = -2} \qquad\qquad \overline{Total = -3}$$

(b) The hydrogenation of acetaldehyde is not only an addition reaction, but it is also a *reduction* because the carbon atom of the C=O group goes from a + 1 to a − 1 oxidation state. The reverse reaction (the *dehydrogenation* of ethanol) is not only an *elimination* reaction, but also an *oxidation*.

Ion-Electron Half-Reaction Method for Balancing Organic Oxidation–Reduction Equations

Only two simple rules are needed:

Rule 1   Electrons (e^-) together with protons (H^+) are arbitrarily considered the reducing agents in the half-reaction for the reduction of the oxidizing agent. Ion charges are balanced by *adding electrons to the left-hand side*. (If the reaction is run in neutral or basic solution, add an equal number of OH^- ions to both sides of the balanced half-reaction to neutralize the H^+, and show the resulting $H^+ + OH^-$ as H_2O.)

Rule 2   Water (H_2O) is arbitrarily taken as the formal source of oxygen for the oxidation of the organic compound, producing *product, protons,* and *electrons* on the right-hand side. (Again, use OH^- to neutralize H^+ in the *balanced* half-reaction in neutral or basic media.)

EXAMPLE 1

Write a balanced equation for the oxidation of RCH_2OH to RCO_2H by $Cr_2O_7^{2-}$ in acid solution.

Reduction half-reaction:

$$Cr_2O_7^{2-} + H^+ + e^- \longrightarrow 2Cr^{3+} + 7H_2O$$

Balancing atoms and charges:

$$Cr_2O_7^{2-} + 14H^+ + 6e^- = 2Cr^{3+} + 7H_2O$$

Oxidation half-reaction:

$$RCH_2OH + H_2O = RCO_2H + 4H^+ + 4e^-$$

The least common multiple of a 6-electron uptake in the reduction step and a 4-electron loss in the oxidation step is 12, so we multiply the first half-reaction by 2 and the second by 3, and add:

$$3RCH_2OH + 3H_2O + 2Cr_2O_7^{2-} + 28H^+ = 3RCO_2H + 12H^+ + 4Cr^{3+} + 14H_2O$$

Canceling common terms, we get:

$$3RCH_2OH + 2Cr_2O_7^{2-} + 16H^+ = 3RCO_2H + 4Cr^{3+} + 11H_2O$$

This shows that the oxidation of 3 mol of a primary alcohol to a carboxylic acid requires 2 mol of dichromate.

EXAMPLE 2

Write a balanced equation for the oxidation of styrene to benzoate ion and carbonate ion by MnO_4^- in alkaline solution.

Reduction:

$$MnO_4^- + 4H^+ + 3e^- = MnO_2 + 2H_2O \text{ (in acid)}$$

Since this reaction is carried out in basic solution, we must add 4 OH^- to neutralize the 4H^+ on the left side, and, of course, 4 OH^- to the right side to maintain a balanced equation.

$$MnO_4^- + 4H^+ + 4\,OH^- + 3e^- = MnO_2 + 2H_2O + 4\,OH^-$$

or, $\quad MnO_4^- + 2H_2O + 3e^- = MnO_2 + 4\,OH^-$

Oxidation:

$$ArCH{=}CH_2 + 5H_2O = ArCO_2^- + CO_3^{2-} + 13H^+ + 10e^-$$

We add 13 OH^- to each side to neutralize the H^+ on the right side,

$$ArCH{=}CH_2 + 5H_2O + 13\,OH^- = ArCO_2^- + CO_3^{2-} + 13H_2O + 10e^-$$

The least common multiple is 30, so we multiply the reduction half-reaction by 10 and the oxidation half-reaction by 3 and add:

$$3ArCH=CH_2 + 39\ OH^- + 10MnO_4^- + 20H_2O = 3ArCO_2^- + 3CO_3^{2-} +$$
$$24H_2O + 10MnO_2 + 40\ OH^-$$

Canceling:

$$3ArCH=CH_2 + 10MnO_4^- = 3ArCO_2^- + 3CO_3^{2-} + 4H_2O + 10MnO_2 + OH^-$$

SAMPLE PROBLEMS

Using the ion–electron half-reaction method, write balanced equations for the following oxidation reactions.

(a) Cyclohexene + MnO_4^- + H^+ $\xrightarrow{\text{(hot)}}$ $HO_2C(CH_2)_4CO_2H$ + Mn^{2+} + H_2O

(b) Cyclopentene + MnO_4^- + H_2O $\xrightarrow{\text{(cold)}}$ cis-1,2-cyclopentanediol + MnO_2 + OH^-

(c) Cyclopentanol + HNO_3 $\xrightarrow{\text{(hot)}}$ $HO_2C(CH_2)_3CO_2H$ + NO_2 + H_2O

(d) 1,2,3-Cyclohexanetriol + HIO_4 $\xrightarrow{\text{(cold)}}$ $OCH(CH_2)_3CHO$ + HCO_2H + HIO_3

SOLUTIONS TO SAMPLE PROBLEMS

(a) Reduction:

$$MnO_4^- + 8\ H^+ + 5\ e^- = Mn^{2+} + 4\ H_2O$$

Oxidation:

The least common multiple is 40:

$$8\ MnO_4^- + 64\ H^+ + 40\ e^- = 8\ Mn^{2+} + 32\ H_2O$$

Adding and canceling:

(b) Reduction:

$$MnO_4^- + 2 H_2O + 3 e^- = MnO_2 + 4 OH^-$$

Oxidation:

$+ 2 OH^- =$ $+ 2 e^-$

The least common multiple is 6:

$$2 MnO_4^- + 4 H_2O + 6 e^- = 2 MnO_2 + 8 OH^-$$

3 $+ 6 OH^- = 3$ $+ 6 e^-$

Adding and cancelling:

3 $+ 2 MnO_4^- + 4 H_2O = 3$ $+ 2 MnO_2 + 2 OH^-$

(c) Reduction:

$$HNO_3 + H^+ + e^- = NO_2 + H_2O$$

Oxidation:

$+ 3 H_2O =$ $+ 8 H^+ + 8 e^-$

The least common multiple is 8:

$$8 HNO_3 + 8 H^+ + 8 e^- = 8 NO_2 + 8 H_2O$$

$+ 3 H_2O =$ $+ 8 H^+ + 8 e^-$

Adding and cancelling:

$+ 8 HNO_3 =$ $+ 8 NO_2 + 5 H_2O$

(d) Reduction:

$$HIO_4 + 2 H^+ + 2 e^- = HIO_3 + H_2O$$

Oxidation:

$$2\ HIO_4 + 4\ H^+ + 4\ e^- = 2\ HIO_3 + 2\ H_2O$$

The least common multiple is 4:

Adding and cancelling:

12.4   (a) $LiAlH_4$

(b) $LiAlH_4$

(c) $NaBH_4$

12.5   (a) [benzene ring] $NH^+\ CrO_3Cl^-$ (PCC)/CH_2Cl_2

(b) $KMnO_4$, OH^-, H_2O, heat; then H_3O^+

(c) H_2CrO_4/acetone

(d) (1) O_3 (2) Zn, HOAc

12.6   (a) $C_6H_5{:}Li$ + $H{-}OH$ $\longrightarrow$ C_6H_6 +   LiOH

Stronger base   Stronger acid   Weaker acid   Weaker base

(b) $C_6H_5{:}Li$ + $H{-}OEt$ $\longrightarrow$ C_6H_6 +   LiOEt

Stronger base   Stronger acid   Weaker acid   Weaker base

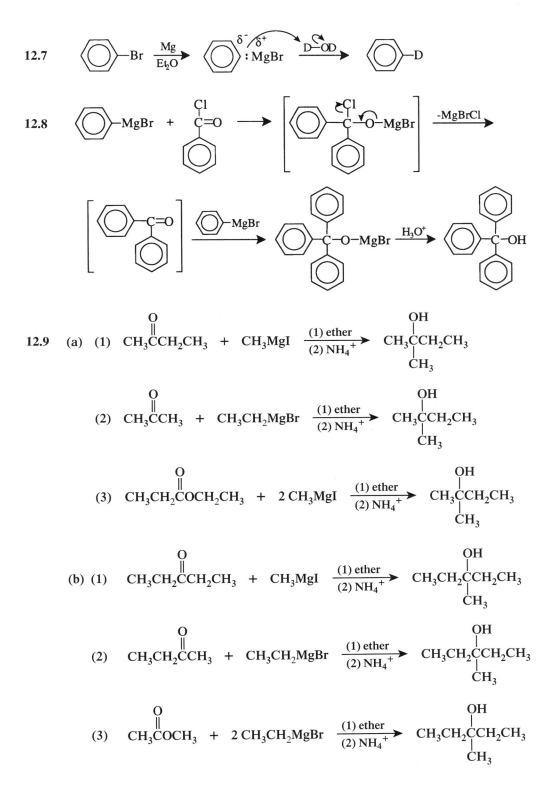

12.7

12.8

12.9 (a) (1) $CH_3CCH_2CH_3$ + CH_3MgI $\xrightarrow[(2)\ NH_4^+]{(1)\ ether}$ $CH_3CCH_2CH_3$

(2) CH_3CCH_3 + CH_3CH_2MgBr $\xrightarrow[(2)\ NH_4^+]{(1)\ ether}$ $CH_3CCH_2CH_3$

(3) $CH_3CH_2COCH_2CH_3$ + $2\ CH_3MgI$ $\xrightarrow[(2)\ NH_4^+]{(1)\ ether}$ $CH_3CCH_2CH_3$

(b) (1) $CH_3CH_2CCH_2CH_3$ + CH_3MgI $\xrightarrow[(2)\ NH_4^+]{(1)\ ether}$ $CH_3CH_2CCH_2CH_3$

(2) $CH_3CH_2CCH_3$ + CH_3CH_2MgBr $\xrightarrow[(2)\ NH_4^+]{(1)\ ether}$ $CH_3CH_2CCH_2CH_3$

(3) CH_3COCH_3 + $2\ CH_3CH_2MgBr$ $\xrightarrow[(2)\ NH_4^+]{(1)\ ether}$ $CH_3CH_2CCH_2CH_3$

(c) (1) $CH_3\overset{O}{\overset{\|}{C}}H$ + $BrMgCHCH_2CH_3$ $\xrightarrow[\text{(2) } H_3O^+]{\text{(1) ether}}$ $CH_3\overset{OH}{\underset{}{C}}HCHCH_2CH_3$

$\underset{CH_3}{\overset{|}{CH_2}}$ $\underset{CH_3}{\overset{|}{CH_2}}$

(2) CH_3MgI + $H\overset{O}{\overset{\|}{C}}CHCH_2CH_3$ $\xrightarrow[\text{(2) } H_3O^+]{\text{(1) ether}}$ $CH_3\overset{OH}{\underset{}{C}}HCHCH_2CH_3$

$\underset{CH_3}{\overset{|}{CH_2}}$ $\underset{CH_3}{\overset{|}{CH_2}}$

(d) (1) Ph$\overset{O}{\overset{\|}{C}}CH_3$ + $CH_3CH_2CH_2MgBr$ $\xrightarrow[\text{(2) } NH_4^+]{\text{(1) ether}}$ Ph$\overset{OH}{\underset{CH_3}{C}}CH_2CH_2CH_3$

(2) Ph$\overset{O}{\overset{\|}{C}}CH_2CH_2CH_3$ + CH_3MgI $\xrightarrow[\text{(2) } NH_4^+]{\text{(1) ether}}$ Ph$\overset{OH}{\underset{CH_3}{C}}CH_2CH_2CH_3$

(3) $CH_3\overset{O}{\overset{\|}{C}}CH_2CH_2CH_3$ + C_6H_5MgBr $\xrightarrow[\text{(2) } NH_4^+]{\text{(1) ether}}$ Ph$\overset{OH}{\underset{CH_3}{C}}CH_2CH_2CH_3$

(e) (1) $C_6H_5\overset{O}{\overset{\|}{C}}OEt$ + $2\,C_6H_5MgBr$ $\xrightarrow[\text{(2) } NH_4^+]{\text{(1) ether}}$ $(C_6H_5)_3COH$

(2) $C_6H_5\overset{O}{\overset{\|}{C}}C_6H_5$ + C_6H_5MgBr $\xrightarrow[\text{(2) } NH_4^+]{\text{(1) ether}}$ $(C_6H_5)_3COH$

12.10 (a) $CH_3CH_2CH_2OH$ $\xrightarrow[CH_2Cl_2]{PCC}$ $CH_3CH_2\overset{O}{\overset{\|}{C}}H$ $\xrightarrow[\text{ether}]{C_6H_5MgBr}$ $CH_3CH_2\overset{OMgBr}{\underset{}{C}}HC_6H_5$

$\xrightarrow[H_2O]{H_3O^+}$ $CH_3CH_2\overset{OH}{\underset{}{C}}HC_6H_5$

(b) C_6H_5MgBr $\xrightarrow[\text{(2) } H_3O^+]{\text{(1) } HCH, \text{ ether}}$ $C_6H_5CH_2OH$ $\xrightarrow[CH_2Cl_2]{PCC}$ $C_6H_5\overset{O}{\overset{\|}{C}}H$

(c) $CH_3CH_2\overset{O}{\overset{||}{C}}OCH_3$ $\xrightarrow[\substack{\text{ether} \\ \text{[from part (a)]}}]{\text{2 } C_6H_5MgBr}$ $C_6H_5\overset{OMgBr}{\underset{C_6H_5}{\overset{|}{C}}}CH_2CH_3$ $\xrightarrow[H_2O]{NH_4^+}$ $C_6H_5\overset{OH}{\underset{C_6H_5}{\overset{|}{C}}}CH_2CH_3$

(d) $CH_3\overset{CH_3}{\underset{|}{C}H}CH_2OH$ $\xrightarrow[CH_2Cl_2]{PCC}$ $CH_3\overset{CH_3}{\underset{|}{C}H}\overset{O}{\overset{||}{C}}H$ $\xrightarrow[\text{ether}]{C_6H_5MgBr}$ $CH_3\overset{CH_3}{\underset{|}{C}H}\overset{OMgBr}{\overset{|}{C}}HC_6H_5$

$\xrightarrow[H_2O]{H_3O^+}$ $CH_3\overset{CH_3}{\underset{|}{C}H}\overset{OH}{\overset{|}{C}}HC_6H_5$

12.11 (a) $(CH_3)_2CHCH_2OH + (CH_3)_2C{=}CH_2$

(b) $(CH_3)_2CHCH_2CN$ (c) $(CH_3)_2C{=}CH_2$

(d) $CH_3\overset{CH_3}{\underset{|}{C}H}CH_2OCH_3 + (CH_3)_2C{=}CH_2$

(e) $(CH_3)_2CHCH_2{-}\overset{OH}{\underset{CH_3}{\overset{|}{C}}}{-}CH_3$ (f) $(CH_3)_2CHCH_2\overset{OH}{\overset{|}{C}}HCH_3$

(g) $(CH_3)_2CHCH_2\overset{OH}{\underset{CH_3}{\overset{|}{C}}}CH_2CH(CH_3)_2$ (h) $(CH_3)_2CHCH_2CH_2CH_2OH$

(i) $(CH_3)_2CHCH_2CH_2OH$ (j) $(CH_3)_2CHCH_3$

(k) $(CH_3)_2CHCH_3 + CH_3C{\equiv}CLi$

12.12 (a) CH_3CH_3 (b) CH_3CH_2D (c) $C_6H_5\overset{OH}{\overset{|}{C}}HCH_2CH_3$

(d) $C_6H_5{-}\overset{OH}{\underset{CH_2CH_3}{\overset{|}{C}}}{-}C_6H_5$ (e) $C_6H_5{-}\overset{OH}{\underset{CH_2CH_3}{\overset{|}{C}}}{-}CH_2CH_3$ (f) $C_6H_5{-}\overset{OH}{\underset{CH_3}{\overset{|}{C}}}{-}CH_2CH_3$

(g) $CH_3CH_3 + CH_3CH_2C{\equiv}C{-}\overset{OH}{\overset{|}{C}}HCH_3$

(h) $CH_3CH_3 + $ $MgBr$

12.13   (a) $(CH_3)_2CHCHCH_2CH_2CH_3$
                        $\overset{OH}{|}$

(b) $(CH_3)_2CHCCH_2CH_2CH_3$
            $\overset{OH}{\underset{CH_3}{|}}$

(c) $CH_3CH_2CH_3$ + $CH_3CH_2CH_2C \equiv C-\overset{OH}{\underset{CH_3}{\overset{|}{C}}}-CH_3$ (d) $CH_3CH_2CH_3$

(e) $CH_3CH_2CH_2CH_2CH=CH_2$

(f) $CH_3CH_2CH_2-\text{⬠}$

(g) $\underset{H}{\overset{CH_3CH_2CH_2}{\diagdown}}C=C\underset{H}{\overset{CH_3}{\diagup}}$

(h) $CH_3CH_2CH_2CH_3$

Note: This variation of the Corey-Posner, Whitesides-House synthesis is stereo-specific.

(i) $CH_3CH_2CH_2D$

12.14   (a) $LiAlH_4$ (d) (1) $KMnO_4$, OH^-, heat,   (2) H_3O^+

(b) $NaBH_4$ (e) PCC/CH_2Cl_2

(c) $LiAlH_4$

12.15   (a) $3\ (CH_3)_2CHOH$ + PBr_3 $\longrightarrow$ $(CH_3)_2CHBr$ + H_3PO_3

$(CH_3)_2CHBr$ + Mg $\xrightarrow{\text{ether}}$ $(CH_3)_2CHMgBr$

$(CH_3)_2CHMgBr$ + $CH_3\overset{O}{\overset{||}{C}}H$ $\xrightarrow[\text{(2) } H_3O^+]{\text{(1) ether}}$ $(CH_3)_2CHCHCH_3$
                                                                                            $\overset{OH}{|}$

(b) $(CH_3)_2CHMgBr$ + $H\overset{O}{\overset{||}{C}}H$ $\xrightarrow[\text{(2) } H_3O^+]{\text{(1) ether}}$ $(CH_3)_2CHCH_2OH$
[from part (a)]

(c) $(CH_3)_2CHMgBr$ + $H_2C\overset{O}{\overset{\diagup\diagdown}{—}}CH_2$ $\xrightarrow[\text{(2) } H_3O^+]{\text{(1) ether}}$ $(CH_3)_2CHCH_2CH_2OH$
[from part (a)]

$(CH_3)_2CHCH_2CH_2Cl$ $\xleftarrow{\text{SOCl}_2}$

(d) $(CH_3)_2CHMgBr$ + $H\overset{O}{\overset{||}{C}}CH(CH_3)_2$ $\xrightarrow[\text{(2) } H_3O^+]{\text{(1) ether}}$ $(CH_3)_2CHCHCH(CH_3)_2$
[from part (a)] $\overset{OH}{|}$

(e) $(CH_3)_2CHMgBr$ + D_2O $\longrightarrow$ $(CH_3)_2CHD$
[from part (a)]

(f) $(CH_3)_2CHBr$ + Li $\longrightarrow$ $(CH_3)_2CHLi$ $\xrightarrow{CuI}$ $[(CH_3)_2CH]_2CuLi$
[from part (a)]

$(CH_3)_2CH-$⬡ $\longleftarrow$ ⬡$-Br$

12.16 (a) $CH_3:Li$ + $H-C{\equiv}C-CH_2CH_3$ $\longrightarrow$ CH_4 + $Li:C{\equiv}CCH_2CH_3$

(b) ⬡$=O$ + $Li:C{\equiv}CCH_2CH_3$ $\longrightarrow$ ⬡$(O^-)(C{\equiv}CCH_2CH_3)$ $\xrightarrow{NH_4^+}$ ⬡$(HO)(C{\equiv}CCH_2CH_3)$

(c) ⬡$(HO)(C{=}C$ with H, H, $CH_2CH_3)$

(d) $CH_3CH_2C{\equiv}C$ ⬡$O-H$ $\xrightarrow{Na:H}$ $CH_3CH_2C{\equiv}C$ ⬡O^-Na^+ $CH_3CH_2-OSO_2CH_3$ $\xrightarrow{}$

$CH_3CH_2C{\equiv}C$ ⬡OCH_2CH_3 + $^-OSO_2CH_3$

(e) $CH_3CH_2\underset{\underset{OH}{|}}{C}HCH_3$

(f) $CH_3CH_2\underset{\underset{OSO_2CH_3}{|}}{C}HCH_3$

(g) $CH_3CH_2CHCH_3$ + $Na^+ \ ^-OCCH_3$ $\longrightarrow$ $CH_3CH_2CHCH_3$
$\quad \quad \quad \underset{OSO_2CH_3}{|}$ $\quad \quad \overset{\overset{O}{||}}{}$ $\quad\quad\quad\quad\quad\quad \underset{\underset{\overset{||}{O}}{OCCH_3}}{|}$

$\quad\quad\quad\quad\quad\quad\quad\quad\quad\quad\quad\quad\quad\quad\quad\quad\quad\quad\quad\quad$ + $Na^+ \ ^-OSO_2CH_3$

(h) $CH_3CH_2\underset{OH}{\underset{|}{C}HCH_3}$ + CH_3CH_2OH

12.17 (a) $CH_3CH_2CH_2CH_2CH_2OH$ $\xrightarrow{PBr_3}$ $CH_3CH_2CH_2CH_2CH_2Br$

(b) $CH_3CH_2CH_2CH_2CH_2Br$ $\xrightarrow[\text{(CH}_3)_3\text{COH}]{\text{(CH}_3)_3\text{COK}}$ $CH_3CH_2CH_2CH=CH_2$
$\quad\quad$ [from (a)]

(c) $CH_3CH_2CH_2CH=CH_2$ $\xrightarrow[\text{(2) NaBH}_4/\text{OH}^-]{\text{(1) Hg(OAc)}_2/\text{THF-H}_2\text{O}}$ $CH_3CH_2CH_2\underset{OH}{\underset{|}{C}HCH_3}$
$\quad\quad$ [from (b)]

(d) $CH_3CH_2CH_2\underset{OH}{\underset{|}{C}HCH_3}$ $\xrightarrow[\text{acetone, H}_2\text{O}]{\text{H}_2\text{CrO}_4}$ $CH_3CH_2CH_2\underset{\overset{||}{O}}{CCH_3}$
$\quad\quad$ [from (c)]

(e) $CH_3CH_2CH_2\underset{OH}{\underset{|}{C}HCH_3}$ $\xrightarrow{PBr_3}$ $CH_3CH_2CH_2\underset{Br}{\underset{|}{C}HCH_3}$
$\quad\quad$ [from (c)]

(f) $CH_3CH_2CH_2CH_2CH_2Br$ $\xrightarrow[\text{Et}_2\text{O}]{\text{Mg}}$ $CH_3CH_2CH_2CH_2CH_2MgBr$ $\xrightarrow[\text{(2) H}_3\text{O}^+]{\text{(1) }H\overset{\overset{O}{||}}{C}H}$
$\quad\quad$ [from (a)]

$\quad\quad\quad\quad\quad\quad\quad\quad\quad\quad\quad\quad\quad\quad\quad\quad\quad\quad$ $CH_3CH_2CH_2CH_2CH_2CH_2OH$

(g) $CH_3CH_2CH_2CH_2CH_2MgBr$ $\xrightarrow[\text{(2) H}_3\text{O}^+]{\text{(1) }H_2C\overset{O}{\overset{\diagup\diagdown}{}}CH_2}$ $CH_3CH_2CH_2CH_2CH_2CH_2CH_2OH$
$\quad\quad$ [from (f)]

(h) $CH_3CH_2CH_2CH_2CH_2OH$ $\xrightarrow[\text{CH}_2\text{Cl}_2]{\text{PCC}}$ $CH_3CH_2CH_2CH_2\overset{\overset{O}{||}}{C}H$

(i) $CH_3CH_2CH_2\underset{OH}{\underset{|}{C}HCH_3}$ $\xrightarrow[\text{acetone, H}_2\text{O}]{\text{H}_2\text{CrO}_4}$ $CH_3CH_2CH_2\underset{\overset{||}{O}}{CCH_3}$
$\quad\quad$ [from (c)]

(j) $CH_3CH_2CH_2CH_2CH_2OH$ $\xrightarrow[\text{(2) H}_3\text{O}^+]{\text{(1) KMnO}_4, \text{OH}^-, \text{heat}}$ $CH_3CH_2CH_2CH_2\overset{\overset{O}{||}}{C}OH$

(k) (1) $CH_3CH_2CH_2CH_2CH_2OH$ $\xrightarrow[\text{140° C}]{\text{H}_2\text{SO}_4}$ $(CH_3CH_2CH_2CH_2CH_2)_2O$

$\quad\quad$ (2) $CH_3CH_2CH_2CH_2CH_2OH$ $\xrightarrow[\text{(-H}_2)]{\text{NaH}}$ $CH_3CH_2CH_2CH_2CH_2ONa$

$\quad\quad\quad\quad$ $\xrightarrow[\text{[from (a)]}]{CH_3CH_2CH_2CH_2CH_2Br}$ $(CH_3CH_2CH_2CH_2CH_2)_2O$

(l) $CH_3CH_2CH_2CH=CH_2$ $\xrightarrow[CCl_4]{Br_2}$ $CH_3CH_2CH_2\underset{Br}{CH}CH_2Br$ $\xrightarrow[heat]{3\ NaNH_2}$
[from (b)]

$CH_3CH_2CH_2C\equiv CNa$ $\xrightarrow{H^+}$ $CH_3CH_2CH_2C\equiv CH$

(m) $CH_3CH_2CH_2C\equiv CH$ $\xrightarrow{HBr}$ $CH_3CH_2CH_2\underset{Br}{C}=CH_2$
[from (l)]

(n) $CH_3CH_2CH_2CH_2CH_2Br$ $\xrightarrow[Et_2O]{Li}$ $CH_3CH_2CH_2CH_2CH_2Li$
[from (a)]

(o) $CH_3CH_2CH_2CH_2CH_2Li$ $\xrightarrow{CuI}$ $(CH_3CH_2CH_2CH_2CH_2)_2CuLi$
[from (n)]

$\xrightarrow[\text{[from (a)]}]{CH_3CH_2CH_2CH_2CH_2Br}$ $CH_3(CH_2)_8CH_3$

(p) $CH_3CH_2CH_2CH_2CH_2MgBr$ $\xrightarrow[(2)\ NH_4^+]{(1)\ CH_3(CH_2)_2\overset{O}{\overset{\|}{C}}CH_3[\text{from (d)}]}$
[from (f)]

$CH_3(CH_2)_4\underset{CH_3}{\overset{OH}{\underset{|}{C}}}(CH_2)_2CH_3$

12.18 (a) $C_6H_5C\equiv CH$ $\xrightarrow[\text{ether }(-CH_4)]{CH_3MgBr}$ $C_6H_5C\equiv CMgBr$ $\xrightarrow[(2)\ NH_4^+]{(1)\ CH_3\overset{O}{\overset{\|}{C}}CH_3}$

$C_6H_5C\equiv CC(OH)(CH_3)_2$

(b) $C_6H_5COCH_3$ $\xrightarrow[OH^-]{NaBH_4}$ $C_6H_5\underset{OH}{\overset{|}{C}H}CH_3$

(c) $C_6H_5C\equiv CH$ $\xrightarrow[Ni_2B\ (P\text{-}2)]{H_2}$ $C_6H_5CH=CH_2$

(d) $C_6H_5CH=CH_2$ $\xrightarrow[(2)\ H_2O_2,\ OH^-]{(1)\ THF:\ BH_3}$ $C_6H_5CH_2CH_2OH$

(e) $C_6H_5CH_2CH_2OH$ $\xrightarrow[(2)\ Mg,\ Et_2O]{(1)\ PBr_3}$ $C_6H_5CH_2CH_2MgBr$ $\xrightarrow[(2)\ H_3O^+]{(1)\ H_2C\overset{O}{\overset{\diagup\diagdown}{-}}CH_2}$

$C_6H_5CH_2CH_2CH_2CH_2OH$

(f) $C_6H_5CH_2CH_2OH$ $\xrightarrow[(-H_2)]{NaH}$ $C_6H_5CH_2CH_2ONa$ $\xrightarrow{CH_3I}$

$C_6H_5CH_2CH_2OCH_3$

12.19   (a) *Partial   Analysis*

Synthesis

(b) *Partial   Analysis*

Synthesis

$$CH_3CH_2CH_2OH \xrightarrow{PBr_3} CH_3CH_2CH_2Br \xrightarrow[Et_2O]{Mg} CH_3CH_2CH_2MgBr$$

(c) *Partial   Analysis*

Synthesis

(d) *Partial Analysis*

Synthesis

(e) *Partial Analysis*

$(CH_3)_2CHCH_2CH_2CO_2H \implies (CH_3)_2CHCH_2{\overset{|}{-}}CH_2CH_2OH \implies$

$(CH_3)_2CHCH_2MgBr + H_2C\overset{O}{\overset{\diagup\diagdown}{-}}CH_2$

Synthesis

$(CH_3)_2CHCH_2OH \xrightarrow{PBr_3} (CH_3)_2CHCH_2Br \xrightarrow[Et_2O]{Mg} (CH_3)_2CHCH_2MgBr$

$\xrightarrow[\text{(2) } H_3O^+]{\text{(1) } H_2C\overset{O}{\overset{\diagup\diagdown}{-}}CH_2} (CH_3)_2CHCH_2CH_2CH_2OH \xrightarrow[\text{(2) } H_3O^+]{\text{(1) } KMnO_4, \ OH^-, \ heat}$

$(CH_3)_2CHCH_2CH_2CO_2H$

(f) *Partial   Analysis*

Synthesis

(g) *Partial   Analysis*

$$CH_3CH_2CH_2\overset{O}{\underset{||}{C}}CH_2CH(CH_3)_2 \implies CH_3CH_2CH_2\overset{}{\underset{OH}{CH}}+CH_2CH(CH_3)_2 \implies$$

$$CH_3CH_2CH_2\overset{O}{\overset{||}{C}}H + BrMgCH_2CH(CH_3)_2$$

Synthesis

$$CH_3CH_2CH_2CH_2OH \xrightarrow[CH_2Cl_2]{PCC} CH_3CH_2CH_2\overset{O}{\overset{||}{C}}H \xrightarrow[(2)\ H_3O^+]{\overset{(1)\ (CH_3)_2CHCH_2MgBr}{[from\ (e)]}}$$

$$CH_3CH_2CH_2\overset{}{\underset{OH}{CH}}CH_2CH(CH_3)_2 \xrightarrow{H_2CrO_4} CH_3CH_2CH_2\overset{O}{\underset{||}{C}}CH_2CH(CH_3)_2$$

(h) *Partial   Analysis*

$$CH_3CH_2\overset{Br}{\underset{C_6H_5}{\overset{|}{C}}}CH_2CH_3 \implies CH_3CH_2\overset{OH}{\underset{C_6H_5}{\overset{|}{C}}}+CH_2CH_3 \implies$$

$$CH_3CH_2MgBr + CH_3CH_2\overset{O}{\overset{||}{C}}C_6H_5 \implies CH_3CH_2\overset{OH}{\underset{|}{CH}}-C_6H_5 \implies$$

$$CH_3CH_2\overset{O}{\overset{||}{C}}H + C_6H_5MgBr$$

Synthesis

12.20   *Analysis*

Synthesis

12.21   The starting compound is a cyclic ester. Addition of two molar equivalents of CH_3MgI will (after acidification) furnish the desired product.

12.22 *Analysis*

$$CH_3-CH_2-\underset{\underset{OH}{|}}{\overset{\overset{CH_3}{|}}{C}}\!\!-\!\!C\equiv CH \implies CH_3-CH_2-\underset{\underset{O}{\|}}{\overset{\overset{CH_3}{|}}{C}} + Na^+ \ ^-\!:\!C\equiv CH$$

Synthesis

$$HC\equiv CH \xrightarrow[\text{liq. } NH_3]{NaNH_2} HC\equiv C\!:^-Na^+ \xrightarrow[\text{(2) } NH_4^+]{\text{(1) } CH_3CH_2\overset{\overset{O}{\|}}{C}CH_3} CH_3-CH_2-\underset{\underset{OH}{|}}{\overset{\overset{CH_3}{|}}{C}}\!-\!C\equiv CH$$

12.23 The three- and four-membered rings are strained, and so they open on reaction with RMgX or RLi. THF possesses an essentially unstrained ring and hence is far more resistant to attack by an organometallic compound.

12.24 (a) $RMgX + C_2H_5-O-\overset{\overset{O}{\|}}{C}-O-C_2H_5 \longrightarrow \left[C_2H_5-O-\underset{\underset{R}{|}}{\overset{\overset{OMgX}{|}}{C}}-O-C_2H_5 \right]$

$$\xrightarrow{-C_2H_5OMgX} \left[R-\overset{\overset{O}{\|}}{C}-O-C_2H_5 \right] \xrightarrow{RMgX} \left[R-\underset{\underset{R}{|}}{\overset{\overset{OMgX}{|}}{C}}-O-C_2H_5 \right] \xrightarrow{-C_2H_5OMgX}$$

$$\left[R-\overset{\overset{O}{\|}}{C}-R \right] \xrightarrow{RMgX} R-\underset{\underset{R}{|}}{\overset{\overset{OMgX}{|}}{C}}-R \xrightarrow{NH_4^+} R-\underset{\underset{R}{|}}{\overset{\overset{OH}{|}}{C}}-R$$

(b) $RMgX + H-\overset{\overset{O}{\|}}{C}-O-C_2H_5 \longrightarrow \left[H-\underset{\underset{R}{|}}{\overset{\overset{OMgX}{|}}{C}}-O-C_2H_5 \right] \xrightarrow{-C_2H_5OMgX}$

$$\left[H-\overset{\overset{O}{\|}}{C}-R \right] \xrightarrow{RMgX} H-\underset{\underset{R}{|}}{\overset{\overset{OMgX}{|}}{C}}-R \xrightarrow{H_3O^+} H-\underset{\underset{R}{|}}{\overset{\overset{OH}{|}}{C}}-R$$

*12.25

Before converting the reactant to a Grignard reagent it is first necessary to mask the alcohol, such as by converting it to a *tert*-butyldimethylsilyl ether. After the Grignard reaction is over the protecting group is removed.

*12.26 2-Phenylethanol, 1,2-diphenylethanol, and 1,1-diphenylethanol are distinct from diphenylacetic acid and benzyl phenylacetate in that they do not have carbonyl groups. IR spectroscopy can be used to segregate these compounds into two groups according to those that do or do not exhibit carbonyl absorptions.

^{1}H NMR can differentiate among all of the compounds. In the case of the alcohols, in the ^{1}H NMR spectrum of 2-phenylethanol there will be two triplets of equal integral value, whereas for 1,2-diphenylethanol there will be a doublet and a triplet in a 2:1 area ratio. The triplet will be downfield of the doublet. 1,1-Diphenylethanol will exhibit a singlet for the unsplit methyl hydrogens.

The broadband proton-decoupled ^{13}C NMR spectrum of 2-phenylethanol should show 6 signals (assuming no overlap), four of which are in the chemical shift region for aromatic carbons. 1,2-Diphenylethanol should exhibit 10 signals (assuming no overlap), 8 of which are in the aromatic region. 1,1-Diphenylethanol should show 6 signals (assuming no overlap), four of which would be in the aromatic region. The DEPT ^{13}C NMR spectra would give direct evidence as to the number of attached hydrogens on each carbon.

Regarding the carbonyl compounds, both diphenylacetic acid and benzyl phenylacetate will show carbonyl absorptions in the IR, but only the former will also have a hydroxyl absorption. The ^{1}H NMR spectrum of diphenylacetic acid should show a broad absorption for the carboxylic acid hydrogen and a sharp singlet for the unsplit hydrogen at C2. Their integral values should be the same, and approximately one tenth the integral value of the signals in the aromatic region. Benzyl phenylacetate will exhibit two singlets, one for each of the unsplit CH$_2$ groups. These signals will have an area ratio of 1:5 with respect to the signal for the 10 aromatic hydrogens. The broadband ^{1}H decoupled ^{13}C NMR spectrum for diphenylacetic acid should show four aromatic carbon signals, whereas that for benzyl phenylacetate (assuming no overlapping signals) would show 10 signals in the aromatic carbon region. Aside from the carbonyl and aromatic carbon signals, benzyl phenylacetate would show two additional signals, whereas diphenylacetic acid would show only one. DEPT ^{13}C NMR spectra for these two compounds would also distinguish them directly.

*12.27   It makes it impossible to distinguish between aldehyde and ketone type sugars (aldoses and ketoses) that had been components of the saccharide. Also, because the R groups of these sugars contain stereocenters, reduction of the ketone carbonyl will be stereoselective. This will complicate the determination of the ratio of sugars differing in configuration at C2.

*12.28   The IR indicates the presence of OH and absence of C=C and C=O. The MS indicates a molecular weight of 116 amu and confirms the presence of hydroxyl. The reaction data indicate X contains 2 protons per molecule that are acidic enough to react with a Grignard reagent, meaning two hydroxyl groups per molecule. (This analytical procedure, the Zerewitinoff determination, was routinely done before the advent of NMR.)

   Thus X has a partial structure like:

   $C_6H_{10}(OH)_2$ with one ring, or
   $C_5H_6O(OH)_2$ with two rings, or (less likely)
   $C_4H_2O_2(OH)_2$ with three rings.

QUIZ

12.1   Which of the following could be employed to transform ethanol into $CH_3CH_2CH_2OH$?

   (a) Ethanol + HBr, then Mg/diethyl ether, then H_3O^+

   (b) Ethanol + HBr, then Mg/diethyl ether, then $\overset{\overset{\displaystyle O}{\|}}{H C H}$, then H_3O^+

   (c) Ethanol + $H_2SO_4/140°$ C

   (d) Ethanol + Na, then $\overset{\overset{\displaystyle O}{\|}}{H C H}$, then H_3O^+

   (e) Ethanol + $H_2SO_4/180°$ C, then $H_2C\overset{\displaystyle O}{\underset{}{-}}CH_2$

12.2   The principal product(s) formed when *1 mol* of methylmagnesium iodide reacts with 1 mol of $CH_3\overset{\overset{\displaystyle O}{\|}}{C}CH_2CH_2OH$.

   (a) $CH_4 + CH_3\overset{\overset{\displaystyle O}{\|}}{C}CH_2CH_2OMgI$

   (b) $CH_3\overset{\overset{\displaystyle OMgI}{|}}{\underset{\underset{\displaystyle CH_3}{|}}{C}}CH_2CH_2OH$

   (c) $CH_3\overset{\overset{\displaystyle O}{\|}}{C}CH_2CH_2OCH_3$

   (d) $CH_3\overset{\overset{\displaystyle CH_3}{|}}{\underset{\underset{\displaystyle OH}{|}}{C}}CH_2CH_2OCH_3$

   (e) None of the above

12.3   Supply the missing reagents.

12.4   Supply the missing reagents and intermediates.

12.5   Supply the missing starting compound.

A

$$\xrightarrow[\text{(2) NH}_4^+]{\text{(1) CH}_3\text{MgBr}}$$

$$\underset{\underset{\text{CH}_3}{|}}{\overset{\overset{\text{CH}_3}{|}}{\text{CH}_3\text{CHCH}_2\text{C}}}-\text{OH}$$

ANSWERS TO
FIRST REVIEW PROBLEM SET

1 (a)

(b)

then,

(c)   The enantiomer of the product given would be formed in an equimolar amount via the following reaction:

The *trans*-1, 2-dibromocyclopentane would be formed as a racemic form via the reaction of the bromonium ion with a bromide ion:

Racemic *trans*-1,2-dibromocyclopentane

And, *trans*-2-bromocyclopentanol (the bromohydrin) would be formed (as a racemic form) via the reaction of the bromonium ion with water.

Racemic *trans*-2-bromocyclopentanol

2 (a) $CHCl_3$ (b)   The cis isomer   (c) CH_3Cl

3   This indicates that the bonds in BF_3 are geometrically arranged so as to cancel each others' polarities in contrast to the case of NF_3. This, together with other evidence, indicates that BF_3 has trigonal planar structure and NF_3 has trigonal pyramidal structure.

4   (a) sp^3
(b)   They are much smaller (60°) than the expected 109.5°.
(c)   That the bent bonds in cyclopropane, like compressed springs, contain stored energy that is available to assist in reactions that open the 3-membered ring.

5   All of these differences can be explained by the contribution to the $CH_2=CHCl$ molecule made by the following **A** and **B** resonance structures.

$$
\begin{array}{ccc}
\underset{\textstyle H}{\overset{\textstyle H}{}}C\!=\!C\underset{\textstyle \ddot{C}l:}{\overset{\textstyle H}{}} & \longleftrightarrow & \underset{\textstyle H}{\overset{\textstyle H}{}}C\!-\!C\underset{\textstyle \overset{+}{C}l:}{\overset{\textstyle H}{}}
\end{array}
$$

A   **B**

(a)   Because of the contribution made to the hybrid by **B,** the C–Cl bond of $CH_2=CH–Cl$ has some double-bond character and is, therefore, shorter than the "pure" single bond of $CH_3CH_2–Cl$.

(b)   The contribution made to the hybrid by **B** imparts some single-bond character to the carbon-carbon double bond of $CH_2=CH–Cl$, causing it to be longer than the "pure" double bond of $CH_2=CH_2$.

(c)   Electronegativity differences would cause a carbon-chlorine bond to be polarized as follows:

$$-\overset{|}{\underset{|}{C}}\overset{\delta^+ \quad \delta^-}{—}Cl$$

And this effect accounts, almost entirely, for the dipole moment of CH_3CH_2Cl.

$$\overset{\delta^+}{CH_3CH_2}\overset{\delta^-}{-Cl} \qquad \mu = 2.05\ D$$

With $CH_2=CH-Cl$, however, the resonance contribution of **B** tends to oppose the polarization of the C–Cl bond caused by electronegativity differences. That is, the resonance effect partially cancels the electronegativity effect, causing the dipole moment to be smaller.

6 $A = CH_3(CH_2)_{11}CH_2C{\equiv}CH$

$B = CH_3(CH_2)_{11}CH_2C{\equiv}CNa$

$C = CH_3(CH_2)_{11}CH_2C{\equiv}CCH_2(CH_2)_6CH_3$

Muscalure =

7

Because catalytic hydrogenation is a syn addition, catalytic hydrogenation of the (Z) isomer would yield a meso compound.

Syn addition of hydrogen to the (E) isomer would yield a racemic form:

(E)   (by addition at one face)   +   (by addition at the other face)

Enantiomers - a racemic form

8   From the molecular formula of **A** and of its hydrogenation product **B,** we can conclude that **A** has two rings and a double bond. (**B** has two rings.)

From the product of strong oxidation with $KMnO_4$ and its stereochemistry (i.e., compound **C**), we can deduce the structure of **A**.

A   (1) $KMnO_4$,OH^-, heat   (2) H_3O^+ CO_2H CO_2H = HO_2C CO_2H

meso-1,3-Cyclopentane-dicarboxylic acid

Compound **B** is bicyclo[2.2.1]heptane and **C** is a glycol.

A H_2 B

A $KMnO_4$, OH^- cold, dilute   OH   OH   H   C   H

Notice that **C** is also a meso compound.

9   (a) $CH_3C\equiv CH$ $\xrightarrow[\text{liq. } NH_3]{NaNH_2}$ $CH_3C\equiv CNa$ $\xrightarrow{CH_3I}$ $CH_3C\equiv CCH_3$

(b) $CH_3C\equiv CCH_3$ $\xrightarrow{\underset{Ni_2B \ (P\text{-}2)}{H_2}}$

$$\underset{H}{\overset{H_3C}{>}}C=C\underset{H}{\overset{CH_3}{<}}$$

[from (a)]

(c) $CH_3C\equiv CCH_3$ $\xrightarrow{\underset{NH_3}{Li}}$

$$\underset{H}{\overset{H_3C}{>}}C=C\underset{CH_3}{\overset{H}{<}}$$

[from (a)]

(d) $CH_3C\equiv CH$ $\xrightarrow{\underset{Ni_2B(P\text{-}2)}{H_2}}$ $CH_3CH=CH_2$ $\xrightarrow[\substack{\text{light} \\ 25°}]{NBS}$ $BrCH_2CH=CH_2$ ⎤

+

CH_3I $\xrightarrow[Et_2O]{Li}$ CH_3Li $\xrightarrow{CuI}$ $(CH_3)_2CuLi$ ⎦

↓

$CH_3CH_2CH=CH_2$

or

$CH_3C\equiv CH$ $\xrightarrow[(2) \ CH_3Br]{(1) \ NaNH_2}$ $CH_3C\equiv CCH_3$ $\xrightarrow[\substack{Ni_2B(P\text{-}2) \\ \text{or } (1) \ Na, \ \text{liq. } NH_3 \\ (2) \ NH_4Cl}]{H_2}$ $CH_3CH=CHCH_3$

↓ HBr

$CH_3CH_2CH=CH_2$ $\xleftarrow[\substack{tert\text{-}BuOH, \\ \text{heat}}]{tert\text{-}BuOK}$ $CH_3CH_2\underset{Br}{CH}CH_3$

(e) $CH_3CH_2CH=CH_2$ $\xrightarrow[CCl_4]{NBS}$ $\begin{matrix} CH_3\underset{Br}{CH}CH=CH_2 \\ + \\ CH_3CH=CHCH_2Br \end{matrix}$ ⎤ $\xrightarrow[(CH_3)_3COH]{(CH_3)_3COK}$

[from (d)]

$CH_2=CH-CH=CH_2$

(f) $CH_3CH_2CH=CH_2$ $\xrightarrow[ROOR]{HBr}$ $CH_3CH_2CH_2CH_2Br$

[from (d)]

(g) $CH_3CH=CHCH_3$ $\xrightarrow[\text{no peroxides}]{HBr}$ $CH_3CH_2\underset{Br}{CH}CH_3$

[from (b) or (c)]

or

$CH_3CH_2CH=CH_2$ $\xrightarrow[\text{no peroxides}]{HBr}$ $CH_3CH_2\underset{Br}{CH}CH_3$

[from (d)]

(h) [from (c)] $\xrightarrow[\text{(anti addition)}]{\text{Br}_2 / \text{CCl}_4}$ (2R, 3S) A meso compound (cf. Section 8.7)

(i) [from (b)] $\xrightarrow[\text{(anti addition)}]{\text{Br}_2 / \text{CCl}_4}$ (2R, 3R) + (2S, 3S) A racemic form

(j) [from (b)] $\xrightarrow[\text{(2) NaHSO}_3]{\text{(1) OsO}_4}$ (syn addition) (cf. Section 8.10)

or [from (c)] $\xrightarrow[\text{(2) H}_3\text{O}^+, \text{heat}]{\text{(1) RCOOH}}$ (anti addition) (cf. Section 11.19)

(k) $CH_3C{\equiv}CCH_3$ $\xrightarrow[\text{CH}_3\text{CO}_2\text{H}]{\text{HBr,Br}^-}$ (cf. Section 8.13)

10 $CH_3CHCH_2CH_3$ with CH_3 $\xrightarrow[hv,\text{heat}]{\text{Br}_2}$ $CH_3CCH_2CH_3$ with Br and CH_3 (cf. Section 10.6)

(a) $CH_3CCH_2CH_3$ with Br and CH_3 $\xrightarrow[\text{heat}]{\text{CH}_3\text{CH}_2\text{ONa} / \text{CH}_3\text{CH}_2\text{OH}}$ $CH_3C{=}CHCH_3$ with CH_3

(b) $CH_3C{=}CHCH_3$ with CH_3 [from (a)] $\xrightarrow[\text{H}_2\text{O}]{\text{H}_3\text{O}^+}$ $CH_3CCH_2CH_3$ with OH and CH_3

B cannot undergo dehydrohalogenation because it has no β hydrogen; however, C and D can, as shown next.

(a meso compound)

13   The eliminations are anti eliminations, requiring an anti periplanar arrangement of the bromine atoms.

14   The eliminations are anti eliminations, requiring an anti periplanar arrangement of the −H and −Br.

(2S,3S)-1,2-Dibromo-1,2-diphenylethane will also give (Z)-1-bromo-1,2-diphenylethene in an anti elimination.

15   In all the following structures, notice that the large *tert*-butyl group is equatorial.

(a)   (bromine addition is anti; cf. Section 8.7)

+ enantiomer
as a racemic form

(b)   (syn hydroxylation; cf. Section 8.10)

+ enantiomer
as a racemic form

(c)   (anti hydroxylation; cf. Section 11.19)

+ enantiomer
as a racemic form

(d)   (syn and anti Markovnikov addition of —H and —OH; cf. Section 11.7)

+ enantiomer
as a racemic form

(e)   (Markovnikov addition of —H and —OH; cf. Section 11.4)

(f)   (anti addition of –Br and –OH, with –Br and –OH placement resulting from the more stable partial carbocation in the intermediate bromonium ion; cf. Section 8.8)

+ enantiomer
as a racemic form

(g)   (anti addition of –I and –Cl, following
                                       Markovnikov's rule; cf. Section 8.2)

+ enantiomer
as a racemic form

(h) $HC(CH_2)_4CC(CH_3)_3$

(i)   (syn addition of deuterium;
                                       cf. Section 7.14)

+ enantiomer
as a racemic form

(j)   (syn, anti Markovnikov addition of −D and
                                       −B− , with −B− being replaced by -T
                                       where it stands; cf. Section 11.7)

+ enantiomer
as a racemic form

16 A = $CH_3C{=}CHCH_2CH_3$ B = $\left(CH_3CHCH\underset{\underset{CH_3}{CH_2}}{} \right)_2 BH$ C = $CH_3CHCHCH_2CH_3$
                                                                                                                    OH

17 (a) The following products are diastereomers. They would have different boiling points
    and would be in separate fractions. Each fraction would be optically active.

(*R*)-3-Methyl-1-pentene   (optically active)   (optically active)

Diastereomers

(b)   Only one product is formed. It is achiral, and, therefore, it would not be optically active.

(optically inactive)

(c)   Two diastereomeric products are formed. Two fractions would be obtained. Each fraction would be optically active.

(optically active)   (optically active)

Diastereomers

(d)   One optically active compound is produced.

(optically active)

(e)   Two diastereomeric products are formed. Two fractions would be obtained. Each fraction would be optically active.

(optically active)   (optically active)

Diastereomers

(f)   Two diastereomeric products are formed. Two fractions would be obtained. Each fraction would be optically active.

19

m/z 120 = M^{+}

105 = M^{+} − 15(CH$_3$ ·) = C$_6$H$_5$—$\overset{+}{\text{C}}$HCH$_3$

77 = M^{+} − 43(i−Pr ·) = C$_6$H$_5$ +

δ 7.2–7.6	5 ring protons
2.95	CH of isopropyl group
1.29	equivalent CH$_3$s of isopropyl group

20   C$_5$H$_{10}$O has IHD = 2

IR absorption indicates C=O

^{13}C NMR spectrum for **X** is consistent with structure

21   (a)

(b)   Isomer **9** is slow to react in an E2 reaction because in its more stable conformation (see following structure) all the chlorine atoms are equatorial and an anti periplanar transition state cannot be achieved. All other isomers **1-8** can have a − Cl axial and thus achieve an anti periplanar transition state.

22 (a) $CH_3CHCH_2CH_3$ $\xrightarrow{F_2}$

(b)   Four fractions. The enantiomeric pairs would not be separated by fractional distillation because enantiomers have the same boiling points.

(c)   All of the fractions would be optically inactive.

(d)   The fraction containing **1** and **2** and the fraction containing **4** and **5.**

23

(R)-2-Fluorobutane   (optically active)   (achiral and, therefore, optically inactive)

1   **2**

3 (optically active)   **4** meso compound (optically inactive)   **5** (optically active)

(b)   Five. Compounds **3** and **4** are diastereomers. All others are constitutional isomers of each other.

(c)   See above.

24

(R)   (S)   (R)   (S)

meso   meso

Each of the two structures just given has a plane of symmetry (indicated by the dashed line), and, therefore, each is a meso compound. The two structures are not superposable one on the other; therefore, they represent molecules of different compounds and are diastereomers.

25   Only a proton or deuteron anti to the bromine can be eliminated; that is, the two groups un-
     dergoing elimination (H and Br or D and Br) must lie in an anti periplanar arrangement. The
     two conformations of *erythro*-2-bromobutane-3-*d* in which a proton or deuteron is anti peri-
     planar to the bromine are **I** and **II.**

Conformation **I** can undergo loss of HBr to yield *cis*-2-butene-2-*d*. Conformation **II** can undergo loss of DBr to yield *trans*-2-butene.

13 CONJUGATED UNSATURATED SYSTEMS

SOLUTIONS TO PROBLEMS

13.1 (a) $^{14}CH_2=CHCH_2X$ and $CH_2=CH\overset{14}{-\!\!-}CH_2X$

(b)   The reaction proceeds through the resonance-stabilized radical.

$$^{14}CH_2=CH-\overset{\bullet}{C}H_2 \quad\longleftrightarrow\quad {}^{14}\overset{\bullet}{C}H_2-CH=CH_2 \quad \text{or} \quad {}^{14}\overset{\delta\bullet}{C}H_2\overset{}{-\!\!-\!\!-}CH\overset{\delta\bullet}{-\!\!-\!\!-}CH_2$$

Thus, attack on X_2 can occur by the carbon atom at either end of the chain since these atoms are equivalent.

(c)   50:50 because attack at the two ends of the chain is equally probable.

13.2 (a)

$$\underset{D}{\overset{4}{CH_3}-\overset{3}{\underset{+}{CH}}\overset{2}{\overset{CH}{\diagup}}\overset{1}{\diagdown}CH_2} \quad\longleftrightarrow\quad \underset{E}{\overset{4}{CH_3}-\overset{3}{CH}\overset{2}{\overset{CH}{\diagup}}\overset{1}{\underset{+}{\diagdown CH_2}}} \quad \text{or} \quad \underset{F}{\overset{4}{CH_3}-\overset{3}{\underset{\delta+}{CH}}\overset{2}{\overset{CH}{\diagup}}\overset{1}{\underset{\delta+}{\diagdown CH_2}}}$$

(b)   We know that the allylic cation is almost as stable as a tertiary carbocation. Here we find not only the resonance stabilization of an allylic cation but also the additional stabilization that arises from contributor **D** in which the plus charge is on a secondary carbon atom.

(c) $CH_3-\overset{\overset{\displaystyle Cl}{|}}{C}H-CH=CH_2$ and $CH_3-CH=CH-CH_2-Cl$, because the Cl will attack the chain at the two positive centers shown in structure **F**.

13.3 (a) $CH_2=\overset{\overset{\displaystyle CH_3}{|}}{C}-\overset{\bullet}{C}H_2 \quad\longleftrightarrow\quad \overset{\bullet}{C}H_2-\overset{\overset{\displaystyle CH_3}{|}}{C}=CH_2$

(b) $CH_2=CH-\underset{+}{CH}-CH=CH_2 \quad\longleftrightarrow\quad {}^+CH_2-CH=CH-CH=CH_2 \quad\longleftrightarrow\quad$

$$CH_2=CH-CH=CH-\overset{+}{C}H_2$$

223

(e) $CH_3CH=CH-CH=\overset{+}{\underset{..}{O}}H$ ⟷ $CH_3CH=CH-\overset{+}{C}H-\overset{..}{O}H$ ⟷

$CH_3\overset{+}{C}H-CH=CH-\overset{..}{O}H$

(f) $CH_2=CH-\overset{..}{\underset{..}{B}}r{:}$ ⟷ ${:}CH_2-CH=\overset{..+}{B}r{:}$

(h) ${:}CH_2-\overset{\overset{..}{O}:}{\underset{\|}{C}}-CH_3$ ⟷ $CH_2=\overset{\overset{..}{\underset{..}{O}:}}{C}-CH_3$

(i) $CH_3-\overset{..}{\underset{..}{S}}-CH_2^+$ ⟷ $CH_3-\overset{+}{\underset{..}{S}}=CH_2$

(j) $CH_3-\overset{+}{N}\overset{\diagup \overset{..}{O}\cdot}{\diagdown \overset{..}{\underset{..}{O}}{:}^-}$ ⟷ $CH_3-\overset{+}{N}\overset{\diagup \overset{..}{\underset{}{O}}{:}^-}{\diagdown \overset{..}{\underset{..}{O}}}$ ⟷ $CH_3-\overset{2+}{N}\overset{\diagup \overset{..}{O}{:}^-}{\diagdown \overset{..}{\underset{..}{O}}{:}^-}$

(minor)

13.4 (a) $CH_3CH_2\overset{\overset{CH_3}{|}}{\underset{+}{C}}-CH=CH_2$ because the positive charge is on a tertiary carbon atom rather than a primary one (rule 8).

(b) because the positive charge is on a secondary carbon atom rather than a primary one (rule 8).

(c) $CH_2=\overset{+}{N}(CH_3)_2$ because all atoms have a complete octet (rule 8b), and there are more covalent bonds (rule 8a).

(d) $CH_3-\overset{\overset{O}{\|}}{C}-OH$ because it has no charge separation (rule 8c).

(e) $CH_2=CH\overset{\bullet}{C}HCH=CH_2$ because the radical is on a secondary carbon atom rather than a primary one (rule 8).

(f) ${:}NH_2-C{\equiv}N{:}$ because it has no charge separation (rule 8c).

13.5 In resonance structures, the positions of the nuclei must remain the same for all structures (rule 2). The keto and enol forms shown differ not only in the positions of their electrons, but also in the position of one of the hydrogen atoms. In the enol form, it is attached to an oxygen atom; in the keto form, it has been moved so that it is attached to a carbon atom.

13.6 (a) *cis*-1,3-Pentadiene, *trans,trans*-2,4-hexadiene, *cis,trans*-2,4-hexadiene, and 1,3-cyclohexadiene are conjugated dienes.

(b) 1,4-Cyclohexadiene and 1,4-pentadiene are isolated dienes.

(c) 1-Penten-4-yne is an isolated enyne.

13.7 The formula, C_6H_8, tells us that **A** and **B** have six hydrogen atoms less than an alkane. This unsaturation may be due to three double bonds, one triple bond and one double bond, or combinations of two double bonds and a ring, or one triple bond and a ring. Since both **A** and **B** react with 2 mol of H_2 to yield cyclohexane, they are either cyclohexyne or cyclohexadienes. The absorption maximum of 256 nm for **A** tells us that it is conjugated. Compound **B,** with no absorption maximum beyond 200 nm, possesses isolated double bonds. We can rule out cyclohexyne because of ring strain caused by the requirement of linearity of the $-C{\equiv}C-$ system. Therefore, **A** is 1,3-cyclohexadiene; **B** is 1,4-cyclohexadiene.

13.8 All three compounds have an unbranched five-carbon chain, because the product of hydrogenation is unbranched pentane. The formula, C_5H_6, suggests that they have one double bond and one triple bond. Compounds **D, E,** and **F** must differ, therefore, in the way the multiple bonds are distributed in the chain. Compounds **E** and **F** have a terminal $-C{\equiv}CH$ [IR absorption at ~3300 cm^{-1}]. The UV absorption maximum near 230 nm for **D** and **E** suggests that in these compounds, the multiple bonds are conjugated. The structures are

$$CH_3-C{\equiv}C-CH{=}CH_2 \qquad HC{\equiv}C-CH{=}CH-CH_3 \qquad HC{\equiv}C-CH_2-CH{=}CH_2$$

$$\textbf{D} \qquad\qquad\qquad \textbf{E} \qquad\qquad\qquad \textbf{F}$$

13.9 (a) Recall that 1,2 and 1,4 addition refer to the conjugated system itself and not the entire carbon chain. $CH_3CH_2\underset{\underset{Cl}{|}}{C}HCH{=}CHCH_3$ and $CH_3CH_2CH{=}CH\underset{\underset{Cl}{|}}{C}HCH_3$

(b) The most stable cation is a hybrid of equivalent forms:

$$CH_3\overset{+}{C}HCH{=}CHCH_3 \quad\longleftrightarrow\quad CH_3CH{=}CH\overset{+}{C}HCH_3$$

Thus, 1,4 and 1,2 addition yield the same product.

$$CH_3\underset{\underset{Cl}{|}}{C}HCH{=}CHCH_3$$

13.10   Addition of the proton gives the resonance hybrid.

(a) $CH_3-\overset{+}{C}H-CH=CH_2$ ⟷ $CH_3-CH=CH-\overset{+}{C}H_2$

           I   **II**

The inductive effect of the methyl group in **I** stabilizes the positive charge on the adjacent carbon. Such stabilization of the positive charge does not occur in **II**. Because **I** contributes more
heavily to the resonance hybrid than does **II**, C2 bears a greater positive charge and reacts faster
with the bromide ion.

(b)   In the 1,4-addition product, the double bond is more highly substituted than in the 1,2-
addition product; hence it is the more stable alkene.

13.11   (a), (c)

(b)   π-Electron
       interaction
       occurs here.

                                      Endo adduct

13.12   (a)

(b)

(c)

                  (major product)   (minor product)

13.13   Use the trans diester because the stereochemistry is retained in the adduct.

13.14

13.15

(Or, in each case, the other face of the dienophile could present itself to the diene, resulting in the respective enantiomer.)

13.16 (a) $BrCH_2CH_2CH_2CH_2Br$ $\xrightarrow[\text{(CH}_3)_3\text{COH}]{\text{(CH}_3)_3\text{COK}}$ $CH_2\!=\!CH\!-\!CH\!=\!CH_2$

(b) $HOCH_2CH_2CH_2CH_2OH$ $\xrightarrow[\text{heat}]{\text{concd H}_2\text{SO}_4}$ $CH_2\!=\!CH\!-\!CH\!=\!CH_2$

(c) $CH_2\!=\!CH\!-\!CH_2CH_2\!-\!OH$ $\xrightarrow[\text{heat}]{\text{concd H}_2\text{SO}_4}$ $CH_2\!=\!CH\!-\!CH\!=\!CH_2$

(d) $CH_2\!=\!CH\!-\!CH_2CH_2\!-\!Cl$ $\xrightarrow[\text{(CH}_3)_3\text{COH}]{\text{(CH}_3)_3\text{COK}}$ $CH_2\!=\!CH\!-\!CH\!=\!CH_2$

(e) $CH_2\!=\!CH\!-\!\underset{\underset{Cl}{|}}{CH}\!-\!CH_3$ $\xrightarrow[\text{(CH}_3)_3\text{COH}]{\text{(CH}_3)_3\text{COK}}$ $CH_2\!=\!CH\!-\!CH\!=\!CH_2$

(f) $CH_2\!=\!CH\!-\!\underset{\underset{OH}{|}}{CH}\!-\!CH_3$ $\xrightarrow[\text{heat}]{\text{concd H}_2\text{SO}_4}$ $CH_2\!=\!CH\!-\!CH\!=\!CH_2$

(g) $CH_2\!=\!CH\!-\!C\!\equiv\!CH$ $+$ H_2 $\xrightarrow{\text{Ni}_2\text{B (P-2)}}$ $CH_2\!=\!CH\!-\!CH\!=\!CH_2$

13.17 $\underset{\underset{CH_3}{|}}{CH_2\!=\!C}\!-\!\underset{\underset{CH_3}{|}}{C\!=\!CH_2}$

13.18 (a) Cl—CH$_2$CHCH=CH$_2$ + Cl–CH$_2$–CH=CH–CH$_2$–Cl
                     |
                     Cl

(b) CH$_2$–CH–CH–CH$_2$ (c) CH$_2$–CH–CH–CH$_2$
        | | | | | | | |
        Cl   Cl   Cl   Cl   Br   Br   Br   Br

(d) CH$_3$–CH$_2$ –CH$_2$ –CH$_3$

(e) Cl–CH$_2$–CH–CH=CH$_2$ + Cl–CH$_2$–CH=CH–CH$_2$–OH
            |
            OH

[+ ClCH$_2$CHCH =CH$_2$ + ClCH$_2$CH =CHCH$_2$Cl]
        |
        Cl

(f) 4 CO$_2$ (*Note:* KMnO$_4$ oxidizes HO$_2$C–CO$_2$H to 2 CO$_2$)

(g) CH$_3$–CH–CH=CH$_2$ + CH$_3$–CH=CH–CH$_2$–OH
          |
          OH

13.19 (a) CH$_2$=CH–CH$_2$–CH$_3$ + [NBS structure] $\xrightarrow{CCl_4}$ CH$_2$=CH–CH–CH$_3$ (Br)

(NBS)

$\left(+ \text{CH}_2\text{–CH=CH–CH}_3 \; (\text{Br}) \right)$

$\xrightarrow[\text{(CH}_3)_3\text{COH}]{\text{(CH}_3)_3\text{COK}}$ CH$_2$=CH—CH=CH$_2$

Note: In the second step, both allylic halides undergo elimination of **HBr** to yield 1,3-butadiene; therefore, separating the mixture produced in the first step is unnecessary. The BrCH$_2$CH=CHCH$_3$ undergoes a 1,4 elimination (the opposite of a 1,4 addition).

(b) CH$_2$=CH–CH$_2$–CH$_2$–CH$_3$ + NBS $\xrightarrow{CCl_4}$ CH$_2$=CH–CH–CH$_2$–CH$_3$ (Br)

$\left(+ \text{CH}_2\text{–CH=CH–CH}_2\text{–CH}_3 \; (\text{Br}) \right)$ $\xrightarrow[\text{(CH}_3)_3\text{COH}]{\text{(CH}_3)_3\text{COK}}$ CH$_2$=CH–CH=CH–CH$_3$

Here again both products undergo elimination of **HBr** to yield 1,3-pentadiene.

(c) $CH_3CH_2CH_2CH_2OH$ $\xrightarrow[\text{heat}]{\text{concd } H_2SO_4}$ $CH_3CH=CHCH_3$ [as in (a)]

$(+ CH_3CH_2CH=CH_2)$

$\underset{\underset{Br}{|}}{CH_2}-CH=CH-\underset{\underset{Br}{|}}{CH_2}$ $\xleftarrow[\text{heat}]{Br_2}$ $CH_2=CH-CH=CH_2$ $\longleftarrow$

(d) $CH_3-CH=CH-CH_3$ + NBS $\xrightarrow{CCl_4}$ $CH_3-CH=CH-CH_2Br$

$+ CH_2=CH-CHBr-CH_3$

(e)

(excess)

(f)

$\left(\text{same as } \bigcirc\!\!\!\!\!\!\!\!\right)$

13.20 $R-\overset{..}{\underset{..}{O}}-\overset{..}{\underset{..}{O}}-R$ $\xrightarrow[\text{or light}]{\text{heat}}$ $2\ R-\overset{..}{\underset{..}{O}}\bullet$

$R-\overset{..}{\underset{..}{O}}\bullet$ + $H-\overset{..}{\underset{..}{Br}}:$ $\longrightarrow$ $R-\overset{..}{\underset{..}{O}}-H$ + $\bullet\overset{..}{\underset{..}{Br}}:$

$CH_2=CH-CH=CH_2$ + $\bullet\overset{..}{\underset{..}{Br}}:$ $\longrightarrow$ $\left[CH_2=CH-\overset{\bullet}{CH}-\underset{\underset{Br}{|}}{CH_2} \longleftrightarrow \overset{\bullet}{CH_2}-CH=CH-\underset{\underset{Br}{|}}{CH_2}\right]$

$\xrightarrow{HBr}$ $CH_2=CH-\underset{\underset{H}{|}}{CH}-\underset{\underset{Br}{|}}{CH_2}$ + $\underset{\underset{H}{|}}{CH_2}-CH=CH-\underset{\underset{Br}{|}}{CH_2}$ + $\bullet\overset{..}{\underset{..}{Br}}:$

(*E* and *Z*)

13.21 Various IR, ^{1}H NMR, ^{13}C NMR, and MS features could be used to differentiate the members of any of these pairs. IR absorptions can be used to indicate whether double or triple bonds are present in a given molecule, or whether a hydroxyl group or halogen is present. Chemical shifts in the proton and carbon NMR spectra indicated can also be used to indicate the hybridization state of carbon atoms. The number of signals in the proton or carbon NMR spectrum for a given compound can suggest whether there is symmetry to the structure or not. Splitting patterns in the proton spectra can give detailed information about the connectivity of atoms. Mass spectra will indicate molecular weight differences when the molecular ions are present. Fragmentation patterns can indicate specific aspects of molecular structure characteristic to one member of a pair or another. UV would distinguish members of a pair where one is conjugated and the other is not. Consult tables with representative spectral information to predict specific aspects of the spectra for each compound.

13.22 (a) Because a highly resonance-stabilized radical is formed:

$$CH_2=CH-\overset{\bullet}{C}H-CH=CH_2 \longleftrightarrow CH_2=CH-CH=CH-\overset{\bullet}{C}H_2 \longleftrightarrow$$

$$\overset{\bullet}{C}H_2-CH=CH-CH=CH_2$$

(b) Because the carbanion is more stable:

$$CH_2=CH-\overset{\bullet\bullet}{\overset{|}{C}H}-CH=CH_2 \longleftrightarrow CH_2=CH-CH=CH-\overset{\bullet\bullet}{\overset{|}{C}H_2} \longleftrightarrow$$

$$\overset{\bullet\bullet}{\overset{|}{C}H_2}-CH=CH-CH=CH_2$$

That is, we can write more reasonance structures of nearly equal energies.

13.23

The resonance hybrid, **I**, has the positive charge, in part, on the tertiary carbon atom; in **II**, the positive charge is on primary and secondary carbon atoms only. Therefore, hybrid **I** is more stable and will be the intermediate carbocation. A 1,4 addition to **I** gives

$$CH_3-\overset{\overset{\displaystyle CH_3}{|}}{C}=CH-CH_2-Cl$$

13.24   (a)

(b)

(c)

(d)

(e)

(f)

13.25   Neither compound can assume the s-cis conformation. 1,3-Butadiyne is linear, and

=CH₂   is held in an s-trans conformation by the requirements of the ring.

13.26   (a)

(b)

13.27

13.28   The endo adduct is less stable than the exo, but is produced at a faster rate at 25°C. At 90°C the Diels-Alder reaction becomes reversible; an equilibrium is established, and the more stable exo adduct predominates.

13.29

13.30 (a)

Norbornadiene

(b)

13.31

Note: The other double bond is less reactive because of the presence of the two chlorine substituents.

Chlordan

allylic
chlorination

Heptachlor

13.32

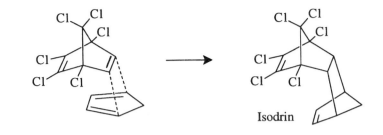

Isodrin

13.33   Protonation of the alcohol and loss of water lead to an allylic cation that can react with a chloride ion at either C1 or C3.

$$CH_3CH=CHCH_2OH \xrightarrow{H^+} CH_3CH=CHCH_2-\overset{\overset{\displaystyle H}{|}}{\underset{+}{O}}-H \xrightarrow{-H_2O}$$

$$CH_3CH=CH\overset{+}{C}H_2 \longleftrightarrow CH_3\overset{+}{C}HCH=CH_2 \xrightarrow{Cl^-}$$

$$CH_3CH=CHCH_2Cl \quad + \quad CH_3\underset{\underset{\displaystyle Cl}{|}}{C}HCH=CH_2$$

13.34   (1) $CH_2=CH-CH=CH_2 \quad + \quad Cl_2 \longrightarrow ClCH_2-\overset{+}{C}H-CH=CH_2$

$$\updownarrow$$

$$ClCH_2-CH=CH-\overset{+}{C}H_2$$

$$\underbrace{}$$

$$\overset{\delta^+}{}\quad\overset{\delta^+}{}$$

$$ClCH_2-CH\text{---}CH\text{---}CH_2$$

(2) $\overset{\delta^+}{}\quad\overset{\delta^+}{}$ $ClCH_2-CH\text{---}CH\text{---}CH_2 \xrightarrow[(-H^+)]{CH_3OH} ClCH_2-\underset{\underset{\displaystyle OCH_3}{|}}{C}H-CH=CH_2$

$$+ \quad ClCH_2-CH=CH-CH_2OCH_3$$

13.35   A six-membered ring cannot accommodate a triple bond because of the strain that would be introduced.

Too highly strained

                                    Br
13.36   The products are $CH_3CH_2\overset{|}{C}HCH=CH_2$ and $CH_3CH_2CH=CHCH_2Br$ (cis and trans).
They are formed from an allylic radical in the following way:

$$Br_2 \longrightarrow 2\ Br\cdot \quad \text{(from NBS)}$$

$$Br\cdot\ +\ CH_3CH_2CH_2CH=CH_2 \longrightarrow CH_3CH_2\overset{\cdot}{C}HCH=CH_2$$

$$\updownarrow \qquad + \text{ HBr}$$

$$CH_3CH_2CH=CH\overset{\cdot}{C}H_2$$

$$\underset{\delta\cdot}{CH_3CH_2CH}=\!=\!\underset{\delta\cdot}{CH}=\!=CH_2\ +\ Br_2 \longrightarrow \underset{\underset{Br}{|}}{CH_3CH_2CHCH}=CH_2$$

$$+ Br\cdot$$

$$+\ CH_3CH_2CH=CHCH_2Br$$
$$\text{(cis and trans)}$$

13.37   (a)   The same carbocation (a resonance hybrid) is produced in the dissociation step:

(b)   Structure **I** contributes more than **II** to the resonance hybrid of the carbocation (rule 8). Therefore, the hybrid carbocation has a larger partial positive charge on the tertiary carbon atom than on the primary carbon atom. Reaction of the carbocation with water will therefore occur more frequently at the tertiary carbon atom.

13.38   (a) Propyne.   (b) Base (: B^-) removes a proton, leaving the anion whose resonance structures are shown:

$$CH_2=C=CH_2\ +\ :B^- \rightleftharpoons H:B\ +\ \underset{\underset{H}{|}}{\overset{\overset{H}{|}}{C}}=C=\overset{..}{\overset{-}{C}}\underset{H}{\diagdown} \longleftrightarrow \underset{\underset{H}{|}}{\overset{\overset{H}{|}}{\overset{..}{C}}}{}^{-}-C\equiv C-H$$

$$\qquad\qquad\qquad\qquad\qquad \textbf{I} \qquad\qquad\qquad\qquad \textbf{II}$$

Reaction with H:B may then occur at the CH_2 carbanion. The overall reaction is

$$CH_2\!\!=\!\!C\!\!=\!\!CH_2 \;+\; :B^- \;\rightleftharpoons\; [CH_2\!\!=\!\!C\!\!=\!\!\overset{\cdot\cdot}{\overset{\cdot}{C}}H \;\longleftrightarrow\; \overset{\cdot\cdot}{\overset{\cdot}{C}}H_2\!\!-\!\!C\!\!\equiv\!\!CH\,] \;+\; H:B$$

$$CH_3\!\!-\!\!C\!\!\equiv\!\!CH \;+\; :B^-$$

13.39 The first crystalline solid is the Diels-Alder adduct below, mp 125°C,

On melting, this adduct undergoes a reverse Diels-Alder reaction, yielding furan (which vaporizes) and maleic anhydride, mp 56°C,

13.40 (a) + enantiomer (b) + enantiomer

(c) + enantiomer (d) + enantiomer

13.41 The product formed when butyl bromide undergoes elimination is 1-butene, a simple monosubstituted alkene. When 4-bromo-1-butene undergoes elimination, the product is 1,3-butadiene, a conjugated diene, and therefore, a more stable product. The transition states leading to the products reflect the relative stabilities of the products. Since the transition state leading to 1,3-butadiene has the lower free energy of activation of the two, the elimination reaction of 4-bromo-1-butene will occur more rapidly.

13.42   The diene portion of the molecule is locked into an s-trans conformation. It cannot, therefore, achieve the s-cis conformation necessary for a Diels-Alder reaction.

13.43

(c)   See above

QUIZ

13.1   Give the 1,4-addition product of the following reaction:

$$CH_3CH=CHCH=CHCH_3 \ + \ HCl \ \longrightarrow \ ?$$

(a) $CH_3CH=CHC=CHCH_3$
   $\quad\quad\quad\quad\quad\ |$
   $\quad\quad\quad\quad\quad Cl$

(b) $CH_3CH_2CHCH=CHCH_3$
   $\quad\quad\quad\quad\ |$
   $\quad\quad\quad\quad Cl$

(c) $CH_2CH=CHCH=CHCH_3$
   $\ |$
   Cl

(d) $CH_3CH_2CH=CHCHCH_3$
   $\quad\quad\quad\quad\quad\quad\ |$
   $\quad\quad\quad\quad\quad\quad Cl$

(e) $CH_3CH_2CHCHCH_2CH_3$
   $\quad\quad\quad\ |\ \ |$
   $\quad\quad\quad Cl\ Cl$

13.2 Which diene and dienophile could be used to synthesize the following compound?

13.3 Which reagent(s) could be used to carry out the following reaction?

(a) NBS/CCl$_4$ $\left(NBS = \begin{array}{c} O \\ \\ NBr \\ \\ O \end{array} \right)$ (b) NBS/CCl$_4$, then Br$_2$/hv

(c) Br$_2$/hv, then (CH$_3$)$_3$COK/(CH$_3$)$_3$COH, then NBS/CCl$_4$

(d) (CH$_3$)$_3$COK/(CH$_3$)$_3$COH, then NBS/CCl$_4$

13.4 Which of the following structures does not contribute to the hybrid for the carbocation formed when 4-chloro-2-pentene ionizes in an S$_N$1 reaction?

(a) CH$_3$CH=CHĊHCH$_3$ (b) CH$_3$ĊHCH=CHCH$_3$ (c) CH$_3$ĊHCH$_2$CH=CH$_2$

(d) All of these contribute to the resonance hybrid.

13.5 Which of the following resonance structures accounts at least in part for the lack of S$_N$2 reactivity of vinyl chloride?

(a) CH$_2$=CH—C̈l: (b) C̈H$_2$—CH=C̈l:$^+$ (c) Neither   (d) Both

SUMMARY OF REACTIONS BY TYPE
CHAPTERS 1–13

I. SUBSTITUTION REACTIONS

Type	Stereochemical Result	Favoring Conditions
S_N2 Chapter 6 See Section 6.15 for a summary of nucleophiles mentioned in early chapters	inversion	1°, 2°, benzylic (1° or 2°), or allylic (1° or 2°) leaving group (e.g., halide, tosylate, mesylate); strong nucleophile; polar aprotic solvent
S_N1 Chapter 6	racemization (via carbocation)	3°, benzylic, or allylic leaving group
Radical substitution Chapter 10 Section 13.2	racemization (via radical)	3°, benzylic, or allylic hydrogen; peroxides, heat/light

II. ELIMINATION REACTIONS

Type	Stereochemical/Regiochemical Result	Favoring Conditions
E2 (dehydrohalogenation) Section 7.6	elimination to form the most substituted alkene (Zaitsev elimination) with small bases formation of the less substituted alkene with use of a bulky base (e.g., *tert*-BuOK/*tert*-BuOH)	strong base (e.g., NaOEt/EtOH, KOH/EtOH, *tert*-BuOK/*tert*-BuOH); 2° or 3° leaving group (e.g., halide, tosylate, mesylate, etc.); heat
El (dehydrohalogenation) Sections 6.18 and 7.10	formation of most substituted alkene; may occur with carbocation rearrangement	3° leaving group; weak base; heat
Dehydration Section 7.7	formation of most substituted alkene; may occur with carbocation rearrangement	catalytic acid (H^+, e.g., concd H_2SO_4, H_3PO_4); heat

III. MECHANISTIC SUMMARY OF ALKENE AND ALKYNE ADDITION REACTIONS

Reactant	Stereochemical Result		Regiochemical Result	
	syn	**anti**	**Markovnikov**	**anti-Markovnikov**
Alkenes	H_2/Pt, Pd or Ni (Section 7.13)	✻ X_2/CCl_4 (Section 8.6)	✻ H_2O, H^+ (hydration) (Section 8.5)	(i) THF:BH_3 (ii) OH^-, H_2O_2 (Section 11.7)
	(i) OsO_4, (ii) Na_2SO_3 (Section 8.10)	✻ X_2,H_2O (Section 8.8)	✻ (i) $Hg(OAc)_2$, H_2O, THF (ii) $NaBH_4$ (Section 11.5)	
	RCO_3H (e.g., MMPP) (Section 11.17)	(i) RCO_3H, (ii) H_3O^+ or OH^- (Section 11.18)	✻ HX (no peroxides) (Section 8.2)	HBr (w/peroxides) (Section 10.9)
	✻ (i) THF:BH_3 (ii) H_2O_2, OH^- (Section 10.7)	✻ X_2/ROH (Section 8.8)	✻ addition of other alkenes	
		✻ X_2, Nucleophile (e.g., RO^-, RMgX) (Section 8.8)		
	:C⟨ Addition of carbenes (Section 8.9)			

✻Shares mechanistic themes with other reactions denoted by the same symbol.

Alkynes	H_2, Pd on $BaCO_3$ w/quinoline Lindlar's cat.) (Section 7.15A)	(i) Li, $EtNH_2$, $-78°C$ (ii) $NH_4^+Cl^-$ (Section 7.15B)	HX (one or two molar equivalents) (Section 8.13)	
	H_2, Ni_2B (P-2) (Section 7.15A)	X_2 (2 equiv.)/ CCl_4 (Section 8.12)		
	H_2(2 equiv.), Pt or Ni (complete hydrogenation) (Section 7.15)			

IV. ALKENE AND ALKYNE CLEAVAGE WITH OXIDATION

Conditions	Reactant		Product(s)
Alkenes			
(i) KMnO$_4$/OH$^-$, heat; (ii) H$_3$O$^+$	tetrasubstituted alkene	$\rightarrow$	two ketones
	trisubstituted alkene	$\rightarrow$	one ketone, one carboxylic acid
(Section 8.11)	monosubstituted alkene	$\rightarrow$	one carboxylic acid and CO$_2$
(i) O$_3$, (ii) Zn, HOAc	tetrasubstituted alkene	$\rightarrow$	two ketones
	trisubstituted alkene	$\rightarrow$	one ketone, one aldehyde
(Section 8.11A)	monosubstituted alkene	$\rightarrow$	one aldehyde and formaldehyde
Alkynes			
(i) KMnO$_4$/OH$^-$, heat; (ii) H$_3$O$^+$	terminal alkyne	$\rightarrow$	a carboxylic acid and formic acid
	internal alkyne	$\rightarrow$	two carboxylic acids
(Section 8.14)			
(i) O$_3$, (ii) Zn, HOAc	terminal alkyne	$\rightarrow$	a carboxylic acid and formic acid
	internal alkyne	$\rightarrow$	two carboxylic acids
(Section 8.14)			

V. CARBON—CARBON BOND-FORMING REACTIONS

(a) Alkylation of alkynide anions (with 1° alkyl halides, epoxides, and aldehydes or ketones) (Sections 4.18C, 4.20B, 7.12 and 12.8D)

(b) Grignard reaction (with aldehydes and ketones, or epoxides) (Sections 12.7B, C, and 12.8)

(c) Reactions of lithium dialkylcuprates (Corey-House, Posner-Whitesides reaction) (Section 12.9)

(d) Carbocation addition to alkenes (e.g., polymerization) (Special Topic A)

(e) Diels-Alder reaction (Section 13.11)

(f) Addition of a carbene to an alkene (Section 8.9)

VI. REDUCTIONS/OXIDATIONS (NOT INCLUDING ALKENES/ALKYNES)

(a) 2 R—X (w/Zn/H$^+$) $\rightarrow$ 2 R—H + ZnX$_2$ (Section 4.18B)

(b) Lithium aluminum hydride (LiAlH$_4$) and sodium borohydride (NaBH$_4$) reduction of carbonyl compounds (Section 12.3)

(c) Oxidation of alcohols with chromic acid, pyridinium chlorochromate (PCC), or hot potassium permanganate (Section 12.4)

VII. MISCELLANEOUS

(a) R—CO—R + PCl$_5$ → R—CCl$_2$—R → alkynes (Section 7.10)

(b) R—COOH + SOCl$_2$ or PCl$_5$ → R—COCl → acyl chlorides for Friedel-Crafts reactions, esters, amides, etc. (Sections 15.7 and 18.5)

(c) Terminal alkynes + NaNH$_2$ in NH$_3$ → alkynide anions (Sections 4.18C, 12.8D)

(d) R—OH + TsCl or MsCl (with pyridine) → R—OTs or R—OMs (Section 11.10)

(e) R—OH + Na (or NaNH$_2$) → R—O$^-$Na$^+$ + H$_2$ (or NH$_3$) (Sections 6.16B, 11.15B)

VIII. CHEMICAL TESTS

(a) Alkenes/Alkynes: Br$_2$/CCl$_4$ (Section 8.6)

(b) Rings/Unsaturation/etc.: Index of Hydrogen Deficiency (Section 7.16)

(c) Position of Unsaturation: KMnO$_4$ (Section 8.11); O$_3$ (Section 8.11A)

METHODS FOR FUNCTIONAL GROUP PREPARATION
CHAPTERS 1–13

I. ALKYL HALIDES

(a) HX addition to alkenes [Markovnikov (Section 8.2) and HBr anti-Markovnikov (Section 10.9)]

(b) X_2 addition to alkenes (Section 8.6)

(c) Radical halogenation

    (i) X_2/light/heat for alkanes (Sections 10.3, 10.4, 10.6)

    (ii) *N*-Bromosuccinimide (NBS)/heat/light for allylic and benzylic substitution (Section 13.2B)

(d) $R-OH + SOCl_2$ (with pyridine) $\rightarrow R-Cl + SO_2 +$ pyridinium hydrochloride (Section 11.14)

(e) $3 R-OH + PBr_3 \rightarrow 3 R-Br + P(OH)_3$ (Section 11.14)

(f) $R-OH + HX \rightarrow R-X + H_2O$ (Sections 11.12, 11.13)

II. ALCOHOLS

(a) Markovnikov addition of H_2O with catalytic H^+ (Section 8.5)

(b) Markovnikov addition of H_2O via (i) $Hg(OAc)_2$, H_2O, THF; (ii) $NaBH_4$ (Section 11.5)

(c) Anti-Markovnikov addition of H_2O via (i) $THF:BH_3$; (ii) OH^-, H_2O_2 (Section 11.7)

(d) OsO_4 addition to alkenes (syn) (Section 8.10)

(e) (i) RCO_3H (peroxycarboxylic acids); (ii) H_3O^+ (Section 11.18)

(f) Cleavage of ethers with one molar equiv. HX (Section 11.16)

(g) Opening of epoxides by a nucleophile (Sections 11.18, 11.19)

(h) Lithium aluminum hydride or sodium borohydride reduction of carbonyl compounds (Section 12.3)

(i) Grignard reaction with aldehydes, ketones, and epoxides (Sections 12.17B, C, 12.8)

(j) Cleavage of silyl ethers (Section 11.15D)

III. ALKENES

(a) E2 Dehydrohalogenation (preferred over E1 for alkene synthesis) (Section 7.6)

(b) Dehydration of alcohols (Section 7.7)

(c) Hydrogenation of alkynes: Z by catalytic hydrogenation, E by dissolving metal reduction (Sections 7.15A, B)

(d) Diels-Alder reaction (forms one new double bond with ring formation) (Section 13.11)

IV. ALKYNES

(a) Alkylation of alkynide anions (Sections 4.18C, 4.20B, 7.12)

(b) Double dehydrohalogenation of vicinal or geminal dihalides (Section 7.10)

V. CARBON—CARBON BONDS

(a) Alkylation of alkynide anions (Sections 4.18C, 4.20B, 7.12, 12.8D)

(b) Organometallic addition to carbonyl compounds and epoxides (e.g., Grignard or RLi reactions) (Sections 12.7B, C, 12.8)

(c) Lithium dialkylcuprate reactions (Corey-House, Posner-Whitesides) (Section 12.9)

(d) Diels-Alder reaction (Section 13.11)

(e) Addition of alkenes to other alkenes (e.g., polymerization) (Section 10.10 and Special Topic A)

(f) Addition of a carbene to an alkene (Section 8.9)

VI. ALDEHYDES

(a) (i) O_3; (ii) Zn, HOAc with appropriate alkenes (Section 8.11A)

(b) Pyridinium chlorochromate (PCC) oxidation of 1° alcohols (Section 12.4A)

VII. KETONES

(a) (i) O_3; (ii) Zn, HOAc with appropriate alkenes (Section 8.11A)

(b) $KMnO_4/OH^-$ cleavage of appropriate alkenes (Section 8.11)

(c) H_2CrO_4 oxidation of 2° alcohols (Section 12.4C)

VIII. CARBOXYLIC ACIDS

(a) (i) $KMnO_4/OH^-$; (ii) H_3O^+ with 1° alcohols (Section 12.4B)

(b) (i) O_3; (ii) Zn, HOAc with alkynes (Section 8.14)

(c) (i) $KMnO_4/OH^-$, heat; (ii) H_3O^+ with alkynes and alkenes (Sections 8.11 and 8.15)

IX. ETHERS (INCLUDING EPOXIDES)

(a) $RO^- + R'X \rightarrow ROR' + X^-$ (Williamson synthesis) (Section 11.15B)

(b) $2\ ROH$ (+cat. H_2SO_4, heat) $\rightarrow ROR + HOH$ (for symmetrical ethers only) (Section 11.15A)

(c) Alkene + RCO_3H (a peroxycarboxylic acid, e.g., MMPP) $\rightarrow$ an epoxide (Section 11.17)

(d) An epoxide + $RO^- \rightarrow$ an α-hydroxy ether (also acid catalyzed) (Section 11.18)

(e) $ROH + ClSiR'_3 \rightarrow ROSiR'_3$ (silyl ether protecting groups, e.g., TBDMS ethers) (Section 11.15D)

14 AROMATIC COMPOUNDS

SOLUTIONS TO PROBLEMS

14.1   Compounds (a) and (b) would yield only one monosubstitution product.

14.2   Resonance structures are defined as being structures that differ *only* in the positions of the electrons. In the two 1,3,5-cyclohexatrienes shown, the carbon atoms are in different positions; therefore, they cannot be resonance structures.

14.3   Inscribing a square in a circle with one corner at the bottom gives the following results:

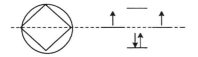

We see, therefore, that cyclobutadiene would be a diradical and would not be aromatic.

14.4   (a)

The cyclopentadienyl anion (above) should be aromatic because it has a closed bonding shell of delocalized π electrons.
(b) and (c) The cyclopentadienyl cation (below) would be a diradical. We would not expect it to be aromatic.

(d) No, 4 is not a Hückel number.

14.5   (a) The cycloheptatrienyl cation (below) would be aromatic because it would have a closed bonding shell of delocalized π electrons.

(b) No, the cycloheptatrienyl anion (below) would be a diradical.

(c) No, 8 is not a Hückel number.

14.6   If the 1,3,5-cycloheptatrienyl anion *were* aromatic, we would expect it to be unusually stable. This would mean that 1,3,5-cycloheptatriene should be unusually acidic. The fact that 1,3,5-cycloheptatriene is not unusually acidic (it is less acidic than 1,3,5-heptatriene) confirms the prediction made in the previous problem, that the 1,3,5-cycloheptatrienyl anion should not be aromatic.

14.7   (a)

Tropylium bromide

These results suggest that the bonding in tropylium bromide is ionic; that is, it consists of a positive tropylium ion and a negative bromide ion.

14.8   It suggests that the cyclopropenyl cation should be aromatic.

14.9   The fact that the cyclopentadienyl cation is antiaromatic means that the following hypothetical transformation would occur with an increase in π-electron energy.

14.10 The cyclopropenyl cation (below).

14.11 The high field signal arises from the six methyl protons of *trans*-15,16-dimethyldihydropyrene, which by virtue of their location are strongly shielded by the magnetic field created by the aromatic ring current (see Figure 14.8).

14.12 Major contributors to the hybrid must be ones that involve separated charges. Contributors like the following would have separated charges, and would have aromatic five- and seven-membered rings.

14.13 (a) (b)

14.14 Because of their symmetries, *p*-dibromobenzene would give two ^{13}C signals, *o*-dibromobenzene would give three, and *m*-dibromobenzene would give four.

Two signals   Three signals   Four signals

14.15

A   **B**   **C**   **D**

A. Strong absorption at 740 cm^{-1} is characteristic of ortho substitution.

B. A very strong absorption peak at 800 cm^{-1} is characteristic of para substitution.

C. Strong absorption peaks at 680 and 760 cm^{-1} are characteristic of meta substitution.

D. Strong absorption peaks at 693 and 765 cm^{-1} are characteristic of a monosubstituted benzene ring.

14.16 (a) [structure: benzene with CO_2H and NO_2]

(b) [structure: benzene with Br and CH_3]

(c) [structure: benzene with two Br (ortho)]

(d) [structure: benzene with two NO_2 (meta)]

(e) [structure: benzene with OH, O_2N, NO_2]

(f) [structure: benzene with CO_2H and NO_2 (para)]

(g) [structure: benzene with OCH_2CH_3 and Cl]

(h) [structure: benzene with SO_3H and Cl]

(i) [structure: benzene with SO_2OCH_3 and CH_3]

(j) [structure: benzene with CH_2Br]

(k) [structure: benzene with NH_2 and NO_2]

(l) [structure: benzene with two CH_3 (ortho)]

(m) [structure: benzene with $C(CH_3)_3$]

(n) [structure: benzene with OH and CH_3 (para)]

(o) [structure: benzene with $C(=O)CH_3$ and Br]

(p) [structure: cyclohexane with OH and C_6H_5]

(q) $HOCH_2CHCHCH_3$ with C_6H_5 and CH_3

(r) [structure: benzene with OCH_3 and Cl]

14.17 (a) [structure: benzene with three Br]

1,2,3-Tribromo-
benzene

[structure: benzene with three Br]

1,2,4-Tribromo-
benzene

[structure: benzene with three Br]

1,3,5-Tribromo-
benzene

(b)

2,3-Dichloro-
phenol

2,4-Dichloro-
phenol

2,5-Dichloro-
phenol

2,6-Dichloro-
phenol

3,4-Dichloro-
phenol

3,5-Dichloro-
phenol

(c)

4-Nitroaniline
(*p*-nitroaniline)

3-Nitroaniline
(*m*-nitroaniline)

2-Nitroaniline
(*o*-nitroaniline)

(d)

4-Methylbenzene-
sulfonic acid
(*p*-toluenesulfonic
acid)

3-Methylbenzene-
sulfonic acid
(*m*-toluenesulfonic
acid)

2-Methylbenzene-
sulfonic acid
(*o*-toluenesulfonic
acid)

(e)

Butylbenzene

Isobutylbenzene

sec-Butylbenzene

tert-Butylbenzene

14.18 Hückel's rule should apply to both pentalene and heptalene. Pentalene's antiaromaticity can be attributed to its having 8 π electrons. Heptalene's lack of aromaticity can be attributed to its having 12 π electrons. Neither 8 nor 12 is a Hückel number.

14.19 (a) The extra two electrons go into the two partly filled (nonbonding) molecular orbitals (Fig. 14.6), causing them to become filled. The dianion, therefore, is not a diradical. Moreover, the cyclooctatetraene dianion has 10 π electrons (a Hückel number), and this apparently gives it the stability of an aromatic compound. (The highest occupied molecular orbitals may become slightly lower in energy and become bonding molecular orbitals.) The stability gained by becoming aromatic is apparently large enough to overcome the extra strain involved in having the ring of the dianion become planar.

(b) The strong base (butyllithium) removes two protons from the compound on the left. This acid-base reaction leads to the formation of the 10 π electron pentalene dianion, an aromatic dianion.

Pentalene dianion

14.20 The bridging –CH$_2$–group causes the 10 π electron ring system (below) to become planar. This allows the ring to become aromatic.

14.21 (a) Resonance contributors that involve the carbonyl group of **I** resemble the *aromatic* cycloheptatrienyl cation and thus stabilize **I**. Similar contributors to the hybrid of **II** resemble the *antiaromatic* cyclopentadienyl cation (see Problem 14.9) and thus destabilize **II**.

(a)

I   **IA**

Contributors like **IA** are exceptionally stable because they resemble an aromatic compound. They therefore make large stabilizing contributions to the hybrid.

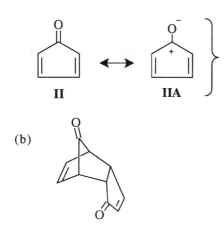

Contributors like **IIA** are exceptionally unstable because they resemble an anti-aromatic compound. Any contribution they make to the hybrid is destabilizing.

(b)

14.22 Ionization of 5-chloro-1,3-cyclopentadiene would produce a cyclopentadienyl cation, and the cyclopentadienyl cation (see Problem 14.9) would be highly unstable because it would be antiaromatic.

Antiaromatic ion
(highly unstable)

14.23 (a) The cycloheptatrienyl anion has 8 π electrons and does not obey Hückel's rule; the cyclononatetraenyl anion with 10 π electrons obeys Hückel's rule.

Cycloheptatrienyl anion
8 π-electrons
Antiaromatic

Cyclononatetraenyl anion
10 π-electrons
Aromatic

(b) By adding 2 π electrons, [16]annulene becomes an 18 π electron system and therefore obeys Hückel's rule.

14.24 As noted in Problem 13.28, furan can serve as the diene component of Diels-Alder reactions, readily losing all aromatic character in the process. Benzene, on the other hand, is so unreactive in a Diels-Alder reaction that it can be used as a nonreactive solvent for Diels-Alder reactions.

14.25

(a)	doublet δ 1.25
(b)	septet δ 2.9
(c)	multiplet δ 7.3

(a)	doublet δ 1.35
(b)	quartet δ 4.1
(c)	singlet δ 1.8
(d)	multiplet δ 7.3

(a)	triplet δ 2.9
(b)	quintet δ 2.1
(c)	multiplet δ 7.1

14.26 A ¹H NMR signal this far upfield indicates that cyclooctatetraene is a cyclic polyene and is not aromatic.

14.27 Compound F is *p*-isopropyltoluene. ¹H NMR assignments are shown in the following spectrum.

We can make the following ^{1}H NMR assignments:

(a)   doublet δ 1.25
(b)   singlet δ 2.3
(c)   septet δ 2.85
(d)   multiplet δ 7.1

14.28   Compound **L** is allylbenzene,

(d) Doublet, δ 3.1 (2H)
(a) or (b) Multiplet, δ 4.8
(a) or (b) Multiplet, δ 5.1
(c) Multiplet, δ 5.8
(e) Multiplet, δ 7.1 (5H)

The following IR assignments can be made.

3035 cm^{-1}, C–H stretching of benzene ring
3020 cm^{-1}, C–H stretching of –CH=CH$_2$ group
2925 cm^{-1} and 2853 cm^{-1}, C–H stretching of –CH$_2$ –group
1640 cm^{-1}, C=C stretching
990 cm^{-1} and 915 cm^{-1}, C–H bendings of –CH=CH$_2$ group
740 cm^{-1} and 695 cm^{-1}, C–H bendings of –C$_6$H$_5$ group

The UV absorbance maximum at 255 nm is indicative of a benzene ring that is not conjugated with a double bond.

14.29   Compound **M** is *m*-ethyltoluene. We can make the following assignments in the spectra.

(a)   triplet δ 1.4
(b)   quartet δ 2.6
(c)   singlet δ 2.4
(d)   multiplet δ 7.05

Meta substitution is indicated by the very strong peaks at 690 and 780 cm^{-1} in the IR spectrum.

14.30 Compound **N** is $C_6H_5CH=CHOCH_3$. The absence of absorption peaks due to O–H or C=O stretching in the IR spectrum of **N** suggests that the oxygen atom is present as part of an ether linkage. The (5H) ^{1}H NMR multiplet between δ 7.1–7.6 strongly suggests the presence of a monosubstituted benzene ring; this is confirmed by the strong peaks at ~690 and ~770 cm^{-1} in the IR spectrum.

We can make the following assignments in the ^{1}H NMR spectrum:

$$(a)\ (b)\ (c)\ (d)$$
$$C_6H_5–CH=CH–OCH_3$$

- *(a)*   Multiplet δ 7.1–7.6
- *(b)*   Doublet δ 6.1
- *(c)*   Doublet δ 5.2
- *(d)*   Singlet δ 3.7

14.31 Compound **X** is *meta*-xylene. The upfield signal at δ 2.3 arises from the two equivalent methyl groups. The downfield signals at δ 6.9 and 7.1 arise from the protons of the benzene ring. Meta substitution is indicated by the strong IR peak at 680 cm^{-1} and very strong IR peak at 760 cm^{-1}.

14.32 The broad IR peak at 3400 cm^{-1} indicates a hydroxy group, and the two bands at 720 and 770 cm^{-1} suggest a monosubstituted benzene ring. The presence of these groups is also indicated by the peaks at δ 4.4 and δ 7.2 in the ^{1}H NMR spectrum. The ^{1}H NMR spectrum also shows a triplet at δ 0.85 indicating a –CH$_3$ group coupled with an adjacent –CH$_2$–group. There is a complex multiplet at δ 1.7 and there is also a triplet at δ 4.5 (1H). Putting these pieces together in the only way possible gives us the following structure for **Y**.

14.33 (a) Four unsplit signals,

(b) Absorptions arising from: =C–H, CH₃, and C=O groups.

*14.34 The vinylic protons of *p*-chlorostyrene should give a spectrum approximately like the following:

δ_H (ppm)

*14.35

*14.36

Na$^+$ ⟶ (a "sandwich compound")

*14.37   Ph—C≡C—C≡C—Ph

QUIZ

14.1   Which of the following reactions is inconsistent with the assertion that benzene is aromatic?

(a)   Br$_2$/CCl$_4$/25°C ⟶ no reaction

(b)   H$_2$/Pt/25°C ⟶ no reaction

(c)   Br$_2$/FeBr$_3$ ⟶ C$_6$H$_5$Br + HBr

(d)   KMnO$_4$/H$_2$O/25°C ⟶ no reaction

(e)   None of the above

14.2   Which is the correct name of the compound shown?

(a)   3-Chloro-5-nitrotoluene   (b)   *m*-Chloro-*m*-nitrotoluene

(c)   1-Chloro-3-nitro-5-toluene   (d)   *m*-Chloromethylnitrobenzene

(e)   More than one of these

14.3   Which is the correct name of the compound shown?

(a) 2-Fluoro-1-hydroxyphenylbenzene (b) 2-Fluoro-4-phenylphenol

(c) *m*-Fluoro-*p*-hydroxybiphenyl   (d) *o*-Fluoro-*p*-phenylphenol

(e) More than one of these

14.4 Which of the following molecules or ions is not aromatic according to Huckel's rule?

(e) All are aromatic.

14.5 Give the structure of a compound with the formula C_7H_7Cl that is capable of undergoing both S_N1 and S_N2 reactions.

14.6 Write the name of an aromatic compound that is isomeric with naphthalene.

15

REACTIONS OF AROMATIC COMPOUNDS

SOLUTIONS TO PROBLEMS

15.1

15.2 The rate is dependent on the concentration of NO_2^+ ion formed from protonated nitric acid.

$$H-\overset{+}{\underset{H}{O}}-NO_2 \quad + \quad HA \quad \longrightarrow \quad NO_2^+ \quad + \quad H_3O^+ \quad + \quad A^-$$

(where HA = HNO_3 or $HOSO_3H$)

Because H_2SO_4 ($HOSO_3H$) is a stronger acid, a mixture of it and HNO_3 will contain a higher concentration of protonated nitric acid than will nitric acid alone.

That is, the reaction,

$$H-O-NO_2 \quad + \quad HOSO_3H \quad \rightleftharpoons \quad H-\overset{+}{\underset{H}{O}}-NO_2 \quad + \quad HSO_4^-$$

Protonated
nitric acid

produces more protonated nitric acid than the reaction,

$$H-O-NO_2 \quad + \quad H-O-NO_2 \quad \rightleftharpoons \quad H-\overset{+}{\underset{H}{O}}-NO_2 \quad + \quad NO_3^-$$

15.3   *Step 1*

15.4

15.5   The carbocation formed by the action of $AlCl_3$ on neopentyl chloride is primary. This carbocation rearranges to the more stable tertiary carbocation before it can react with the benzene ring:

15.6 $CH_3CH_2CH_2-OH + BF_3 \rightleftharpoons CH_3CH_2CH_2^+ + HOBF_3$

The propyl cation can rearrange into an isopropyl cation:

$$CH_3CH_2CH_2^+ \xrightarrow[\text{shift}]{\text{hydride}} CH_3\overset{+}{C}HCH_3$$

Both carbocations can then attack the ring.

15.7 (a)

15.8 If the methyl group had no directive effect on the incoming electrophile, we would expect to obtain the products in purely statistical amounts. Since there are two ortho hydrogen atoms, two meta hydrogen atoms, and one para hydrogen, we would expect to get 40% ortho (2/5), 40% meta (2/5), and 20% para (1/5). Thus, we would expect that only 60% of the mixture of mononitrotoluenes would have the nitro group in the ortho or para position. And we would expect to obtain 40% of *m*-nitrotoluene. In actuality, we get 96% of combined *o*- and *p*-nitrotoluene and only 4% *m*-nitrotoluene. This shows the ortho-para directive effect of the methyl group.

15.9 (a)

15.10 As the following structures show, attack at the ortho and para positions of phenol leads to arenium ions that are more stable (than the one resulting from meta attack) because they are hybrids of four resonance structures, one of which is relatively stable. Only three resonance structures are possible for the meta arenium ion, and none is relatively stable.

Ortho attack

Relatively stable

Meta attack

Para attack

Relatively stable

15.11 (a) The atom (an oxygen atom) attached to the benzene ring has an unshared electron pair that it can donate to the arenium ions formed from ortho and para attack stabilizing them. (The arenium ions are analogous to the previous answer with a $-COCH_3$ group replacing the $-H$).

(b) Structures such as the following compete with the benzene ring for the oxygen electrons, making them less available to the benzene ring.

This effect makes the benzene ring of phenyl acetate less electron rich and, therefore, less reactive.

(c) Because the acetamido group has an unshared electron pair on the nitrogen atom that it can donate to the benzene ring it is an ortho-para director.

(d) Structures such as the following compete with the benzene ring for the nitrogen electrons, making them less available to the benzene ring.

15.12 The electron-withdrawing inductive effect of the chlorine of chloroethene makes its double bond less electron rich than that of ethene. This causes the rate of reaction of chloroethene with an electrophile (i.e., a proton) to be slower than the corresponding reaction of ethene.

When chloroethene adds a proton, the orientation is governed by a resonance effect. In theory, two carbocations can form:

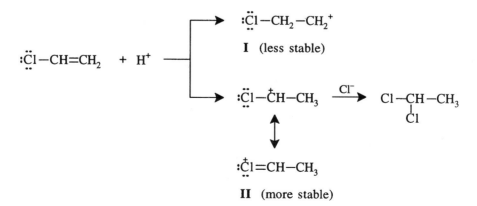

Carbocation **II** is more stable than **I** because of the resonance contribution of the extra structure just shown in which the chlorine atom donates an electron pair (see Section 15.11D).

15.13 *Ortho attack*

Relatively stable

Para attack

Relatively stable

15.14 The phenyl group, as the following resonance structures show, can act as an electron-releasing group and can stabilize the arenium ions formed from ortho and para attack.

Ortho attack

Para   attack

15.15

I   leads to 1-chloro-1-phenylpropane

II   leads to 2-chloro-1-phenylpropane

III   leads to 1-chloro-3-phenylpropane

The major product is 1-chloro-1-phenylpropane because **I** is the most stable radical. It is a benzylic radical and therefore is stabilized by resonance.

15.17   The addition of hydrogen bromide to 1-phenylpropene proceeds through a benzylic radical in the presence of peroxides, and through a benzylic cation in their absence (cf., a and b as follows).

(a)   Hydrogen bromide addition in the presence of peroxides.

Chain   Initiation

Step 1   R—O—O—R ⟶ 2 R—O•

Step 2   RO• + H—Br ⟶ R—O—H + Br•

Step 3   Br• + C₆H₅CH=CHCH₃ ⟶ C₆H₅ĊH—CHCH₃
                                                                                    |
                                                                                   Br

A benzylic radical

Chain   Propagation

Step 4   C₆H₅ĊHCHCH₃ + H—Br ⟶ C₆H₅CH₂CHCH₃ + Br•
                          | |
                         Br   Br

2-Bromo-1-phenylpropane

REACTIONS OF AROMATIC COMPOUNDS 267

The mechanism for the addition of hydrogen bromide to 1-phenylpropene in the presence of peroxides is a chain mechanism analogous to the one we discussed when we described anti-Markovnikov addition in Section 10.9. The step that determines the orientation of the reaction is the first chain-propagating step. Bromine attacks the second carbon atom of the chain because by doing so the reaction produces a more stable benzylic radical. Had the bromine atom attacked the double bond in the opposite way, a less stable secondary radical would have been formed.

$$C_6H_5CH=CHCH_3 \quad + \quad Br\cdot \quad \xrightarrow{X} \quad C_6H_5CH-\overset{\cdot}{C}HCH_3$$

$$Br$$

A secondary radical

(b) Hydrogen bromide addition in the absence of peroxides.

$$C_6H_5CH=CHCH_3 \quad + \quad HBr \quad \longrightarrow \quad C_6H_5\overset{+}{C}HCH_2CH_3 \quad + \quad Br^-$$

A benzylic cation

$$C_6H_5CHCH_2CH_3$$

$$Br$$

1-Bromo-1-phenylpropane

In the absence of peroxides, hydrogen bromide adds through an ionic mechanism. The step that determines the orientation in the ionic mechanism is the first, where the proton attacks the double bond to give the more stable benzylic cation. Had the proton attacked the double bond in the opposite way, a less stable secondary cation would have been formed.

$$C_6H_5CH=CHCH_3 \quad + \quad HBr \quad \xrightarrow{X} \quad C_6H_5CH-\overset{+}{C}HCH_3 \quad + \quad Br^-$$

$$H$$

A secondary cation

15.18 (a) because the more stable carbocation intermediate is the benzylic carbocation, which then reacts with a chloride ion.

(b) because the more stable intermediate is a benzylic cation, which then reacts with H_2O.

15.19 (a) The first method would fail because introducing the chlorine substituent first would introduce an ortho-para directing group. Consequently, the subsequent Friedel-Crafts reaction would not then take place at the desired meta position.

   The second method would fail for essentially the same reasons. Introducing the ethyl group first would introduce an ortho-para director, and subsequent ring chlorination would not take place at the desired meta position.

(b) If we introduce an acetyl group first, which we later convert to an ethyl group, we install a meta director. This allows us to put the chlorine atom in the desired position. Conversion of the acetyl group to an ethyl group is then carried out using the Clemmensen reduction.

15.20 (a) [structure: benzene ring with OH, NO₂, CF₃]   (b) [structure: benzene ring with CN, O₂N, SO₃H]   (c) [structure: benzene ring with OCH₃, NO₂, NO₂]   +   [structure: benzene ring with OCH₃, O₂N, NO₂]

15.21 (a) In concentrated base and ethanol (a relatively nonpolar solvent), the S_N2 reaction is favored. Thus, the rate depends on the concentration of both the alkyl halide and $NaOC_2H_5$. Since no carbocation is formed, the only product is

$$CH_3CH=CHCH_2OCH_2CH_3$$

(b) When the concentration of $C_2H_5O^-$ ion is small or zero, the reaction occurs through the S_N1 mechanism. The carbocation that is produced in the first step of the S_N1 mechanism is a resonance hybrid.

$$CH_3CH=CHCH_2Cl \quad \rightleftharpoons \quad \left[\begin{array}{c} CH_3CH=CH\overset{+}{C}H_2 \\ \updownarrow \\ CH_3\overset{+}{C}H-CH=CH_2 \end{array} \right] + Cl^-$$

This ion reacts with the nucleophile ($C_2H_5O^-$ or C_2H_5OH) to produce two isomeric ethers

$$CH_3CH=CHCH_2-OCH_2CH_3 \quad \text{and} \quad CH_3\overset{\displaystyle OCH_2CH_3}{\underset{|}{C}H}CH=CH_2$$

(c)   In the presence of water, the first step of the S_N1 reaction occurs. The reverse of this reaction produces two compounds because the positive charge on the carbocation is distributed over carbon atoms one and three

15.22 (a)   The carbocation that is produced in the S_N1 reaction is exceptionally stable because one resonance contributor is not only allylic but also tertiary.

A 3° allylic carbocation

(b)   CH$_3$C=CHCH$_2$OH   +   CH$_3$CCH=CH$_2$
         | |
         CH$_3$ CH$_3$
                                              OH

15.23 Compounds that undergo reactions by an S_N1 path must be capable of forming relatively stable carbocations. Primary halides of the type ROCH$_2$X form carbocations that are stabilized by resonance:

$$R-\overset{..}{\underset{..}{O}}-\overset{+}{C}H_2 \quad \longleftrightarrow \quad R-\overset{+}{\underset{..}{O}}=CH_2$$

15.24 The relative rates are in the order of the relative stabilities of the carbocations:

$$C_6H_5\overset{+}{C}H_2 \quad < \quad C_6H_5\overset{+}{C}HCH_3 \quad < \quad (C_6H_5)_2\overset{+}{C}H \quad < \quad (C_6H_5)_3\overset{+}{C}$$

The solvolysis reaction involves a carbocation intermediate.

15.25

15.26 (a)

(b)

(e) $\langle \bigcirc \rangle$—CHCH$_2$CH$_2$CH$_3$
          |
          OH

(f) $\langle \bigcirc \rangle$—CH$_2$CH$_2$CH$_2$CH$_3$

(g) $\langle \bigcirc \rangle$—$\overset{\overset{\text{O}}{\|}}{\text{C}}$—OH

15.29 (a) benzene + CICHCH$_3$ (with CH$_3$) →(AlCl$_3$) isopropylbenzene (CHCH$_3$ with CH$_3$)

(b) benzene + ClCCH$_3$ (with CH$_3$, CH$_3$) →(AlCl$_3$) tert-butylbenzene (CCH$_3$ with CH$_3$, CH$_3$)

(c) benzene + Cl—CCH$_2$CH$_3$ (C=O) →(AlCl$_3$) phenyl ketone $\overset{\overset{\text{O}}{\|}}{\text{C}}CH_2CH_3$ →(Zn(Hg) / HCl, reflux) $\bigcirc$—CH$_2$CH$_2$CH$_3$

(*Note:* The use of Cl-CH$_2$CH$_2$CH$_3$ in a Friedel-Crafts synthesis gives mainly the rearranged product, isopropylbenzene.)

(d) benzene + Cl—$\overset{\overset{\text{O}}{\|}}{\text{C}}CH_2CH_2CH_3$ →(AlCl$_3$) $\bigcirc$—$\overset{\overset{\text{O}}{\|}}{\text{C}}CH_2CH_2CH_3$ →(Zn(Hg) | HCl, reflux) $\bigcirc$—CH$_2$CH$_2$CH$_2$CH$_3$

(e) $\bigcirc$—C(CH$_3$)$_3$ + Cl$_2$ →(FeCl$_3$ / dark) Cl—$\bigcirc$—C(CH$_3$)$_3$

[from (b)]

(i)

(j)

(+ ortho)

15.32 (a)

(b)

(c) [from (b)]

(d) [from (b)]

(e)

15.33 (a) Step (2) will fail because a Friedel-Crafts reaction will not take place on a ring that bears an $-NO_2$ group (or any meta director).

(b) The last step will first brominate the double bond.

15.34 (a) Electrophilic aromatic substitution will take place as follows:

(b) The ring directly attached to the oxygen atom is activated toward electrophilic attack because the oxygen atom can donate an unshared electron pair to it and stabilize the intermediate arenium ion when attack occurs at the ortho or para position.

15.35 (a)

(b)

(c)

15.36

15.37 This problem serves as another illustration of the use of a sulfonic acid group as a block-ing group in a synthetic sequence. Here we are able to bring about nitration between two meta substituents.

(b) 1,2 Addition.

(c) Yes. The carbocation given in (a) is a hybrid of *secondary allylic and benzylic* con-tributors and is therefore more stable than any other possibility; for example,

$$C_6H_5CH=CH-CH=CH_2 \xrightarrow{H^+} C_6H_5CH_2-\overset{+}{C}H-CH=CH_2$$

$$C_6H_5CH_2-CH=CH-\overset{+}{C}H_2$$

A hybrid of allylic contri-butors only

(d) Since the reaction produces only *the more stable isomer*—that is, the one in which the double bond is conjugated with the benzene ring—the reaction is likely to be under equilibrium control:

15.40

Toluene

Succinic
anhydride

A
(C₁₁H₁₂O₃)

B
(C₁₁H₁₄O₂)

C
(C₁₁H₁₃ClO)

D
(C₁₁H₁₂O)

E
(C₁₁H₁₄O)

F
(C₁₁H₁₂)

G
(C₁₁H₁₁Br)

2-Methylnaphthalene

15.41

15.42 (a)

rearrangement

−HA

(b) CH_3—C=CH_2

15.43 (a)

(b)

15.44 (a) Large ortho substituents prevent the two rings from becoming coplanar and prevent rotation about the single bond that connects them. If the correct substitution patterns are present, the molecule as a whole will be chiral. Thus, enantiomeric forms are possible even though the molecules do not have a stereocenter. The compound with 2-NO_2, 6-CO_2H, 2'-NO_2, 6'-CO_2H is an example,

These molecules are nonsuperposable mirror images and, thus, are enantiomers.

(b) Yes

(c) This molecule has a plane of symmetry.

The plane of the page is a plane of symmetry.

15.45 (a)

15.47   (a and b)   The *tert*-butyl group is easily introduced by any of the variations of the Friedel-Crafts alkylation reaction, and, because of the stability of the *tert*-butyl cation, it is easily removed under acidic conditions.

(c)   In contrast to the -SO$_3$H group often used as a blocking group, –C(CH$_3$)$_3$ activates the ring to further electrophilic substitution.

15.48   At the lower temperature, the reaction is kinetically controlled, and the usual o/p directive effects of the –CH$_3$ group are observed. At higher temperatures, the reaction is thermodynamically controlled. At reaction times long enough for equilibrium to be reached, the most stable isomer, *m*-toluenesulfonic acid, is the principal product.

15.49   The evidence indicates that the mechanistic step in which the C–H bond is broken is *not* rate determining. (In the case cited, it makes no difference kinetically if a C–H or C–D bond is broken in electrophilic aromatic substitution.) This evidence is consistent with the two-step mechanism given in Section 15.2. The step in which the aromatic compound reacts with the electrophile (NO$_2$$^+$) is the slow rate-determining step. Proton (or deuteron) loss from the arenium ion to return to an aromatic system is a rapid step and has no effect on the overall rate.

15.50 (a) $C_6H_5CH_2Br \xrightarrow[\substack{DMF \\ (-NaBr)}]{NaCN} C_6H_5CH_2CN$

(b) $C_6H_5CH_2Br \xrightarrow[\substack{CH_3OH \\ (-NaBr)}]{CH_3ONa} C_6H_5CH_2OCH_3$

(c) $C_6H_5CH_2Br \xrightarrow[\substack{CH_3CO_2H \\ (-NaBr)}]{CH_3CO_2Na} C_6H_5CH_2O_2CCH_3$

(d) $C_6H_5CH_2Br \xrightarrow[\substack{acetone \\ (-NaBr)}]{NaI} C_6H_5CH_2I$

(e) $CH_2{=}CHCH_2Br \xrightarrow[\substack{acetone \\ (-NaBr)}]{NaN_3} CH_2{=}CHCH_2N_3$

(f) $CH_2{=}CHCH_2Br \xrightarrow[\substack{(CH_3)_2CHCH_2OH \\ (-NaBr)}]{NaOCH_2CH(CH_3)_2} CH_2{=}CHCH_2OCH_2CH(CH_3)_2$

15.51 $A = $ (benzene ring) $B = $ (bromobenzene ring, Br) $C = $ (toluene ring, CH_3)

15.52 $(C_6H_5)_3C{-}O{-}H \underset{-H^+}{\overset{+H^+}{\rightleftarrows}} (C_6H_5)_3C{-}\overset{+}{O}H_2 \underset{+H_2O}{\overset{-H_2O}{\rightleftarrows}} (C_6H_5)_3C^+ \underset{-EtOH}{\overset{+EtOH}{\rightleftarrows}}$

Very stable carbocation

$(C_6H_5)_3\overset{+}{C}{-}\underset{\underset{H}{|}}{O}{-}Et \underset{+H^+}{\overset{-H^+}{\rightleftarrows}} (C_6H_5)_3C{-}O{-}Et$

15.53 (a) $CH_3CH_2CH{=}CHCH_2Br$ would be the most reactive in an S_N2 reaction because it is a 1° allylic halide. There would, therefore, be less steric hindrance to the attacking nucleophile.

(b) $CH_2{=}CHCBr(CH_3)_2$ would be the most reactive in an S_N1 reaction because it is a 3° allylic halide. The carbocation formed in the rate-determining step, being both 3° and allylic, would be the most stable.

15.54 The final product is *o*-nitroaniline. (The reactions are given in Section 15.14.) The presence of six signals in the ^{13}C NMR spectrum confirms that the substitution in the final product is *ortho* and not *para*. A final product with para substitution (i.e., *p*-nitroaniline) would have given only four signals in the ^{13}C NMR spectrum.

*15.55 $CH_2OHCHOHCH_2OH$ (Glycerol) After sodium borohydride reduction of the aldehyde, ozonolysis oxidatively degrades the aromatic ring, leaving only the polyhydroxy side chain. Water is an alternative to HOAc sometimes used to work up ozonolysis reactions.

*15.56

$$
\begin{array}{c}
\text{COOH} \\
\text{HO—C—H} \\
\text{H—C—OH} \\
\text{CH}_2\text{OH}
\end{array}
\quad \textbf{(Threonic acid)}
$$

Ozonolysis oxidatively degrades the aromatic rings, leaving only the carboxyl carbon as a remnant of the alkyl substituted benzene ring. Water is an alternative to HOAc used to work up the ozonolysis reaction.

QUIZ

15.1 Which of the following compounds would be most reactive toward ring bromination?

15.2 Which of the following is *not* a meta directing substituent when present on a benzene ring?

(a) $-C_6H_5$ (b) $-NO_2$ (c) $-N(CH_3)_3{}^+$ (d) $-C\equiv N$ (e) $-CO_2H$

15.3 The major product(s), **C**, of the reaction,   **C**, would be

(a)   (b)   (c)

(d)   Equal amounts of (a) and (b)   (e)   Equal amounts of (a) and (c)

15.4 Complete the following syntheses.

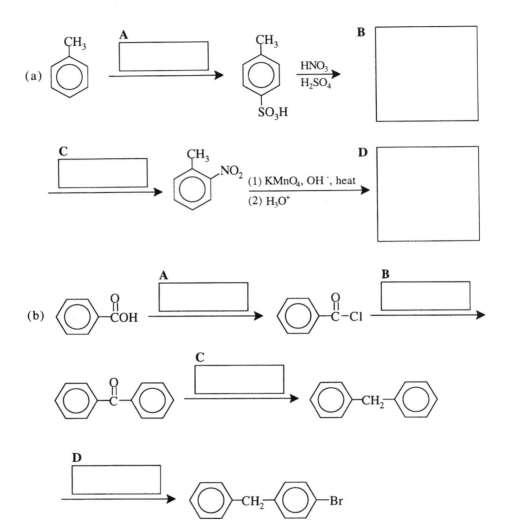

16

ALDEHYDES AND KETONES I. NUCLEOPHILIC ADDITION TO THE CARBONYL GROUP

SOLUTIONS TO PROBLEMS

16.1 (a)

$$\underset{\text{Pentanal}}{CH_3CH_2CH_2CH_2\overset{\displaystyle O}{\overset{\|}{C}}H}$$

$$\underset{\substack{\displaystyle | \\ CH_3}}{\underset{\text{2-Methylbutanal}}{CH_3CH_2CH\overset{\displaystyle O}{\overset{\|}{C}}H}}$$

$$\underset{\substack{\displaystyle | \\ CH_3}}{\underset{\text{3-Methylbutanal}}{CH_3CHCH_2\overset{\displaystyle O}{\overset{\|}{C}}H}}$$

2,2-Dimethylpropanal

$$\underset{\text{2-Pentanone}}{CH_3CH_2CH_2\overset{\displaystyle O}{\overset{\|}{C}}CH_3}$$

$$\underset{\text{3-Pentanone}}{CH_3CH_2\overset{\displaystyle O}{\overset{\|}{C}}CH_2CH_3}$$

3-Methyl-2-butanone

(b)

Acetophenone or
methyl phenyl ketone

Phenylethanal or
phenylacetaldehyde

2-Methylbenzaldehyde
(*o*-tolualdehyde)

3-Methylbenzaldehyde
(*m*-tolualdehyde)

4-Methylbenzaldehyde
(*p*-tolualdehyde)

16.2 (a) 1-Pentanol, because its molecules form hydrogen bonds to each other.

(b) 2-Pentanol, because its molecules form hydrogen bonds to each other.

(c) Pentanal, because its molecules are more polar.

(d) 2-Phenylethanol, because its molecules form hydrogen bonds to each other.

(e) Benzyl alcohol because its molecules form hydrogen bonds to each other.

16.3 (a) $CH_3CH_2CH_2OH \xrightarrow[CH_2Cl_2]{PCC} CH_3CH_2\overset{\overset{\displaystyle O}{\|}}{C}H$

(b) $CH_3CH_2CO_2H \xrightarrow{SOCl_2} CH_3CH_2\overset{\overset{\displaystyle O}{\|}}{C}Cl \xrightarrow[\text{diethyl ether}]{LiAlH[OC(CH_3)_3]_3} CH_3CH_2\overset{\overset{\displaystyle O}{\|}}{C}H$

16.4 (a) ⬡ $\xrightarrow[Fe]{Br_2}$ ⬡—Br $\xrightarrow[Et_2O]{Mg}$ ⬡—MgBr $\xrightarrow[(2) H_3O^+]{(1) HCHO}$

⬡—CH_2OH $\xrightarrow[CH_2Cl_2]{PCC}$ ⬡—CHO

(b) ⬡—CH_3 $\xrightarrow[(2) H_3O^+]{(1) KMnO_4, OH^-, heat}$ ⬡—CO_2H $\xrightarrow{SOCl_2}$

⬡—COCl $\xrightarrow[\text{diethyl ether}]{LiAlH[OC(CH_3)_3]_3}$ ⬡—CHO

(c) $CH_3CH_2Br \xrightarrow{HC\equiv CNa} CH_3CH_2C\equiv CH \xrightarrow[H_2O]{H_3O^+, Hg^{2+}} CH_3CH_2\overset{\overset{\displaystyle O}{\|}}{C}CH_3$

(d) $CH_3C\equiv CCH_3 \xrightarrow[H_2O]{H_3O^+, Hg^{2+}} CH_3CH_2\overset{\overset{\displaystyle O}{\|}}{C}CH_3$

(e) ⬡—$\underset{OH}{\overset{}{C}HCH_3}$ $\xrightarrow{H_2CrO_4}$ ⬡—$\underset{\underset{\displaystyle O}{\|}}{C}CH_3$

(f) ⬡ $\xrightarrow[AlCl_3]{CH_3COCl}$ ⬡—$\underset{\underset{\displaystyle O}{\|}}{C}CH_3$

(g) ⬡—$\underset{\underset{\displaystyle O}{\|}}{C}Cl$ $\xrightarrow{(CH_3)_2CuLi}$ ⬡—$\underset{\underset{\displaystyle O}{\|}}{C}CH_3$

(h)

(i)

(j) $C_6H_5CH_2CN$ $\xrightarrow[\text{(2) } H_3O^+]{\text{(1) } i\text{-Bu}_2\text{AlH}}$ $C_6H_5CH_2\overset{\overset{\displaystyle O}{\|}}{C}H$

(k) $CH_3(CH_2)_4CO_2CH_3$ $\xrightarrow[\text{(2) } H_3O^+]{\text{(1)} i\text{-Bu}_2\text{AlH}}$ $CH_3(CH_2)_4\overset{\overset{\displaystyle O}{\|}}{C}H$

16.5 (a) The nucleophile is the negatively charged carbon of the Grignard reagent *acting as a carbanion.*

(b) The magnesium portion of the Grignard reagent acts as a Lewis acid and accepts an electron pair of the carbonyl oxygen. This acid-base interaction makes the carbonyl carbon even more positive and, therefore, even more susceptible to nucleophilic attack.

(c) The product that forms initially (above) is a magnesium derivative of an alcohol.

(d) On addition of water, the organic product that forms is an alcohol.

16.6 The nucleophile is a hydride ion.

16.7

16.8 *Acid-Catalyzed Reaction*

Base-Catalyzed Reaction

$$OH^- \;+\; H_2{}^{18}O \;\rightleftharpoons\; H_2O \;+\; {}^{18}OH^-$$

16.9

Sucrose　　　　　　　Sucrose

16.10

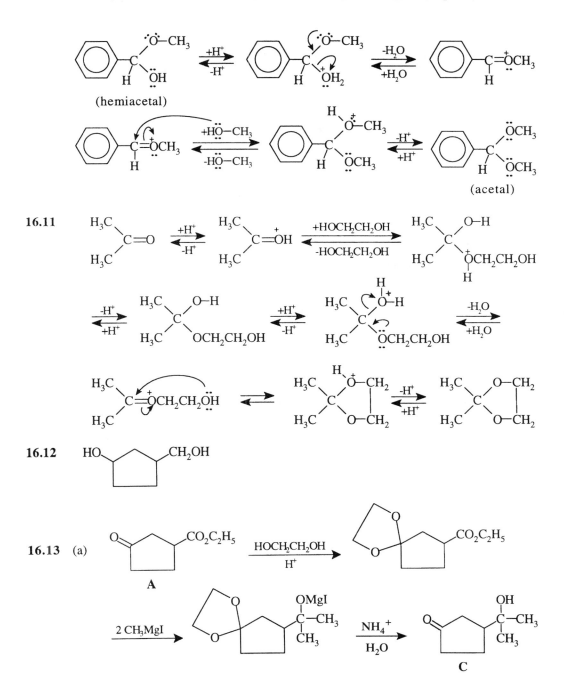

16.11

16.12

16.13 (a)

A

C

(b)   Addition would take place at the ketone group as well as at the ester group. The product (after hydrolysis) would be.

16.14   (a)

(b)   Tetrahydropyranyl ethers are acetals; thus, they are stable in aqueous base and hydrolyze readily in aqueous acid.

5-Hydroxypentanal

(c) $HOCH_2CH_2CH_2CH_2Cl$

$(+ HOCH_2CH_2CH_2CH_2CH)$

16.15 (a)

16.B1

16.B2

16.16 (a) $CH_3\overset{O}{\overset{\|}{C}}H$ $\xrightarrow{HCN}$ $CH_3\overset{OH}{\overset{|}{C}}HCN$ $\xrightarrow[\text{reflux}]{HCl, H_2O}$ $CH_3\overset{OH}{\overset{|}{C}}HCO_2H$

Lactic acid

(b) A racemic form

16.17 (a) CH_3I $\xrightarrow[(2)\text{ RLi}]{(1)(C_6H_5)_3P}$ $:\overset{-}{C}H_2-\overset{+}{P}(C_6H_5)_3$ $\xrightarrow{C_6H_5\overset{O}{\overset{\|}{C}}CH_3}$ $C_6H_5\overset{}{C}=CH_2$
$\overset{|}{C}H_3$

(b) CH_3CH_2Br $\xrightarrow[(2)\text{ RLi}]{(1)(C_6H_5)_3P}$ $CH_3\overset{..}{\overset{-}{C}}H-\overset{+}{P}(C_6H_5)_3$ $\xrightarrow{C_6H_5\overset{O}{\overset{\|}{C}}CH_3}$ $C_6H_5\overset{}{C}=CHCH_3$
$\overset{|}{C}H_3$

(c) $:\overset{-}{C}H_2-\overset{+}{P}(C_6H_5)_3$ $\xrightarrow{CH_3\overset{O}{\overset{\|}{C}}CH_3}$ $\overset{H_3C}{\underset{H_3C}{}}C=CH_2$
[from part (a)]

(d) $:\overset{-}{C}H_2-\overset{+}{P}(C_6H_5)_3$ [from part (a)] $\longrightarrow$ (see structure)

(e) $CH_3CH_2CH_2Br$ $\xrightarrow[(2)\text{ RLi}]{(1)(C_6H_5)_3P}$ $CH_3CH_2\overset{..}{\overset{-}{C}}H-\overset{+}{P}(C_6H_5)_3$ $\xrightarrow{CH_3\overset{O}{\overset{\|}{C}}CH_2CH_3}$

$\overset{CH_3}{\overset{|}{}}$
$CH_3CH_2CH=CCH_2CH_3$

(f) $CH_2{=}CHCH_2Br$ $\xrightarrow[(2)\text{ RLi}]{(1)(C_6H_5)_3P}$ $CH_2{=}CH\overset{..}{\overset{-}{C}}H-\overset{+}{P}(C_6H_5)_3$ $\xrightarrow{C_6H_5\overset{O}{\overset{\|}{C}}H}$

$C_6H_5CH{=}CHCH{=}CH_2$

(g) $C_6H_5CH_2Br$ $\xrightarrow[(2)\text{ RLi}]{(1)(C_6H_5)_3P}$ $C_6H_5\overset{..}{\overset{-}{C}}H-\overset{+}{P}(C_6H_5)_3$ $\xrightarrow{C_6H_5\overset{O}{\overset{\|}{C}}H}$

$C_6H_5CH{=}CHC_6H_5$

16.21 The product is a lactone, formed as follows:

(a lactone)

16.22

CH$_3$C—O—CHCH$_3$ The isopropyl group has a greater migratory aptitude than

the methyl group. The mechanism is as follows:

16.23 (a) HCHO   Methanal

(b) CH$_3$CHO   Ethanal

(c) C$_6$H$_5$CH$_2$CHO   Phenylethanal

(d) CH$_3$COCH$_3$ Propanone

(e) CH$_3$COCH$_2$CH$_3$ Butanone

(f) CH$_3$COC$_6$H$_5$ 1-Phenylethanone or methyl phenyl ketone

(g) C$_6$H$_5$COC$_6$H$_5$ Diphenylmethanone or diphenyl ketone

(h)   2-Hydroxybenzaldehyde or *o*-hydroxybenzaldehyde

(i)   4-Hydroxy-3-methoxybenzaldehyde

(j) CH$_3$CH$_2$COCH$_2$CH$_3$ 3-Pentanone

(k) $CH_3CH_2COCH(CH_3)_2$ 2-Methyl-3-pentanone

(l) $(CH_3)_2CHCOCH(CH_3)_2$ 2,4-Dimethyl-3-pentanone

(m) $CH_3(CH_2)_3CO(CH_2)_3CH_3$ 5-Nonanone

(n) $CH_3(CH_2)_2CO(CH_2)_2CH_3$ 4-Heptanone

(o) $C_6H_5CH{=}CHCHO$ 3-Phenyl-2-propenal

16.24

(a) $CH_3CH_2CH_2OH$

(b) $CH_3CH_2CHOHC_6H_5$

(c) $CH_3CH_2CH_2OH$

(d) $CH_3CH_2\overset{\overset{\displaystyle O}{\|}}{C}{-}O^-$

(e) $CH_3CH_2CH{=}CH_2$

(f) $CH_3CH_2CH_2OH$

(g) $CH_3CH_2CH\overset{\displaystyle O-CH_2}{\underset{\displaystyle O-CH_2}{\big\langle}}$

(h) $CH_3CH_2CH{=}CHCH_3$

(i) $CH_3CH_2CHOHCH_2CO_2C_2H_5$

(j) $CH_3CH_2CO_2^-\,NH_4^+ + Ag\!\downarrow$

(k) $CH_3CH_2CH{=}NOH$

(l) $CH_3CH_2CH{=}NNHCONH_2$

(m) $CH_3CH_2CH{=}NNHC_6H_5$

(n) $CH_3CH_2CO_2H$

(o) $CH_3CH_2CH\overset{\displaystyle S-CH_2}{\underset{\displaystyle S-CH_2}{\big\langle}}$

(p) $CH_3CH_2CH_3 + CH_3CH_3 + NiS$

16.25

(a) $CH_3CHOHCH_3$

(b) $C_6H_5\underset{\displaystyle CH_3}{\overset{\displaystyle |}{C}}OHCH_3$

(c) $CH_3CHOHCH_3$

(d) No reaction

(e) $CH_3\underset{\displaystyle CH_3}{\overset{\displaystyle CH_3}{\overset{\displaystyle |}{C}}}{=}CH_2$

(f) $CH_3CHOHCH_3$

(g)

(h) $CH_3CH{=}C(CH_3)_2$

(i) $CH_3\underset{\displaystyle CH_3}{\overset{\displaystyle OH}{\overset{\displaystyle |}{C}}}CH_2CO_2C_2H_5$

(j) No reaction

(k) $CH_3\underset{\displaystyle CH_3}{\overset{\displaystyle |}{C}}{=}NOH$

(l) $CH_3\underset{\displaystyle CH_3}{\overset{\displaystyle |}{C}}{=}NNHCONH_2$

(m) $CH_3\underset{\displaystyle CH_3}{\overset{\displaystyle |}{C}}{=}NNHC_6H_5$

(n) No reaction

(o)

(p) $CH_3CH_2CH_3 + CH_3CH_3 + NiS$

16.26 (a)

(b)

(c)

(d)

(e)

16.27 (a)

(b)

16.28 (a)

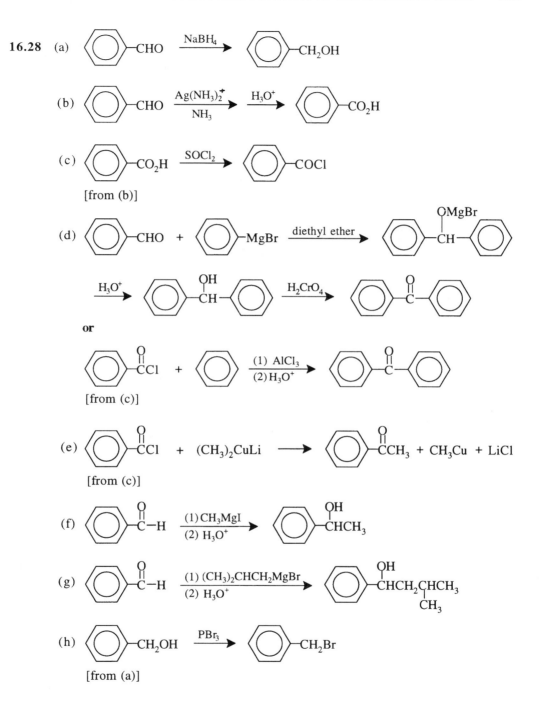

(b)

(c)

[from (b)]

(d)

or

[from (c)]

(e)

[from (c)]

(f)

(g)

(h)

[from (a)]

(i) C_6H_5–CH$_2$Br $\xrightarrow[\text{CH}_3\text{CO}_2\text{H}]{\text{Zn}}$ C_6H_5–CH$_3$

[from (h)]

or

C_6H_5–CHO $\xrightarrow[\text{BF}_3]{\text{HSCH}_2\text{CH}_2\text{SH}}$ (dithiolane) $\xrightarrow[\text{(H}_2)]{\text{Raney Ni}}$ C_6H_5–CH$_3$

(j) C_6H_5–CHO $\xrightarrow[\text{H}^+]{\text{CH}_3\text{OH}}$ C_6H_5–CH(OCH$_3$)$_2$

(k) C_6H_5–CHO $\xrightarrow[\text{H}_3{}^{18}\text{O}^+]{\text{H}_2{}^{18}\text{O}}$ C_6H_5–C^{18}(O)H (See problem 16.8 for the mechanism)

(l) C_6H_5–CHO $\xrightarrow[\text{(2) H}_3\text{O}^+]{\text{(1) NaBD}_4}$ C_6H_5–CHDOH

(m) C_6H_5–CHO $\xrightarrow{\text{HCN}}$ C_6H_5–CH(OH)CN (a cyanohydrin)

(n) C_6H_5–CHO $\xrightarrow{\text{NH}_2\text{OH}}$ C_6H_5–CH=NOH (an oxime)

(o) C_6H_5–CHO + H$_2$NNHC$_6$H$_5$ $\xrightarrow[\text{CH}_3\text{CO}_2\text{H}]{\text{H}_3\text{O}^+}$ C_6H_5–CH=NNHC$_6$H$_5$ (a phenylhydrazone)

(p) C_6H_5–CHO + H$_2$NNHCONH$_2$ $\longrightarrow$ C_6H_5–CH=NNHCONH$_2$ (a semicarbazone)

(q) C_6H_5–CHO + (C$_6$H$_5$)$_3\overset{+}{\text{P}}$–$\overset{\cdot\cdot}{\text{C}}$HCH=CH$_2$ $\longrightarrow$ C_6H_5–CH=CHCH=CH$_2$
(a Wittig reagent)

16.29 (a)

16.30 (a)

16.33

The compound $C_7H_6O_3$ is 3,4-dihydroxybenzaldehyde. The reaction involves hydrolysis of the acetal of formaldehyde.

16.36 (a) $(CH_3)_2SO_4$, NaOH, or CH_3I, NaOH

(b)   PCC/CH_2Cl_2

(c)   Zn, BrCHCO$_2$Et, then H_3O^+
$\overset{CH_3}{|}$

(d)   LiAlH$_4$, then H$_2$O   Intermediate is CH$_3$O—⟨ ⟩—CHO

16.37

The product would be racemic as no chiral reagents were used.

16.38 (R)-3-Phenyl-2-pentanone

16.39

16.40 (a) $Ag(NH_3)_2{}^+OH^-$ (positive test with benzaldehyde)

(b) $Ag(NH_3)_2{}^+OH^-$ (positive test with hexanal)

(c) Concentrated H_2SO_4 (2-hexanone is soluble)

(d) CrO_3 in H_2SO_4 (positive test with 2-hexanol)

(e) Br_2 in CCl_4 (decolorization with $C_6H_5CH{=}CHCOC_6H_5$)

(f) $Ag(NH_3)_2{}^+OH^-$ (positive test with pentanal)

(g) Br_2 in CCl_4 (immediate decolorization occurs with enol form)

(h) $Ag(NH_3)_2{}^+OH^-$ (positive test with cyclic hemiacetal)

16.41 Compound **W** is

Compound **X** is

triplet δ 3.1

16.42 Each ¹H NMR spectrum (Figs. 16.4 and 16.5) has a five-hydrogen peak near δ 7.2, suggesting the **Y** and **Z** each has a C_6H_5—group. The IR spectrum of each compound shows a strong peak near 1710 cm⁻¹. This absorption indicates that each compound has a C=O group not adjacent to the phenyl group. We have, therefore, the following pieces,

If we subtract the atoms of these pieces from the molecular formula,

$$C_{10}H_{12}O$$
$$-C_7H_5O \quad (C_6H_5 + C=O)$$

We are left with, C_3H_7

In the ¹H NMR spectrum of **Y**, we see an ethyl group [triplet, δ 1.0 (3H) and quartet, δ 2.45 (2H)] and an unsplit —CH_2—group [singlet, δ 3.7 (2H)]. This means that **Y** must be

1-Phenyl-2-butanone

In the ¹H NMR spectrum of **Z**, we see an unsplit —CH_3 group [singlet, δ 2.1 (3H)] and two triplets at δ 2.7 and 2.9. This means **Z** must be

4-Phenyl-2-butanone

16.43 That compound **A** forms a phenylhydrazone, gives a negative Tollens' test, and gives an IR band near $1710\ cm^{-1}$ indicates that **A** is a ketone. The ^{13}C spectrum of **A** contains only four signals indicating that A has a high degree of symmetry. The information from the DEPT ^{13}C NMR spectra enables us to conclude that **A** is diisobutyl ketone:

$$
\underset{(d)}{\underset{(a)\quad (b)\ (c)}{(CH_3)_2CHCH_2\overset{\overset{\textstyle O}{\|}}{C}CH_2CH(CH_3)_2}}
$$

Assignments:

(a) δ 22.6

(b) δ 24.4

(c) δ 52.3

(d) δ 210.0

16.44 That the ^{13}C spectrum of **B** contains only three signals indicates that **B** has a highly symmetrical structure. The information from DEPT spectra indicates the presence of equivalent methyl groups (CH_3 at δ 18.8), equivalent $-\overset{|}{\underset{|}{C}}-$ groups (at δ 70.4), and equivalent $\overset{\diagdown}{\diagup}C{=}O$ groups (at δ 215.0). These features allow only one possible structure for **B**:

Assignments:

(a) δ 18.8

(b) δ 70.4

(c) δ 215.0

16.45 The two nitrogen atoms of semicarbazide that are adjacent to the C=O group bear partial positive charges because of resonance contributions made by the second and third structures below.

Only this nitrogen is nucleophilic.

16.46 Hydrolysis of the acetal linkage of multistriatin produces the ketodiol below.

16.47

The *gem*-diol formed in the alkaline hydrolysis step readily loses water to form the aldehyde.

16.48

CH$_3$C—C—CH$_2$CH$_3$ or its enantiomer

16.49 The general formula for an oxime is

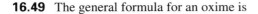

Both carbon and nitrogen are sp^2 hybridized; the electron pair on nitrogen occupies one sp^2 orbital. Aldoximes and ketoximes can exist in either of these two stereoisomeric forms:

This type of stereoisomerism is also observed in the case of other compounds that possess the C=N group, for example, phenylhydrazones and semicarbazones.

*16.52   (a)   O—H stretch at about 3300 cm^{-1}; C=O stretch at about 1710 cm^{-1}

(b)   (Intramolecular hemiacetal from *A*)

*16.53

QUIZ

16.1   Which Wittig reagent could be used to synthesize $C_6H_5CH=CHCH_2CH_3$? (Assume any other needed reagents are available.)

(a) $C_6H_5\overset{..}{C}H\overset{+}{P}(C_6H_5)_3$ (d)   More than one of these

(b) $C_6H_5CH=CH\overset{..}{C}H\overset{+}{P}(C_6H_5)_3$ (e)   None of these

(c) $CH_3CH_2\overset{..}{C}H\overset{+}{P}(C_6H_5)_3$

16.2   Which compound is an acetal?

(a) $C_6H_5\underset{\overset{|}{OH}}{C}HOCH_3$ (d)   More than one of these

(e)   None of these

(b)

(c)

16.3   Which reaction sequence could be used to convert $C_6H_{13}C\equiv CH$ to $C_6H_{13}\overset{\overset{\text{O}}{||}}{C}CH_3$?

(a) O_3, then Zn, HOAc, then AlCl$_3$, the CH$_3$CO$_2$H

(b)   H$_2$SO$_4$, HgSO$_4$, H$_2$O, heat

(c)   HCl, then CH$_3$CO$_2$H

(d)   O$_3$, then Zn, HOAc, then H$_2$SO$_4$, HgSO$_4$, H$_2$O, heat

(e)   CH$_3$CO$_2$H, then H$_2$O$_2$, OH$^-$/H$_2$O

16.4 Complete the following syntheses. If more than one step is required for a transformation, list them as (1), (2), (3), and so on.

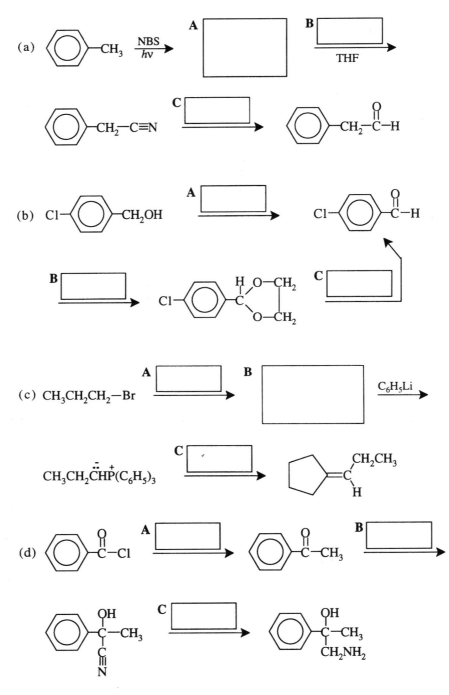

17

ALDEHYDES AND KETONES II. ALDOL REACTIONS

SOLUTIONS TO PROBLEMS

17.1

2,4-Cyclohexadien-1-one
(keto form)

Phenol
(enol form)

The enol form is aromatic, and it is therefore stabilized by the resonance energy of the benzene ring.

17.2   No.

does not have a hydrogen attached to its α-carbon

atom (which is a stereocenter) and thus enol formation involving the stereocenter

is not possible. With the stereocenter is a β carbon

and thus enol formation does not affect it.

17.3   In OD⁻/D₂O:

In D₃O⁺/D₂O:

17.4

cis-Decalone

Protonation on
the bottom face

trans-Decalone

A large group is
axial in *cis*-decalone.

No bulky groups are
axial in *trans*-decalone.

17.5   The reaction is said to be "base promoted" because base is consumed as the reaction takes place. A catalyst is, by definition, not consumed.

17.6   (a)   The slow step in base-catalyzed racemization is the same as that in base-promoted halogenation—*the formation of an enolate ion.* (Formation of an enolate ion from *sec*-butyl phenyl ketone leads to racemization because the enolate ion is achiral. When it accepts a proton, it yields a racemic form.) The slow step in acid-catalyzed racemization is also the same as that in acid-catalyzed halogenation—*the formation of an enol.* (The enol, like the enolate ion, is achiral and tautomerizes to yield a racemic form of the ketone.)

(b)   According to the mechanism given, the slow step for acid-catalyzed iodination (formation of the enol) is the same as that for acid-catalyzed bromination. Thus, we would expect both reactions to occur at the same rate.

(c)   Again, the slow step for both reactions (formation of the enolate ion) is the same, and consequently, both reactions take place at the same rate.

17.7   (a)

(b) For $CH_3CH_2\overset{\underset{\displaystyle |}{OH}}{C}HCH_2CH_2\overset{\underset{\displaystyle}{\overset{\displaystyle O}{\|}}}{C}H$ to form, a hydroxide ion would have to remove a β proton in the first step. This does not happen because the anion that would be produced, that is, $\overset{..}{:}CH_2CH_2CHO$, cannot be stabilized by resonance.

(c) $CH_3CH_2CH{=}\overset{\underset{\displaystyle |}{\underset{\displaystyle CH_3}{}}}{\overset{\displaystyle O}{\overset{\displaystyle \|}{C}}}CH$

17.8 $CH_3CHO \xrightarrow[5°C]{10\% \text{ NaOH}} CH_3\overset{\underset{\displaystyle |}{OH}}{C}HCH_2CHO \xrightarrow{\text{heat}} CH_3CH{=}CHCHO \xrightarrow[Ni]{H_2}$
                                                                (aldol)

$CH_3CH_2CH_2CH_2OH$

17.9 (a) $2 CH_3CH_2CH_2CHO \xrightarrow[H_2O]{OH^-} CH_3CH_2CH_2\overset{\underset{\displaystyle |}{OH}}{C}H\overset{\underset{\displaystyle |}{\underset{\displaystyle CH_2}{\underset{\displaystyle |}{\underset{\displaystyle CH_3}{}}}}}{C}HCHO$

(b) Product of (a) $\xrightarrow[-H_2O]{H^+} CH_3CH_2CH_2CH{=}\overset{\underset{\displaystyle |}{\underset{\displaystyle CH_2}{\underset{\displaystyle |}{\underset{\displaystyle CH_3}{}}}}}{C}CHO \xrightarrow[(2) H_2O]{(1) \text{ LiAlH}_4, \text{ Et}_2O}$

$CH_3CH_2CH_2CH{=}\overset{\underset{\displaystyle |}{\underset{\displaystyle CH_2}{\underset{\displaystyle |}{\underset{\displaystyle CH_3}{}}}}}{C}CH_2OH$

(c) Product of (b) $\xrightarrow[Pt]{H_2} CH_3CH_2CH_2CH_2\overset{\underset{\displaystyle |}{\underset{\displaystyle CH_2}{\underset{\displaystyle |}{\underset{\displaystyle CH_3}{}}}}}{C}HCH_2OH$

(d) Product of (a) $\xrightarrow{\text{NaBH}_4} CH_3CH_2CH_2\overset{\underset{\displaystyle |}{OH}}{C}H\overset{\underset{\displaystyle |}{\underset{\displaystyle CH_2}{\underset{\displaystyle |}{\underset{\displaystyle CH_3}{}}}}}{C}HCH_2OH$

17.10

(shown in text)

2,6-Dimethyl-2,5-hepta-
dien-4-one

17.11   Drawing the molecules as they will appear in the final product helps to visualize the necessary steps:

Mesitylene

The two molecules that lead to mesitylene are shown as follows:

This molecule (4-methyl-3-penten-2-one) is formed by an acid-catalyzed condensation between two molecules of acetone as shown in the text.

The mechanism is,

17.12

17.13

$$\text{C}_6\text{H}_5\text{—}\overset{\overset{\text{O}}{\|}}{\text{C}}\text{H} \; + \; \text{CH}_3\text{CHO} \; \xrightarrow[5°C]{10\% \text{ NaOH}} \; \xrightarrow{\text{heat}} \; \text{C}_6\text{H}_5\text{—CH}=\text{CHCHO}$$

17.14   Three successive aldol additions occur.

First Aldol Addition

$$\text{CH}_3\overset{\overset{\text{O}}{\|}}{\text{C}}\text{H} \; + \; \text{OH}^- \;\; \rightleftharpoons \;\; {}^-\text{:CH}_2\overset{\overset{\text{O}}{\|}}{\text{C}}\text{H} \; + \; \text{H}_2\text{O}$$

$$\text{H}\overset{\overset{\text{O}}{\|}}{\text{C}}\text{H} \; + \; {}^-\text{:CH}_2\overset{\overset{\text{O}}{\|}}{\text{C}}\text{H} \;\; \rightleftharpoons \;\; {}^-\text{OCH}_2\text{CH}_2\overset{\overset{\text{O}}{\|}}{\text{C}}\text{H}$$

$$ {}^-\text{OCH}_2\text{CH}_2\overset{\overset{\text{O}}{\|}}{\text{C}}\text{H} \; + \; \text{H}_2\text{O} \;\; \rightleftharpoons \;\; \text{HOCH}_2\text{CH}_2\overset{\overset{\text{O}}{\|}}{\text{C}}\text{H} \; + \; \text{OH}^-$$

Second Aldol Addition

$$\text{HOCH}_2\text{CH}_2\overset{\overset{\text{O}}{\|}}{\text{C}}\text{H} \; + \; \text{OH}^- \;\; \rightleftharpoons \;\; \text{HOCH}_2\overset{..}{\text{C}}\text{H}\overset{\overset{\text{O}}{\|}}{\text{C}}\text{H} \; + \; \text{H}_2\text{O}$$

$$\text{H}\overset{\overset{\text{O}}{\|}}{\text{C}}\text{H} \; + \; \text{HOCH}_2\overset{..}{\text{C}}\text{H}\overset{\overset{\text{O}}{\|}}{\text{C}}\text{H} \;\; \rightleftharpoons \;\; \text{HOCH}_2\underset{}{\overset{\overset{\text{CH}_2\text{O}^-}{|}}{\text{C}}}\text{HCHO}$$

$$\text{HOCH}_2\underset{}{\overset{\overset{\text{CH}_2\text{O}^-}{|}}{\text{C}}}\text{HCHO} \; + \; \text{H}_2\text{O} \;\; \rightleftharpoons \;\; \text{HOCH}_2\underset{}{\overset{\overset{\text{CH}_2\text{OH}}{|}}{\text{C}}}\text{HCHO} \; + \; \text{OH}^-$$

Third Aldol Addition

$$\text{HOCH}_2\underset{}{\overset{\overset{\text{CH}_2\text{OH}}{|}}{\text{C}}}\text{HCHO} \; + \; \text{OH}^- \;\; \rightleftharpoons \;\; \text{HOCH}_2\underset{..}{\overset{\overset{\text{CH}_2\text{OH}}{|}}{\text{C}}}\text{—CHO} \; + \; \text{H}_2\text{O}$$

$$\text{H}\overset{\overset{\text{O}}{\|}}{\text{C}}\text{H} \; + \; \text{HOCH}_2\underset{..}{\overset{\overset{\text{CH}_2\text{OH}}{|}}{\text{C}}}\text{—CHO} \;\; \rightleftharpoons \;\; \text{HOCH}_2\text{—}\underset{\overset{|}{\text{CH}_2\text{O}^-}}{\overset{\overset{\text{CH}_2\text{OH}}{|}}{\text{C}}}\text{—CHO}$$

$$\text{HOCH}_2\text{—}\underset{\overset{|}{\text{CH}_2\text{O}^-}}{\overset{\overset{\text{CH}_2\text{OH}}{|}}{\text{C}}}\text{—CHO} \; + \; \text{H}_2\text{O} \;\; \rightleftharpoons \;\; \text{HOCH}_2\text{—}\underset{\overset{|}{\text{CH}_2\text{OH}}}{\overset{\overset{\text{CH}_2\text{OH}}{|}}{\text{C}}}\text{—CHO} \; + \; \text{OH}^-$$

17.15 (a) CH_3CO_2H + BF_3 $\rightleftharpoons$ $CH_3CO_2BF_3^-$ + H^+

Pseudoionone

α-Ionone

β-Ionone

(b)   In β-ionone both double bonds and the carbonyl group are conjugated; thus it is more stable.

(c)   β-Ionone, because it is a fully conjugated unsaturated system.

17.16 (a)

17.17 (a)

17.18

17.19 (a)   (b)   (c)

Notice that starting compounds are drawn so as to indicate which atoms are involved in the cyclization reaction.

17.20 It is necessary for conditions to favor the intramolecular reaction rather than the inter-molecular one. One way to create these conditions is to use very dilute solutions when we carry out the reaction. When the concentration of the compound to be cyclized is very low (i.e., when we use what we call a "high dilution technique"), the probability is greater that one end of a molecule will react with the other end of that same molecule rather than with a different molecule.

17.21

(a)

Kinetic enolate

(b) $CH_3CH_2\overset{O}{\overset{\|}{C}}CH_3$ $\xrightarrow{LDA}$ $CH_3CH_2\overset{O^-Li^+}{\overset{|}{C}}=CH_2$ $\xrightarrow[(2)\,H_2O]{(1)\,C_6H_5\overset{O}{\overset{\|}{C}}H}$ $CH_3CH_2\overset{O}{\overset{\|}{C}}CH_2\overset{OH}{\overset{|}{C}}HC_6H_5$

Kinetic enolate

(c) $CH_3\overset{CH_3}{\underset{|}{C}}H\overset{O}{\overset{\|}{C}}CH_3$ $\xrightarrow{LDA}$ $CH_3\overset{CH_3}{\underset{|}{C}}H\overset{O^-Li^+}{\overset{|}{C}}=CH_2$ $\xrightarrow[(2)\,H_2O]{(1)\,CH_3CH_2\overset{O}{\overset{\|}{C}}H}$ $CH_3\overset{CH_3}{\underset{|}{C}}H\overset{O}{\overset{\|}{C}}CH_2\overset{OH}{\overset{|}{C}}HCH_2CH_3$

Kinetic enolate

(d) $CH_3CH=CH\overset{O}{\overset{\|}{C}}CH_3$ $\xrightarrow{LDA}$ $CH_3CH=CH\overset{O^-Li^+}{\overset{|}{C}}=CH_2$ $\xrightarrow[(2)\,H_2O]{(1)\,CH_3\overset{O}{\overset{\|}{C}}H}$

$CH_3CH=CH\overset{O}{\overset{\|}{C}}CH_2\overset{OH}{\overset{|}{C}}HCH_3$

17.22 (a) $CH_3\overset{CH_3}{\overset{|}{C}}=CH\overset{O}{\overset{\|}{C}}CH_3$ $\xrightarrow{LDA}$ $CH_3\overset{CH_3}{\overset{|}{C}}=CH\overset{O^-Li^+}{\overset{|}{C}}=CH_2$ $\xrightarrow[(2)\,H_2O]{(1)\,CH_3-}$

α-Bisabolanone

(b) $CH_3\overset{CH_3}{\overset{|}{C}}=CH\overset{O^-Li^+}{\overset{|}{C}}=CH_2$ $\xrightarrow[(2)\,H_2O]{(1)\,CH_3-}$

[prepared in (a)]

Ocimenone

17.B1

Step 1

Step 2

Step 3

17.23 (a)

(b)

17.24

Kinetic enolate

17.25 (a)

(b)   2-Methyl-1,3-cyclohexanedione is more acidic because its enolate ion is stabilized by an additional resonance structure.

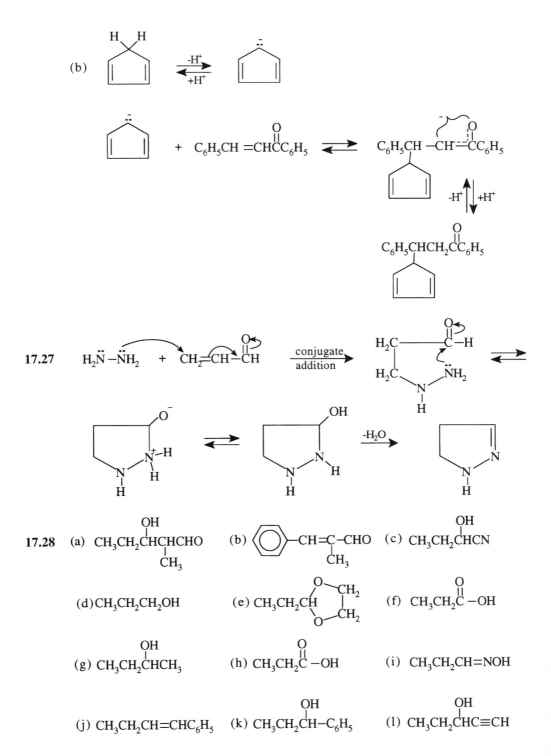

17.28 (a) CH$_3$CH$_2$CHCHCHO with OH and CH$_3$ substituents

(b) C$_6$H$_5$—CH=C—CHO with CH$_3$

(c) CH$_3$CH$_2$CHCN with OH

(d) CH$_3$CH$_2$CH$_2$OH

(e) CH$_3$CH$_2$CH with O—CH$_2$ / O—CH$_2$ ring

(f) CH$_3$CH$_2$C—OH with O

(g) CH$_3$CH$_2$CHCH$_3$ with OH

(h) CH$_3$CH$_2$C—OH with O

(i) CH$_3$CH$_2$CH=NOH

(j) CH$_3$CH$_2$CH=CHC$_6$H$_5$

(k) CH$_3$CH$_2$CH—C$_6$H$_5$ with OH

(l) CH$_3$CH$_2$CHC≡CH with OH

(m) $CH_3CH_2CH_3$

(n) $CH_3CH_2\overset{\overset{\displaystyle OH}{|}}{CH}-\underset{\underset{\displaystyle CH_2CH_3}{|}}{CH}CO_2Et$

17.29 (a) $CH_3\overset{\overset{\displaystyle CH_3}{|}}{\underset{\underset{\displaystyle OH}{|}}{C}}-CH_2\overset{\overset{\displaystyle O}{||}}{C}CH_3$

(b) $C_6H_5CH=CH\overset{\overset{\displaystyle O}{||}}{C}CH_3$

(c) $CH_3\overset{\overset{\displaystyle OH}{|}}{\underset{\underset{\displaystyle CN}{|}}{C}}CH_3$

(d) $CH_3\overset{\overset{\displaystyle OH}{|}}{C}HCH_3$

(e) $\underset{H_3C}{\overset{H_3C}{>}}C\underset{O-CH_2}{\overset{O-CH_2}{<}}$

(f) No reaction

(g) $CH_3\overset{\overset{\displaystyle OH}{|}}{\underset{\underset{\displaystyle CH_3}{|}}{C}}CH_3$

(h) No reaction

(i) $CH_3\overset{\overset{\displaystyle NOH}{||}}{C}CH_3$

(j) $\underset{H_3C}{\overset{H_3C}{>}}C=CHC_6H_5$

(k) $CH_3\overset{\overset{\displaystyle OH}{|}}{\underset{\underset{\displaystyle CH_3}{|}}{C}}C_6H_5$

(l) $CH_3\overset{\overset{\displaystyle OH}{|}}{\underset{\underset{\displaystyle CH_3}{|}}{C}}C\equiv CH$

(m) $CH_3CH_2CH_3$

(n) $CH_3\overset{\overset{\displaystyle HO}{|}}{\underset{\underset{\displaystyle CH_3}{|}}{C}}-\overset{\overset{\displaystyle CH_2CH_3}{|}}{CH}CO_2Et$

17.30 (a) $H_3C-\langle\bigcirc\rangle-CH=CHCHO$

(b) $H_3C-\langle\bigcirc\rangle-\overset{\overset{\displaystyle OH}{|}}{CH}C\equiv CCH_3$

(c) $H_3C-\langle\bigcirc\rangle-\overset{\overset{\displaystyle OH}{|}}{CH}CH_2CH_3$

(d) $H_3C-\langle\bigcirc\rangle-\overset{\overset{\displaystyle O}{||}}{C}-OH$

(e) $HO-\overset{\overset{\displaystyle O}{||}}{C}-\langle\bigcirc\rangle-\overset{\overset{\displaystyle O}{||}}{C}-OH$

(f) $H_3C-\langle\bigcirc\rangle-CH=CH_2$

(g) $H_3C-\langle\bigcirc\rangle-CH=CH-\overset{\overset{\displaystyle O}{||}}{C}-\langle\bigcirc\rangle$

(h) $H_3C-\langle\bigcirc\rangle-\overset{\overset{\displaystyle OH}{|}}{CH}-CH_2\overset{\overset{\displaystyle O}{||}}{C}-OEt$

17.31 (a)

17.32

17.33

17.34 (a) The conjugate base is a hybrid of the following structures:

$$^-:CH_2-CH=CH-\overset{\overset{\displaystyle O}{\|}}{C}H \longleftrightarrow CH_2=CH-\overset{..}{C}H-\overset{\overset{\displaystyle O}{\|}}{C}H \longleftrightarrow \underbrace{CH_2=CH-CH=\overset{\overset{\displaystyle O^-}{}}{C}H}$$

This structure is especially stable because the negative charge is on the oxygen atom.

(b) $CH_3CH=CHCHO \underset{+H^+}{\overset{-H^+}{\rightleftarrows}} {}^-:CH_2CH=CHCHO$

$C_6H_5CH=CH\overset{\overset{\displaystyle O}{\|}}{C}H \; + \; {}^-:CH_2CH=CHCHO \rightleftarrows$

$C_6H_5CH=CH\overset{\overset{\displaystyle O^-}{|}}{C}H-CH_2CH=CHCHO \underset{-H^+}{\overset{+H^+}{\rightleftarrows}} C_6H_5CH=CH\overset{\overset{\displaystyle OH}{|}}{C}H-CH_2CH=CHCHO$

$\xrightarrow{-H_2O} C_6H_5CH=CHCH=CHCH=CHCHO$

17.35 (a)

(b)

(c)

(d)

17.36 (a) In simple addition, the carbonyl peak (1665–1780-cm^{-1} region) does not appear in the product; in conjugate addition it does.

(b) As the reaction takes place, the long-wavelength absorption arising from the conjugated system should disappear. One could follow the rate of the reaction by following the rate at which this absorption peak disappears.

17.37 (a) Compound **U** is ethyl phenyl ketone: (b) Compound **V** is benzyl methyl ketone:

17.38 A is $CH_3\overset{O}{\overset{\|}{C}}CH_2CH(OCH_3)_2$

(a) $\overset{O}{\overset{\|}{CH_3-C}}-CH_2-CH(OCH_3)_2$ (b) (c) (d)

(a) Singlet δ 2.1   (c) Triplet δ 4.7

(b) Doublet δ 2.6   (d) Singlet δ 3.2

17.39 Abstraction of an α hydrogen at the ring junction yields an enolate ion that can then accept a proton to form either *trans*-1-decalone or *cis*-1-decalone. Since *trans*-1-decalone is more stable, it predominates at equilibrium.

(95%)
trans-1-Decalone
(more stable)

(5%)
cis-1-Decalone
(less stable)

17.40 (a) $CH_3OCH_2Br + (C_6H_5)_3P \xrightarrow{(2) \; RLi} CH_3OCH=P(C_6H_5)_3$

(b) Hydrolysis of the ether yields a hemiacetal that then goes on to form an aldehyde.

17.41   (a)   The hydrogen atom that is added to the aldehyde carbon atom in the reduction must come from the other aldehyde rather than from the solvent. It must be transferred as a hydride ion and directly from molecule to molecule, since if it were ever a free species it would react immediately with the solvent. A possible mechanism is the following:

(b)   Although an aldol reaction occurs initially, the aldol reaction is reversible. The Cannizzaro reaction, though slower, is irreversible. Eventually, all the product is in the form of the alcohol and the carboxylate ion.

17.42   This difference in behavior indicates that, for acetaldehyde, the capture of a proton from the solvent (the reverse of the reaction by which the enolate ion is formed) occurs much more slowly than the attack by the enolate ion on another molecule.

When acetone is used, the equilibrium for the formation of the enolate ion is unfavorable, but more importantly, enolate attack on another acetone molecule is disfavored due to steric hindrance. Here proton capture (actually deuteron capture) competes very well with the aldol reaction.

17.43

17.44   **X** is:

*17.45

QUIZ

17.1 Supply formulas for the missing reagents and intermediates in the following synthesis.

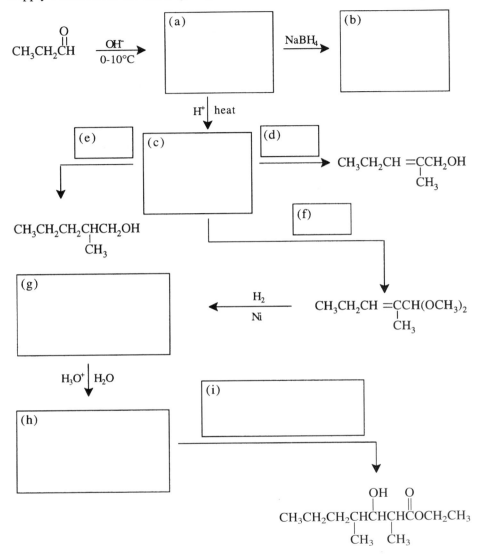

17.2 Supply formulas for the missing reagents and intermediates in the following synthesis.

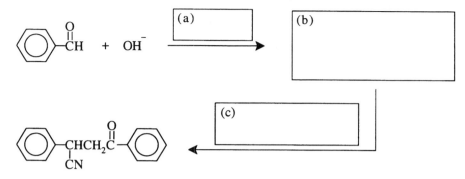

17.3 Supply formulas for the missing reagents and intermediates in the following synthesis.

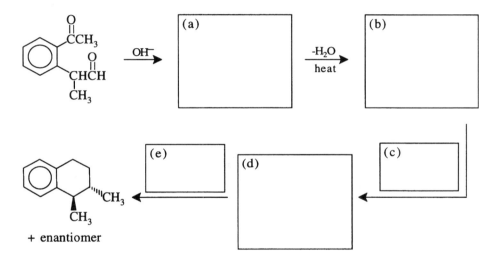

+ enantiomer

17.4 Which would be formed in the following reaction?

$$CH_3\overset{O}{\overset{\|}{C}}H \ + \ CH_3CH_2\overset{O}{\overset{\|}{C}}H \ \xrightarrow[25°C]{OH^-} \ ?$$

(a) $CH_3\overset{OH}{\underset{|}{C}}HCH_2\overset{O}{\overset{\|}{C}}H$ (b) $CH_3CH_2\overset{OH}{\underset{|}{C}}HCH\overset{O}{\overset{\|}{C}}H$ (c) $CH_3\overset{OH}{\underset{|}{C}}HCH\overset{O}{\overset{\|}{C}}H$
$\underset{CH_3}{|}\underset{CH_3}{|}$

(d) $CH_3CH_2\overset{OH}{\underset{|}{C}}HCH_2\overset{O}{\overset{\|}{C}}H$ (e) All of these will be formed.

17.5 What would be the major product of the following reaction?

$$C_6H_5\overset{O}{\overset{\|}{C}}\underset{\underset{CH_3}{|}}{C}HCH_3 \ + \ Br_2 \ + \ OH^- \ \longrightarrow$$

(a) $C_6H_5\overset{\overset{\displaystyle O}{\|}}{C}\underset{\underset{\displaystyle CH_3}{|}}{C}BrCH_3$

(b) $C_6H_5\overset{\overset{\displaystyle O}{\|}}{C}\underset{\underset{\displaystyle CH_3}{|}}{C}HCH_2Br$

(c) $C_6H_5\overset{\overset{\displaystyle O}{\|}}{C}\underset{\underset{\displaystyle CH_3}{|}}{C}HCH_2OH$

(d) $C_6H_5CBr_2\underset{\underset{\displaystyle CH_3}{|}}{\overset{\overset{\displaystyle CH_3}{|}}{C}}HCH_3$

(e)   None of these

18

CARBOXYLIC ACIDS AND THEIR DERIVATIVES: NUCLEOPHILIC ADDITION-ELIMINATION AT THE ACYL CARBON

SOLUTIONS TO PROBLEMS

18.1 (a) 2-Methylbutanoic acid

(b) 3-Pentenoic acid

(c) Sodium 4-bromobutanoate

(d) 5-Phenylpentanoic acid

(e) 3-Methyl-3-pentenoic acid

18.2 Acetic acid, in the absence of solvating molecules, exists as a dimer owing to the formation of two intermolecular hydrogen bonds:

At temperatures much above the boiling point, the dimer dissociates into the individual molecules.

18.3 (a) CH_2FCO_2H (F– is more electronegative than H-)

(b) CH_2FCO_2H (F– is more electronegative than Cl-)

(c) CH_2ClCO_2H (Cl– is more electronegative than Br-)

(d) $CH_3CHFCH_2CO_2H$ (F– is closer to –CO_2H)

(e) $CH_3CH_2CHFCO_2H$ (F– is closer to –CO_2H)

(f) $(CH_3)_3\overset{+}{N}$—⟨O⟩—CO_2H [$(CH_3)_3\overset{+}{N}$- is more electronegative than H-]

(g) CF_3—⟨O⟩—CO_2H (CF_3– is more electronegative than CH_3-)

18.4 (a) The carboxyl group is an electron-withdrawing group; thus, in a dicarboxylic acid such as those in Table 18.3, one carboxyl group increases the acidity of the other.

(b) As the distance between the carboxyl groups increases the acid-strengthening, inductive effect decreases.

333

18.5 (a) $CH_3CH_2\overset{O}{\underset{\|}{C}}-OCH_3$

(b) $O_2N-\langle\bigcirc\rangle-\overset{O}{\underset{\|}{C}}-OCH_2CH_3$

(c) $CH_3O-\overset{O}{\underset{\|}{C}}CH_2\overset{O}{\underset{\|}{C}}-OCH_3$

(d) $\langle\bigcirc\rangle-\overset{O}{\underset{\|}{C}}-N(CH_3)_2$

(e) $CH_3CH_2CH_2CH_2C\equiv N$

(f) benzene ring with $\overset{O}{\underset{\|}{C}}-OCH_3$ and $\overset{O}{\underset{\|}{C}}-OCH_3$

(g) $\begin{array}{c} H-C-\overset{O}{\underset{\|}{C}}-OCH_2CH_2CH_3 \\ \| \\ H-C-\overset{}{\underset{\overset{\|}{O}}{C}}-OCH_2CH_2CH_3 \end{array}$

(h) $H-\overset{O}{\underset{\|}{C}}-N(CH_3)_2$

(i) $CH_3\underset{\underset{Br}{|}}{C}H\overset{O}{\underset{\|}{C}}-Br$

(j) $\begin{array}{c} H_2C-\overset{O}{\underset{\|}{C}}-OCH_2CH_3 \\ | \\ H_2C-\underset{\overset{\|}{O}}{C}-OCH_2CH_3 \end{array}$

18.6 (a) $\langle\bigcirc\rangle-CH_2CH_3$ $\xrightarrow[\text{(2) H}_3O^+]{\text{(1) KMnO}_4,\ OH^-,\ \text{heat}}$ $\langle\bigcirc\rangle-\overset{O}{\underset{\|}{C}}OH$ + CO_2

(b) $\langle\bigcirc\rangle-Br$ $\xrightarrow[\text{Et}_2O]{Mg}$ $\langle\bigcirc\rangle-MgBr$ $\xrightarrow{CO_2}$ $\langle\bigcirc\rangle-\overset{O}{\underset{\|}{C}}OMgBr$ $\xrightarrow{H_3O^+}$ $\langle\bigcirc\rangle-\overset{O}{\underset{\|}{C}}OH$

(c) $\langle\bigcirc\rangle-\overset{O}{\underset{\|}{C}}-CH_3$ $\xrightarrow[\text{(2) H}_3O^+]{\text{(1)Cl}_2/\text{NaOH}}$ $\langle\bigcirc\rangle-\overset{O}{\underset{\|}{C}}OH$ + $CHCl_3$

(d) $\langle\bigcirc\rangle-CH=CH_2$ $\xrightarrow[\text{(2) H}_3O^+]{\text{(1) KMnO}_4,\ OH^-,\ \text{heat}}$ $\langle\bigcirc\rangle-\overset{O}{\underset{\|}{C}}OH$

(e) $\langle\bigcirc\rangle-CH_2OH$ $\xrightarrow[\text{(2) H}_3O^+]{\text{(1) KMnO}_4,\ OH^-,\ \text{heat}}$ $\langle\bigcirc\rangle-\overset{O}{\underset{\|}{C}}OH$

(f) $\langle\bigcirc\rangle-\overset{O}{\underset{\|}{C}}H$ $\xrightarrow[\text{(2) H}_3O^+]{\text{(1) KMnO}_4,\ OH^-}$ $\langle\bigcirc\rangle-\overset{O}{\underset{\|}{C}}OH$

18.7 These syntheses are easy to see if we work backward.

(a) $C_6H_5CH_2CO_2H$ $\xleftarrow[\text{(2) H}^+]{\text{(1) CO}_2}$ $C_6H_5CH_2MgBr$ $\xleftarrow[\text{Et}_2O]{\text{Mg}}$ $C_6H_5CH_2Br$

(b) $CH_3CH_2CH_2\overset{\overset{\displaystyle CH_3}{|}}{\underset{\underset{\displaystyle CH_3}{|}}{C}}CO_2H$ $\xleftarrow[\text{(2) H}^+]{\text{(1) CO}_2}$ $CH_3CH_2CH_2\overset{\overset{\displaystyle CH_3}{|}}{\underset{\underset{\displaystyle CH_3}{|}}{C}}MgBr$ $\xleftarrow[\text{Et}_2O]{\text{Mg}}$ $CH_3CH_2CH_2\overset{\overset{\displaystyle CH_3}{|}}{\underset{\underset{\displaystyle CH_3}{|}}{C}}Br$

(c) $CH_2{=}CHCH_2CO_2H$ $\xleftarrow[\text{(2) H}^+]{\text{(1) CO}_2}$ $CH_2{=}CHCH_2MgBr$ $\xleftarrow[\text{Et}_2O]{\text{Mg}}$ $CH_2{=}CHCH_2Br$

(d) $H_3C{-}\langle\bigcirc\rangle{-}CO_2H$ $\xleftarrow[\text{(2) H}^+]{\text{(1) CO}_2}$ $H_3C{-}\langle\bigcirc\rangle{-}MgBr$ $\xleftarrow[\text{Et}_2O]{\text{Mg}}$ $H_3C{-}\langle\bigcirc\rangle{-}Br$

(e) $CH_3CH_2CH_2CH_2CH_2CO_2H$ $\xleftarrow[\text{(2) H}^+]{\text{(1) CO}_2}$ $CH_3CH_2CH_2CH_2CH_2MgBr$ $\xleftarrow[\underset{\displaystyle Mg}{}]{\text{Et}_2O}$ $CH_3CH_2CH_2CH_2CH_2Br$

18.8 (a) $C_6H_5CH_2CO_2H$ $\xleftarrow[\text{(2) H}^+,\ \text{H}_2\text{O, heat}]{\text{(1) CN}^-}$ $C_6H_5CH_2Br$

$CH_2{=}CHCH_2CO_2H$ $\xleftarrow[\text{(2) H}^+,\ \text{H}_2\text{O, heat}]{\text{(1) CN}^-}$ $CH_2{=}CHCH_2Br$

$CH_3CH_2CH_2CH_2CH_2CO_2H$ $\xleftarrow[\text{(2) H}^+,\ \text{H}_2\text{O, heat}]{\text{(1) CN}^-}$ $CH_3CH_2CH_2CH_2CH_2Br$

(b) A nitrile synthesis. Preparation of a Grignard reagent from $HOCH_2CH_2CH_2CH_2Br$ would not be possible because of the presence of the acidic hydroxyl group.

18.9 Since maleic acid is a cis dicarboxylic acid, dehydration occurs readily:

Maleic acid   Maleic anhydride

Being a trans dicarboxylic acid, fumaric acid must undergo isomerization to maleic acid first. This isomerization requires a higher temperature.

Fumaric acid

18.10 The labeled oxygen atom should appear in the carboxyl group of the acid. (Follow the reverse steps of the mechanism in Section 18.7A of the text using $H_2^{18}O$.)

(b)   Method (3) should give a higher yield of **F** than method (4). Since the hydroxide ion is a strong base and since the alkyl halide is secondary, method (4) is likely to be accompanied by considerable elimination. Method (3), on the other hand, employs a weaker base, acetate ion, in the S_N2 step and is less likely to be complicated by elimination. Hydrolysis of the ester **E** that results should also proceed in high yield.

18.13 (a)   Steric hindrance presented by the di-ortho methyl groups of methyl mesitoate prevents formation of the tetrahedral intermediate that must accompany attack at the acyl carbon.

(b)   Carry out hydrolysis with labeled $^{18}OH^-$ in labeled $H_2{}^{18}O$. The label should appear in the methanol.

18.15 (a) $(CH_3)_3CCO_2H$ $\xrightarrow{SOCl_2}$ $(CH_3)_3CCOCl$ $\xrightarrow{NH_3}$ $(CH_3)_3CCONH_2$

$\xrightarrow[\text{heat}]{P_4O_{10}}$ $(CH_3)_3CC\equiv N$

(b) An elimination reaction would take place because CN^- is a strong base.

(e)

$$\xrightarrow[\text{CH}_3\text{CO}_2\text{H}]{\text{HBr}} \quad \text{H}_3\overset{+}{\text{N}}\text{CH}_2\text{CO}_2\text{H} \quad + \quad \text{CO}_2 \quad +$$

(f) $\text{H}_2\text{N}-\overset{\overset{\displaystyle O}{\|}}{\text{C}}-\text{NH}_2 \quad \xrightarrow[\text{heat}]{\text{OH}^-,\,\text{H}_2\text{O}} \quad 2\,\text{NH}_3 \;+\; \text{CO}_3^{2-}$

18.17 (a) By decarboxylation of a β-keto acid:

$$\text{CH}_3(\text{CH}_2)_3\overset{\overset{\displaystyle O}{\|}}{\text{C}}\text{CH}_2\overset{\overset{\displaystyle O}{\|}}{\text{C}}\text{OH} \quad \xrightarrow{100\text{-}150°\text{C}} \quad \text{CH}_3(\text{CH}_2)_3\overset{\overset{\displaystyle O}{\|}}{\text{C}}\text{CH}_3 \quad + \quad \text{CO}_2$$

(b) By decarboxylation of a substituted malonic acid:

$$\underset{\underset{\displaystyle\text{CH}_3}{|}}{\overset{\overset{\displaystyle\text{CO}_2\text{H}}{|}}{\text{CH}_3\text{CH}_2\text{C}}}-\text{CO}_2\text{H} \quad \xrightarrow{100\text{-}150°\text{C}} \quad \underset{\underset{\displaystyle\text{CH}_3}{|}}{\text{CH}_3\text{CH}_2\text{CH}}\text{CO}_2\text{H} \quad + \quad \text{CO}_2$$

(c) By decarboxylation of a β-keto acid:

$$\xrightarrow{100\text{-}150°\text{C}} \qquad + \quad \text{CO}_2$$

(d) By decarboxylation of a substituted malonic acid:

$$\underset{\underset{\displaystyle\text{CO}_2\text{H}}{|}}{\overset{\overset{\displaystyle\text{CO}_2\text{H}}{|}}{\text{CH}_3\text{CH}_2\text{CH}_2\text{CH}}} \quad \xrightarrow{100\text{-}150°\text{C}} \quad \text{CH}_3\text{CH}_2\text{CH}_2\text{CH}_2\text{CO}_2\text{H} \quad + \quad \text{CO}_2$$

18.18 (a) The oxygen-oxygen bond of the diacyl peroxide has a low homolytic bond dissociation energy ($DH° \simeq 146$ kJ mol^{-1}). This allows the following reaction to occur at a moderate temperature.

$$\text{R}-\overset{\overset{\displaystyle O}{\|}}{\text{C}}-\text{O}-\text{O}-\overset{\overset{\displaystyle O}{\|}}{\text{C}}-\text{R} \quad \longrightarrow \quad 2\,\text{R}-\overset{\overset{\displaystyle O}{\|}}{\text{C}}-\text{O}\bullet \qquad \Delta H° \simeq 146 \text{ kJ mol}^{-1}$$

(b) By decarboxylation of the carboxylate radical produced in part (a).

$$\text{R}-\overset{\overset{\displaystyle O}{\|}}{\text{C}}-\text{O}\bullet \quad \longrightarrow \quad \text{R}\bullet \;+\; \text{CO}_2$$

(c) *Chain Initiation*

$$\text{Step 1} \quad R-\overset{\overset{\displaystyle O}{\|}}{C}-O-O-\overset{\overset{\displaystyle O}{\|}}{C}-R \quad \xrightarrow{\text{heat}} \quad 2\ R-\overset{\overset{\displaystyle O}{\|}}{C}-O\bullet$$

$$\text{Step 2} \quad R-\overset{\overset{\displaystyle O}{\|}}{C}-O\bullet \quad \longrightarrow \quad R\bullet \ + \ CO_2$$

Chain Propagation

$$\text{Step 3} \quad R\bullet \ + \ CH_2{=}CH_2 \quad \longrightarrow \quad RCH_2CH_2\bullet$$

$$\text{Step 4} \quad RCH_2CH_2\bullet \ + \ CH_2{=}CH_2 \quad \longrightarrow \quad RCH_2CH_2CH_2CH_2\bullet$$

Step 3, 4, 3, 4, and so on.

18.19 (a) $CH_3(CH_2)_4CO_2H$

(b) $CH_3(CH_2)_4CONH_2$

(c) $CH_3(CH_2)_4CONHC_2H_5$

(d) $CH_3(CH_2)_4CON(C_2H_5)_2$

(e) $CH_3CH_2CH{=}CHCH_2CO_2H$

(f) $CH_3CH{=}CHCH_2\underset{\underset{\displaystyle CH_3}{|}}{C}HCO_2H$

(g) $HO_2CCH_2CH_2CH_2CH_2CO_2H$

(h)

(i)

(j)

(k) $C_2H_5O_2C{-}CO_2C_2H_5$

(l) $C_2H_5O_2C(CH_2)_4CO_2C_2H_5$

(m) $CH_3CH_2CO_2CH_2CH(CH_3)_2$

(n)

(o)

(p) $HO_2CCHOHCH_2CO_2H$

(q)

(r) $HO_2CCH_2CH_2CO_2H$

(s)

(t) $HO_2CCH_2CO_2H$

(u) $C_2H_5O_2CCH_2CO_2C_2H_5$

18.20 (a) Benzoic acid

(b) Benzoyl chloride

(c) Benzamide

(d) Benzoic anhydride

(e) Benzyl benzoate

(f) Phenyl benzoate

(g) Isopropyl acetate or 1-methylethyl ethanoate

(h) *N,N*-Dimethylacetamide or *N,N*-dimethylethanamide

(i) Acetonitrile or ethanenitrile

18.21 (a)

(d) Cl—⟨benzene ring⟩—CH₂CO₂H $\xrightarrow[\text{(2)LiAlH[OC(CH}_3)_3]_3]{\text{(1) SOCl}_2}$ Cl—⟨benzene ring⟩—CH₂CHO $\xrightarrow{\text{HCN}}$

[from (b)]

Cl—⟨benzene ring⟩—CH₂CHCN with OH $\xrightarrow[\text{heat}]{\text{H}_3\text{O}^+,\text{H}_2\text{O}}$ [Cl—⟨benzene ring⟩—CH₂CHCO₂H with OH] ⟶

Cl—⟨benzene ring⟩—CH=CHCO₂H

18.22 (a) H₃C—CO₂H $\xrightarrow[\text{P}]{\text{Br}_2}$ CH₂Br—CO₂H $\xrightarrow{\text{NaOH}}$ CH₂Br—CO₂Na $\xrightarrow{\text{NaCN}}$

CH₂—CO₂Na with CN $\xrightarrow[\text{heat}]{\text{H}_3\text{O}^+,\text{H}_2\text{O}}$ HO₂C—CH₂—CO₂H

(b) HO—CH₂—CH₂—CH₂—CH₂—OH $\xrightarrow[\text{(2) H}_3\text{O}^+]{\text{(1) KMnO}_4,\text{ OH}^-,\text{ heat}}$ HO—C(=O)—CH₂—CH₂—C(=O)—OH

(c) ⟨cyclohexane ring⟩—OH $\xrightarrow[\text{(2) H}_3\text{O}^+]{\text{(1)KMnO}_4,\text{ OH}^-,\text{ heat}}$ ⟨structure with C—OH and C—OH groups⟩

18.23 (a) CH₃CH₂CH₂CH₂CH₂OH $\xrightarrow[\text{(2) H}_3\text{O}^+]{\text{(1) KMnO}_4,\text{ OH}^-,\text{ heat}}$ CH₃CH₂CH₂CH₂CO₂H

(b) CH₃CH₂CH₂CH₂Br $\xrightarrow[\text{(2) CO}_2]{\text{(1) Mg, Et}_2\text{O}}$ CH₃CH₂CH₂CH₂CO₂MgBr $\xrightarrow{\text{H}_3\text{O}^+}$

CH₃CH₂CH₂CH₂CO₂H

CH₃CH₂CH₂CH₂Br $\xrightarrow{\text{CN}^-}$ CH₃CH₂CH₂CH₂CN $\xrightarrow[\text{heat}]{\text{H}_3\text{O}^+,\text{ H}_2\text{O}}$

CH₃CH₂CH₂CH₂CO₂H

(c) CH₃(CH₂)₃CH=CH(CH₂)₃CH₃ $\xrightarrow[\text{(2) H}_3\text{O}^+]{\text{(1) KMnO}_4,\text{ OH}^-,\text{ heat}}$ 2 CH₃(CH₂)₃CO₂H

(d) CH₃CH₂CH₂CH₂CHO $\xrightarrow[\text{(2) H}_3\text{O}^+]{\text{(1) Ag(NH}_3)_2{}^+\text{OH}^-}$ CH₃CH₂CH₂CH₂CO₂H

18.24   (a) $CH_3CO_2H + HCl$

(b) $CH_3\overset{\displaystyle OH}{\underset{|}{C}}(CH_2CH_3)_2$

(c) $CH_3CO_2CH_2(CH_2)_2CH_3$

(d) CH_3CONH_2

(e) $+$ CH_3——$\overset{\displaystyle O}{\overset{||}{C}}CH_3$ (f) CH_3CHO

(g) CH_3COCH_3

(h) CH_3CO_2Na

(i) $CH_3CONHCH_3$

(j) $CH_3CONHC_6H_5$

(k) $CH_3CON(CH_3)_2$

(l) $CH_3CO_2CH_2CH_3$

(m) $(CH_3CO)_2O$

(n) $(CH_3CO)_2O$

(o) $CH_3CO_2C_6H_5$

(p) $BrCH_2COCl$

18.25   (a) $CH_3CONH_2 + CH_3CO_2^-\ NH_4^+$

(b) $2\ CH_3CO_2H$

(c) $CH_3CO_2CH_2CH_2CH_3 + CH_3CO_2H$

(d) $C_6H_5COCH_3 + CH_3CO_2H$

(e) $CH_3CONHCH_2CH_3 + CH_3CO_2^-\ CH_3CH_2NH_3^+$

(f) $CH_3CON(CH_2CH_3)_2 + CH_3CO_2^-\ (CH_3CH_2)_2NH_2^+$

18.26   (a) $\overset{\displaystyle CONH_2}{\underset{\displaystyle \underset{CO_2^-\ NH_4^+}{\overset{|}{CH_2}}}{\overset{|}{CH_2}}}$

(b) $\overset{\displaystyle CO_2H}{\underset{\displaystyle \underset{CO_2H}{\overset{|}{CH_2}}}{\overset{|}{CH_2}}}$

(c) $\overset{\displaystyle CO_2CH_2CH_2CH_3}{\underset{\displaystyle \underset{CO_2H}{\overset{|}{CH_2}}}{\overset{|}{CH_2}}}$

(d)

(e) $\overset{\displaystyle CONHCH_2CH_3}{\underset{\displaystyle \underset{CO_2^-\ CH_3CH_2NH_3^+}{\overset{|}{CH_2}}}{\overset{|}{CH_2}}}$

(f) $\overset{\displaystyle CON(CH_2CH_3)_2}{\underset{\displaystyle \underset{CO_2^-\ (CH_3CH_2)_2NH_2^+}{\overset{|}{CH_2}}}{\overset{|}{CH_2}}}$

18.27   See text, p. 677.

18.28   (a)

(b)

(c)

18.29   (a)   CH$_3$CH$_2$CO$_2$H + CH$_3$CH$_2$OH

(b)   CH$_3$CH$_2$CO$_2^-$ + CH$_3$CH$_2$OH

(c)   CH$_3$CH$_2$CO$_2$(CH$_2$)$_7$CH$_3$ + CH$_3$CH$_2$OH

(d)   CH$_3$CH$_2$CONHCH$_3$ + CH$_3$CH$_2$OH

(e)   CH$_3$CH$_2$CH$_2$OH + CH$_3$CH$_2$OH

(f)   CH$_3$CH$_2$$\overset{\overset{\displaystyle C_6H_5}{|}}{\underset{\underset{\displaystyle OH}{|}}{C}}$—C$_6H_5$ + CH$_3$CH$_2$OH

18.30 (a) $CH_3CH_2CO_2H + NH_4^+$

(b) $CH_3CH_2CO_2^- + NH_3$

(c) CH_3CH_2CN

18.31 See the mechanisms in Section 18.8F, where $R = CH_2CH_3$ for propanamide.

18.32 (a)

(b) $CH_3CH=CHCO_2H$

(c) CH_3CH_2CH ... $CHCH_2CH_3$

(d)

(e)

(f)

18.33 (a)

(R)-$(-)$-2-butanol

$\xrightarrow[\text{(retention)}]{\text{TsCl, pyridine}}$ **A**

$\xrightarrow[\text{(inversion)}]{CN^-}$ **B**

$\xrightarrow[\text{(retention)}]{H_2SO_4, H_2O}$ $(+)$ **C**

$\xrightarrow[\substack{(2)\ H_2O \\ \text{(retention)}}]{(1)\ LiAlH_4}$ $(-)$ **D**

(b) (R)-(−)-2-butanol →[PBr₃, pyridine (inversion)] **E** →[CN⁻ (inversion)] **F**

→[H₂SO₄, H₂O (retention)] (−) **C** →[(1) LiAlH₄, (2) H₂O (retention)] (+) **D**

(c) **A** →[CH₃CO₂⁻ (inversion)] **G** →[OH⁻ (retention)] (+) **H**

(S)-(+)-2-butanol

(d) (−) **D** →[PBr₃ (retention)] **J** →[Mg, diethyl ether (retention)]

K →[(1) CO₂, (2) H⁺ (retention)] **L**

(e) (R)-(+)-Glyceraldehyde →[HCN] **M** + **N**

(f)   **M** $\xrightarrow[\text{heat}]{\text{H}_2\text{SO}_4,\ \text{H}_2\text{O}}$ **P** $\xrightarrow[\text{HNO}_3]{[\text{O}]}$ *meso*-Tartaric acid

(g)   **N** $\xrightarrow[\text{heat}]{\text{H}_2\text{SO}_4,\ \text{H}_2\text{O}}$ **Q** $\xrightarrow[\text{HNO}_3]{[\text{O}]}$ (−)-Tartaric acid

18.34 (a)

18.35

An interpretation of the ¹H NMR spectral data for phenacetin is as follows:

(*a*)   triplet δ 1.4

(*b*)   singlet δ 2.1

(*c*)   quartet δ 3.95

(*d*)   multiplet δ 6.8-7.4

(*e*)   broad singlet δ 9.0

18.36 (a) $$CH_3CH_2-O-\overset{\overset{\displaystyle O}{\|}}{C}-CH_2CH_2-\overset{\overset{\displaystyle O}{\|}}{C}-O-CH_2CH_3$$
$\quad\quad\quad$ (*a*) (*c*) $\quad\quad$ (*b*) (*b*) $\quad\quad$ (*c*) (*a*)

Interpretation:

(*a*) Triplet δ 1.2 (6H)$\quad\quad$ $2 -\overset{\overset{\displaystyle O}{\|}}{C}-O-$, 1740 cm⁻¹ (ester)
(*b*) Singlet δ 2.5 (4H)
(*c*) Quartet δ 4.1 (4H)

(b) $$\text{C}_6\text{H}_5-\overset{\overset{\displaystyle O}{\|}}{C}-O-CH_2-\overset{\overset{\displaystyle CH_3\ (a)}{|}}{CH}-CH_3$$
$\quad\quad\quad\quad\quad\quad\quad\quad$ (*c*) (*b*) $\quad\quad$ (*a*)

(*d*)

Interpretation:

(*a*) Doublet δ 1.0 (6H)$\quad\quad$ $-\overset{\overset{\displaystyle O}{\|}}{C}-O-$, 1720 cm⁻¹ (ester)
(*b*) Multiplet δ 2.1 (1H)
(*c*) Doublet δ 4.1 (2H)
(*d*) Multiplet δ 7.8 (5H)

(c) $\langle \bigcirc \rangle$—CH$_2$—$\overset{\overset{\textstyle O}{\|}}{C}$—O—CH$_2CH_3$
       (b)   (c) (a)
 (d)

Interpretation:

(a) Triplet δ 1.2 (3H) $-\overset{\overset{\textstyle O}{\|}}{C}-O-$, 1740 cm^{-1} (ester)
(b) Singlet δ 3.5 (2H)
(c) Quartet δ 4.1 (2H)
(d) Multiplet δ 7.3 (5H)

(d) Cl—$\overset{\overset{\textstyle Cl}{|}}{CH}$—CO$_2$H
        (a)   (b)

Interpretation: —OH , 2500-2700 cm^{-1}

(a) Singlet δ 6.0 $\overset{\overset{\textstyle O}{\|}}{C}$
(b) Singlet δ 11.7 $-\overset{\overset{\textstyle O}{\|}}{C}-O-$, 1705 cm^{-1} (acid)

(e) Cl—CH$_2$—$\overset{\overset{\textstyle O}{\|}}{C}$—O—CH$_2CH_3$
        (b)   (c) (a)

Interpretation:

(a) Triplet δ 1.3 $-\overset{\overset{\textstyle O}{\|}}{C}-O-$, 1745 cm^{-1} (ester)
(b) Singlet δ 4.0
(c) Quartet δ 4.2

18.37

18.38 Alkyl groups are electron releasing; they help disperse the positive charge of an alkyl-ammonium salt and thereby help to stabilize it.

$$R\ddot{N}H_2 \; + \; H_3O^+ \; \longrightarrow \; R{\rightarrow}NH_3{}^+ \; + \; H_2O$$

Stabilized by
electron-releasing
alkyl group

Consequently, alkylamines are somewhat stronger bases than ammonia.

Amides, on the other hand, have acyl groups, , attached to nitrogen, and acyl groups are electron withdrawing. They are especially electron withdrawing because of resonance contributions of the kind shown here,

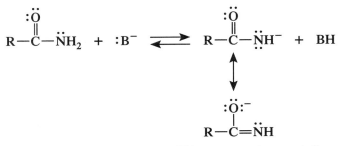

This kind of resonance also *stabilizes* the amide. The tendency of the acyl group to be electron withdrawing, however, *destabilizes* the conjugate acid of an amide, and reactions such as the following do not take place to an appreciable extent.

$$R-\overset{\overset{\displaystyle O}{\|}}{C}-\ddot{N}H_2 \; + \; H_3O^+ \; \underset{\longrightarrow}{\longleftarrow} \; R-\overset{\overset{\displaystyle O}{\|}}{C}-NH_3{}^+ \; + \; H_2O$$

Stabilized   *Destabilized by*
by   *electron-withdrawing*
resonance   *acyl group*

18.39 (a)   The conjugate base of an amide is stabilized by resonance.

$$R-\overset{\overset{\displaystyle :\ddot{O}}{\|}}{C}-\ddot{N}H_2 \; + \; :B^- \; \rightleftharpoons \; R-\overset{\overset{\displaystyle :\ddot{O}}{\|}}{C}-\ddot{N}H^- \; + \; BH$$

$$R-\overset{\overset{\displaystyle :\ddot{O}:^-}{|}}{C}=\ddot{N}H$$

This structure is especially
stable because the negative
charge is on oxygen.

(b)   The conjugate base of an imide is stabilized by an additional resonance structure,

18.40   That compound **X** does not dissolve in aqueous sodium bicarbonate indicates that **X** is not a carboxylic acid. That **X** has an IR absorption peak at 1740 cm^{-1} indicates the presence of a carbonyl group, probably that of an ester (Table 18.5). That the molecular formula of **X** ($C_7H_{12}O_4$) contains four oxygen atoms suggests that **X** is a diester.

   The ^{13}C spectrum shows only four signals indicating a high degree of symmetry for **X.** The single signal at δ 166.7 is that of an ester carbonyl carbon, indicating that both ester groups of **X** are equivalent.

   Putting these observations together with the information gathered from DEPT ^{13}C spectra and the molecular formula leads us to the conclusion that **X** is diethyl malonate. The assignments are

$$
\begin{array}{c}
\quad\quad\quad O \quad\quad O \\
\quad\quad\quad || \quad\quad || \\
CH_3CH_2OCCH_2COCH_2CH_3 \\
(a) \ (c) \ \ (d)(b) \ (d) \ (c) \ (a)
\end{array}
$$

(a) δ 14.2
(b) δ 41.6
(c) δ 61.3
(d) δ 166.7

18.41   (a)   *Chain   Initiation*

   Step 1   RO—OR $\longrightarrow$ 2 RO•

   Step 2 $CH_3\overset{\overset{\displaystyle O}{||}}{C}SH$ +   RO• $\longrightarrow$ $CH_3\overset{\overset{\displaystyle O}{||}}{C}S$•   +   ROH

   Chain   Propagation

   Step 3 $CH_3\overset{\overset{\displaystyle O}{||}}{C}S$•   + $CH_2{=}CHR$ $\longrightarrow$ $CH_3\overset{\overset{\displaystyle O}{||}}{C}SCH_2\overset{\displaystyle \bullet}{C}HR$

   Step 4 $CH_3\overset{\overset{\displaystyle O}{||}}{C}SCH_2\overset{\displaystyle \bullet}{C}HR$ + $CH_3\overset{\overset{\displaystyle O}{||}}{C}SH$ $\longrightarrow$ $CH_3\overset{\overset{\displaystyle O}{||}}{C}SCH_2CH_2R$ +$CH_3\overset{\overset{\displaystyle O}{||}}{C}S$•

18.42 *cis*-4-Hydroxycyclohexanecarboxylic acid can assume a boat conformation that permits lactone formation.

Neither of the chair conformations nor the boat form of *trans*-4-hydroxycyclohexanecarboxylic acid places the –OH group and the –CO$_2$H group close enough together to permit lactonization.

18.43

18.44 (a)

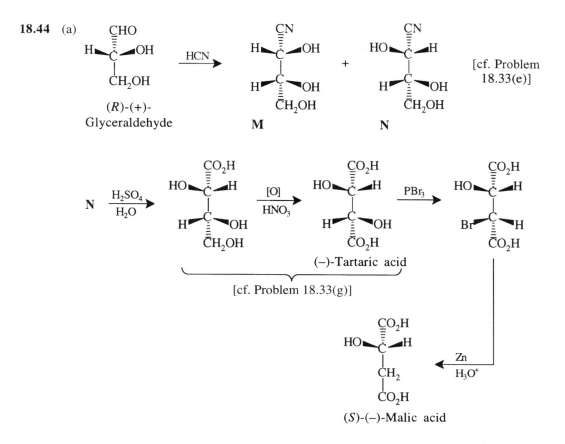

(R)-(+)-
Glyceraldehyde   **M**   **N**   [cf. Problem 18.33(e)]

(−)-Tartaric acid

[cf. Problem 18.33(g)]

(S)-(−)-Malic acid

(b)   Replacement of either alcoholic –OH by a reaction that proceeds with inversion produces the same stereoisomer.

(c)   Two. The stereoisomer given in (b) and the one given next, below.

(d)   It would have made no difference because treating either isomer (or both together) with zinc and acid produces (−) malic acid.

(−)-Malic acid

18.45 (a) $CH_3O_2C\text{—}C\equiv C\text{—}CO_2CH_3$. This is a Diels-Alder reaction.

(b) H_2, Pd. The disubstituted double bond is less hindered than the tetrasubstituted double bond and hence is more reactive.

(c) $CH_2=CH\text{—}CH=CH_2$. Another Diels-Alder reaction.

(d) $LiAlH_4$

(e) $CH_3\text{—}\overset{\overset{O}{\parallel}}{\underset{\underset{O}{\parallel}}{S}}\text{—}Cl$ and pyridine

(f) $CH_3CH_2S^-$

(g) OsO_4

(h)   Raney Ni

(i)   Base. This is an aldol condensation.

(j) C_6H_5Li (or C_6H_5MgBr) followed by H_3O^+

(k) H_3O^+. This is an acid-catalyzed rearrangement of an allylic alcohol.

(l) $CH_3\overset{\overset{O}{\parallel}}{C}Cl$, pyridine

(m) O_3, followed by oxidation

(n)   Heat

18.46 (a)

Furan   Dimethylmaleic
              anhydride

Cantharidin

(b)   Cantharidin apparently undergoes dehydrogenation to the Diels-Alder adduct shown here, and then the adduct spontaneously decomposes through a reverse Diels-Alder reaction to furan and dimethylmaleic anhydride. These results suggest that the attempted Diels-Alder synthesis fails because the position of equilibrium favors reactants rather than products.

18.47 The very low hydrogen content of the molecular formula of **Y** ($C_8H_4O_3$) indicates that **Y** is highly unsaturated. That **Y** dissolves slowly in warm aqueous $NaHCO_3$ suggests that **Y** is a carboxylic acid anhydride that hydrolyzes and dissolves because it forms a carboxylate salt:

The infrared absorption peaks at 1779 and 1854 cm^{-1} are consistent with those of an aromatic carboxylic anhydride (Table 18.5).

That only four signals appear in the ^{13}C spectrum of **Y** indicates a high degree of symmetry for **Y**. Three of the signals occur in the aromatic region (δ 120–δ 140) and one signal is downfield (δ 163.1)

These signals and the information from the DEPT ^{13}C NMR spectra lead us to conclude that **Y** is phthalic anhydride. The assignments are

(a) δ 124.3
(b) δ 131.1
(c) δ 136.1
(d) δ 163.1

Z is phthalic acid and **AA** is ethyl hydrogen phthalate.

18.48   (a)   Ethyl acetate   (b)   Acetic anhydride   (c)   *N*-Ethylacetamide.

18.49   In the first instance, nucleophilic attack by the amine occurs preferentially at the less hindered carbon of the formyl group. (Recall that aldehydes are more reactive than ketones toward nucleophiles for the same reason.) In the second case, $CF_3CO_2^-$ is a better leaving group than $CH_3CO_2^-$ since the former is the conjugate base of the stronger acid.

18.50

18.51 $C_6H_6 + (CH_3)_2CHCCl$ $\xrightarrow{AlCl_3}$ (phenyl)-$CCH(CH_3)_2$ $\xrightarrow[\text{Wolff-Kishner}]{\text{Clemmensen or}}$

(phenyl)-$CH_2CH(CH_3)_2$ $\xrightarrow[\text{HCl, ZnCl}_2]{\text{HCHO}}$ $(CH_3)_2CHCH_2$-(phenyl)-CH_2Cl

$\xrightarrow{KCN}$ $(CH_3)_2CHCH_2$-(phenyl)-CH_2CN $\xrightarrow[\text{CH}_3\text{I}]{\text{NaNH}_2}$ $(CH_3)_2CHCH_2$-(phenyl)-$\overset{\displaystyle |}{\underset{\displaystyle CH_3}{CHCN}}$

$\xrightarrow[\text{heat}]{H_3O^+}$ $(CH_3)_2CHCH_2$-(phenyl)-$\overset{\displaystyle |}{\underset{\displaystyle CH_3}{CHCOOH}}$

18.52 $A =$ (phenyl)-$CCH(CH_3)_2$ $C =$ $(CH_3)_2CHCH_2$-(phenyl)-CCH_3

$B =$ (phenyl)-$CH_2CH(CH_3)_2$ $D =$ $(CH_3)_2CHCH_2$-(phenyl)-$\overset{\displaystyle |}{\underset{\displaystyle CH_3}{C(OH)CN}}$

In the last step, HI/red P accomplishes both the reduction of $-OH$ to $-H$ and the hydrolysis of the nitrile function.

18.53 (a) The signal at δ 193.8 is consistent with the carbonyl carbon of an aldehyde and shows that the PCC reaction produced cinnamaldehyde.

(b) The signal at δ 164.5 is consistent with the carbonyl carbon of a carboxylic acid, and suggests that the oxidation with $K_2Cr_2O_7$ in sulfuric acid produced cinnamic acid.

*18.54

A

(IR absorption at 2550 cm^{-1} is S—H stretch)

B

(IR absorption at 3020 cm^{-1} is C—H stretch for C—H of 3-membered ring)

*18.55

QUIZ

18.1   Which of the following would be the strongest acid?

(a)   Benzoic acid   (b)   4-Nitrobenzoic acid   (c)   4-Methylbenzoic acid

(d)   4-Methoxybenzoic acid   (e)   4-Ethylbenzoic acid

18.2   Which of the following would yield (S)-2-butanol?

(a)   (R)-2-Bromobutane + $CH_3CO_2^- Na^+$ ⟶ product $\xrightarrow[\text{heat}]{OH^-, H_2O}$

(b)   (R)-2-Bromobutane $\xrightarrow[\text{heat}]{OH^-, H_2O}$

(c)   (S)-2-Butyl acetate $\xrightarrow[\text{heat}]{OH^-, H_2O}$

(d)   All of the above

(e)   None of the above

18.3 Which reagent would serve as the basis for a simple chemical test to distinguish between hexanoic acid and hexanamide?

(a) Cold dilute NaOH   (b) Cold dilute $NaHCO_3$

(c) Cold concd. H_2SO_4 (d) More than one of these

(e) None of these

18.4 Give an acceptable name for:

A

B

C

18.5 Complete the following syntheses.

B

SPECIAL TOPIC
Step-Growth Polymers

SOLUTIONS TO PROBLEMS

B.1 (a) cyclohexanone $\xrightarrow[\text{H}_2\text{Cr}_2\text{O}_7]{[\text{O}]}$ $HO_2C(CH_2)_4CO_2H$

(b) $HO_2C(CH_2)_4CO_2H + 2\ NH_3 \longrightarrow NH_4O_2C(CH_2)_4CO_2NH_4 \xrightarrow[-2H_2O]{\text{heat}}$

$H_2N\overset{\text{O}}{\underset{\|}{C}}(CH_2)_4\overset{\text{O}}{\underset{\|}{C}}NH_2 \xrightarrow[\text{catalyst}]{350°C} N{\equiv}C(CH_2)_4C{\equiv}N \xrightarrow[\text{catalyst}]{4\ H_2}$

$H_2NCH_2(CH_2)_4CH_2NH_2$

(c) $CH_2{=}CH{-}CH{=}CH_2 \xrightarrow{Cl_2} ClCH_2CH{=}CHCH_2Cl \xrightarrow{2\ NaCN}$

$N{\equiv}CCH_2CH{=}CHCH_2C{\equiv}N \xrightarrow[\text{Ni}]{H_2} N{\equiv}C(CH_2)_4C{\equiv}N \xrightarrow[\text{catalyst}]{4\ H_2}$

$H_2NCH_2(CH_2)_4CH_2NH_2$

(d) tetrahydrofuran $\xrightarrow{2\ HCl}$ $ClCH_2CH_2CH_2CH_2Cl \xrightarrow{2\ NaCN} N{\equiv}C(CH_2)_4C{\equiv}N$

$\xrightarrow[\text{catalyst}]{4\,H_2}$ $H_2NCH_2(CH_2)_4CH_2NH_2$

B.2 (a) $HOCH_2CH_2OH + {:}B^- \rightleftarrows HOCH_2CH_2O^- + HB$

$RO\overset{\text{O}}{\underset{\|}{C}}{-}\bigcirc{-}\overset{\text{O}}{\underset{\|}{C}}OCH_3 + {}^-OCH_2CH_2OH \rightleftarrows$

$RO\overset{\text{O}}{\underset{\|}{C}}{-}\bigcirc{-}\underset{\underset{OCH_3}{|}}{\overset{\overset{O^-}{|}}{C}}{-}OCH_2CH_2OH \rightleftarrows RO\overset{\text{O}}{\underset{\|}{C}}{-}\bigcirc{-}\overset{\text{O}}{\underset{\|}{C}}OCH_2CH_2OH$

$+ CH_3O^-$

$CH_3O^- + HB \rightleftarrows CH_3OH + {:}B^-$

[R = CH$_3$- or HOCH$_2$CH$_2$-]

(b)

[R=CH₃- or HOCH₂CH₂-]

B.3 (a)

(b) By high-pressure catalytic hydrogenation

B.4

B.5

Lexan

B.6 (a)   The resin is probably formed in the following way. Base converts the bisphenol A to a phenoxide ion that attacks a carbon atom of the epoxide ring of epichlorohydrin:

(b)   The excess of epichlorohydrin limits the molecular weight and ensures that the resin has epoxy ends.

(c)   Adding the hardener brings about cross linking by reacting at the terminal epoxide groups of the resin:

$H_2NCH_2CH_2NHCH_2CH_2\overset{..}{N}H_2$ + $H_2C{-}CHCH_2{-}$[polymer]${-}CH_2CH{-}CH_2$ $\longrightarrow$

${-}CH_2{-}CHCH_2{-}NCH_2CH_2{-}N{-}CH_2CH_2{-}N{-}CH_2CHCH_2$[polymer]$CH_2CHCH_2{-}$etc.
      OH   H   CH$_2$ H   OH   OH

                                   CHOH
                                   CH$_2$
                              [polymer]
                                   CH$_2$
                                   CHOH
                                   CH$_2$ etc.
                     N${-}$CH$_2$CH$_2$N${-}$CH$_2$CH$_2${-}N{-}CH_2$CHCH$_2$[polymer]CH$_2$CHCH$_2$
                     H   H   H   OH   OH

B.7

(a)

(b)   To ensure that the polyester chain has –CH₂OH end groups.

B.8   Because the para position is occupied by a methyl group, cross-linking does not occur and the resulting polymer remains thermoplastic (See Section B.4.)

B.9

19

SYNTHESIS AND REACTIONS OF β-DICARBONYL COMPOUNDS: MORE CHEMISTRY OF ENOLATE ANIONS

SUMMARY OF ACETOACETIC ESTER AND MALONIC ESTER SYNTHESES

A. Acetoacetic Ester Synthesis

B. Malonic Ester Synthesis

SOLUTIONS TO PROBLEMS

19.1 (a) *Step 1*

Step 2   CH$_3$CH$_2$C$\overset{O}{\underset{OC_2H_5}{\parallel}}$ + $\ddot{:}$CHCOC$_2$H$_5$ $\overset{O}{\parallel}$ CH$_3$ $\rightleftharpoons$ CH$_3$CH$_2$C$\overset{O^-}{-}$CH$-$COC$_2$H$_5$ C$_2$H$_5$O CH$_3$

C$_2$H$_5$O$^-$ + CH$_3$CH$_2$C$\overset{O}{\parallel}-$CH$-$COC$_2$H$_5$ CH$_3$

Step 3   CH$_3$CH$_2$C$\overset{O}{\parallel}-$C$\overset{H}{-}$C$\overset{O}{\parallel}$OC$_2$H$_5$ CH$_3$ + $^-$OC$_2$H$_5$ $\rightleftharpoons$ CH$_3$CH$_2$C$\overset{O}{=}$C$=$COC$_2$H$_5$ CH$_3$ +C$_2$H$_5$OH

(b) CH$_3$CH$_2$C$\overset{O}{\parallel}-$CH$-$COC$_2$H$_5$ CH$_3$ + CH$_3$CH$_2$C$\overset{OH}{=}$C$-$COC$_2$H$_5$ CH$_3$

19.2 (a)

![cyclohexanone with COC$_2$H$_5$ substituent]

(b) To undergo a Dieckmann condensation, diethyl glutarate would have to form a highly strained four-membered ring.

19.3   CH$_3$COC$_2$H$_5$$\overset{O}{\parallel}$ + C$_2$H$_5$O$^-$ $\rightleftharpoons$ $\ddot{:}$CH$_2$COC$_2$H$_5$$\overset{O}{\parallel}$ + C$_2$H$_5$OH

C$_6$H$_5$COC$_2$H$_5$$\overset{O}{\parallel}$ + $\ddot{:}$CH$_2$COC$_2$H$_5$$\overset{O}{\parallel}$ $\rightleftharpoons$ C$_6$H$_5$C$\overset{O^-}{-}$CH$_2$COC$_2$H$_5$$\overset{O}{\parallel}$ OC$_2$H$_5$

C$_6$H$_5$C$\overset{O}{\parallel}-$CH$_2$COC$_2$H$_5$$\overset{O}{\parallel}$ + C$_2$H$_5$O$^-$ $\rightleftharpoons$ C$_6$H$_5$C$\overset{O}{=}$CH$=$COC$_2$H$_5$ + C$_2$H$_5$OH

$\xrightarrow{H_3O^+}$ C$_6$H$_5$C$\overset{O}{\parallel}-$CH$_2$COC$_2$H$_5$$\overset{O}{\parallel}$

C$_6$H$_5$CH$_2$COC$_2$H$_5$$\overset{O}{\parallel}$ + C$_2$H$_5$O$^-$ $\rightleftharpoons$ C$_6$H$_5$CHCOC$_2$H$_5$$\overset{^-\;\;O}{\parallel}$ + C$_2$H$_5$OH

19.6   CH$_3$CCH$_2$CH$_2$CH$_2$CH$_2$COC$_2$H$_5$ +   C$_2$H$_5$O$^-$ ⇌ (-C$_2$H$_5$OH)

19.7   The partially negative oxygen atom of sodioacetoacetic ester acts as the nucleophile.

$$CH_3\overset{O}{\underset{\|}{C}}-\overset{..}{\underset{}{C}}H-\overset{O}{\underset{\|}{C}}-OC_2H_5 \quad \longleftrightarrow \quad CH_3\overset{O^-}{\underset{}{C}}=CH-\overset{O}{\underset{\|}{C}}-OC_2H_5$$

19.8   Again, working backward,

(a)   CH$_3$CCH$_2$CH$_2$CH$_3$ ←(heat, -CO$_2$)   CH$_3$CCH—COH with CH$_2$CH$_3$ ←(1) dil. NaOH, heat (2) H$_3$O$^+$

CH$_3$CCHCOC$_2$H$_5$ with CH$_2$CH$_3$ ←(1) NaOC$_2$H$_5$ (2) CH$_3$CH$_2$Br   CH$_3$CCH$_2$COC$_2$H$_5$

(b)   CH$_3$CCHCH$_2$CH$_2$CH$_3$ with CH$_2$CH$_2$CH$_3$ ←(heat, -CO$_2$)   CH$_3$C—C—CO$_2$H   ←(1) dil. NaOH, heat (2) H$_3$O$^+$

19.9 (a) Reactivity is the same as with any S_N2 reaction. With primary halides substitution is highly favored, with secondary halides elimination competes with substitution, and with tertiary halides elimination is the exclusive course of reaction.

(b) Acetoacetic ester and 2-methylpropene

(c) Bromobenzene is unreactive to nucleophilic substitution.

19.10

19.11 The carboxyl group that is lost more readily is the one that is β to the keto group (cf. Section 18.11 of the text).

19.12

19.13 CH$_3$CCH$_2$CC$_6$H$_5$ ←$\frac{\text{heat}}{-CO_2}$ CH$_3$CCHCOH ←$\frac{(1)OH^-, H_2O, heat}{(2) H_3O^+}$

19.14 (a) One molar equivalent of NaNH$_2$ converts acetoacetic ester to its anion,

and one molar equivalent of NaNH$_2$ converts bromobenzene to benzyne (cf. Section 21.11B):

Then the anion of acetoacetic ester adds to the benzyne as it forms in the mixture.

This is the end product of the addition.

(b)   1-phenyl-2-propanone, as follows:

(c)   By treating bromobenzene with diethyl malonate and two molar equivalents of NaNH$_2$ to form diethyl phenylmalonate.

[The mechanism for this reaction is analogous to that given in part (a).]
    Then hydrolysis and decarboxylation will convert diethyl phenylmalonate to phenylacetic acid.

19.15 Here we alkylate the dianion,

$$CH_3\overset{O}{\overset{\|}{C}}-CH_2-\overset{O}{\overset{\|}{C}}OC_2H_5 \xrightarrow[\text{liq. NH}_3]{2\ KNH_2} \ ^-:CH_2\overset{O}{\overset{\|}{C}}-\overset{..}{C}H-\overset{O}{\overset{\|}{C}}OC_2H_5 \xrightarrow[\text{(2) NH}_4Cl]{\text{(1) C}_6H_5CH_2Cl}$$

$$C_6H_5CH_2CH_2\overset{O}{\overset{\|}{C}}CH_2\overset{O}{\overset{\|}{C}}OC_2H_5$$

19.16 Working backward,

(a) $CH_3CH_2CH_2CH_2CO_2H \xleftarrow[-CO_2]{\text{heat}} CH_3CH_2CH_2\underset{\overset{|}{CO_2H}}{\overset{\overset{|}{CO_2H}}{CH}} \xleftarrow[\text{(2) H}_3O^+]{\text{(1) OH}^-,\ H_2O,\ \text{heat}}$

$CH_3CH_2CH_2\underset{\overset{|}{CO_2C_2H_5}}{\overset{\overset{|}{CO_2C_2H_5}}{CH}} \xleftarrow[\text{(2) CH}_3CH_2CH_2Br]{\text{(1) NaOC}_2H_5} \underset{\overset{|}{CO_2C_2H_5}}{\overset{\overset{|}{CO_2C_2H_5}}{CH_2}}$

(b) $CH_3CH_2CH_2\underset{\overset{|}{CH_3}}{CH}CO_2H \xleftarrow[-CO_2]{\text{heat}} CH_3CH_2CH_2\underset{\overset{|}{CO_2H}}{\overset{\overset{|}{CO_2H}}{C}}CH_3 \xleftarrow[\text{(2) H}_3O^+]{\text{(1) OH}^-,\ H_2O,\ \text{heat}}$

$CH_3CH_2CH_2\underset{\overset{|}{CO_2C_2H_5}}{\overset{\overset{|}{CO_2C_2H_5}}{C}}CH_3 \xleftarrow[\text{(CH}_3)_3COK]{\text{CH}_3I} CH_3CH_2CH_2\underset{\overset{|}{CO_2C_2H_5}}{\overset{\overset{|}{CO_2C_2H_5}}{CH}}$

$\xleftarrow[\text{(2) CH}_3CH_2CH_2Br]{\text{(1) NaOC}_2H_5} \underset{\overset{|}{CO_2C_2H_5}}{\overset{\overset{|}{CO_2C_2H_5}}{CH_2}}$

(c) $CH_3\underset{\overset{|}{CH_3}}{CH}CH_2CH_2CO_2H \xleftarrow[-CO_2]{\text{heat}} CH_3\underset{\overset{|}{CH_3}}{CH}CH_2\underset{\overset{|}{CO_2H}}{\overset{\overset{|}{CO_2H}}{CH}} \xleftarrow[\text{(2) H}_3O^+]{\text{(1) OH}^-,\ H_2O,\ \text{heat}}$

$CH_3\underset{\overset{|}{CH_3}}{CH}CH_2\underset{\overset{|}{CO_2C_2H_5}}{\overset{\overset{|}{CO_2C_2H_5}}{CH}} \xleftarrow[\text{(2) CH}_3CHCH_2Br \atop \overset{|}{CH_3}]{\text{(1) NaOC}_2H_5} \underset{\overset{|}{CO_2C_2H_5}}{\overset{\overset{|}{CO_2C_2H_5}}{CH_2}}$

19.17 $2\ H_3C-CH_2-CH_2-Br + H_2C\overset{\overset{\displaystyle CN}{|}}{\underset{\overset{|}{COEt}\atop\overset{\|}{O}}{}} \xrightarrow[-2\ HBr]{\text{base}}$

Valproic acid

19.18   (a)   Formaldehyde, $H-\overset{O}{\overset{\|}{C}}-H$

(b)

(c)

19.19   By treating the thioketal with Raney nickel.

19.20   (a)

(c)

19.23 These syntheses are easier to see if we work backward.

(b) See Section 19.3.

(b) $CH_3\overset{O}{\overset{\|}{C}}CH_2CH_2CH_2CH_3$ $\xleftarrow[-CO_2]{heat}$ $CH_3\overset{O}{\overset{\|}{C}}-\underset{\underset{\underset{CH_3}{|}}{\underset{CH_2}{|}}{\underset{|}{CH}}-\overset{O}{\overset{\|}{C}}OH$ $\xleftarrow[(2)\ H_3O^+]{(1)\ OH^-,\ H_2O,\ heat}$

$CH_3\overset{O}{\overset{\|}{C}}-\underset{\underset{\underset{CH_3}{|}}{\underset{CH_2}{|}}{\underset{|}{CH}}-\overset{O}{\overset{\|}{C}}OC_2H_5$ $\xleftarrow[(2)\ CH_3CH_2CH_2Br]{(1)\ NaOC_2H_5}$ $CH_3\overset{O}{\overset{\|}{C}}CH_2\overset{O}{\overset{\|}{C}}OC_2H_5$

(c) $CH_3\overset{O}{\overset{\|}{C}}CH_2CH_2\overset{O}{\overset{\|}{C}}CH_3$ $\xleftarrow[-CO_2]{heat}$ $CH_3\overset{O}{\overset{\|}{C}}-\underset{\underset{\underset{CH_3}{|}}{\underset{C=O}{|}}{\underset{CH_2}{\underset{|}{CH}}}}-\overset{O}{\overset{\|}{C}}OH$ $\xleftarrow[(2)\ H_3O^+]{(1)\ OH^-,\ H_2O,\ heat}$

$CH_3\overset{O}{\overset{\|}{C}}-\underset{\underset{\underset{CH_3}{|}}{\underset{C=O}{|}}{\underset{CH_2}{\underset{|}{CH}}}}-\overset{O}{\overset{\|}{C}}OC_2H_5$ $\xleftarrow[(2)CH_3COCH_2Br]{(1)\ NaOC_2H_5}$ $CH_3\overset{O}{\overset{\|}{C}}CH_2\overset{O}{\overset{\|}{C}}OC_2H_5$

(d) $CH_3\underset{\underset{OH}{|}}{CH}CH_2CH_2CO_2H$ $\xleftarrow{NaBH_4}$ $CH_3\overset{O}{\overset{\|}{C}}CH_2CH_2\overset{O}{\overset{\|}{C}}OH$ $\xleftarrow[-CO_2]{heat}$

$CH_3\overset{O}{\overset{\|}{C}}-\underset{\underset{CO_2H}{|}}{\underset{CH_2}{\underset{|}{CH}}}-\overset{O}{\overset{\|}{C}}OH$ $\xleftarrow[(2)\ H_3O^+]{(1)\ OH^-,\ H_2O,\ heat}$ $CH_3\overset{O}{\overset{\|}{C}}-\underset{\underset{\underset{O}{\|}}{\underset{COC_2H_5}{|}}}{\underset{CH_2}{\underset{|}{CH}}}-\overset{O}{\overset{\|}{C}}OC_2H_5$

$\xleftarrow[(2)\ BrCH_2CO_2C_2H_5]{(1)\ NaOC_2H_5}$ $CH_3\overset{O}{\overset{\|}{C}}CH_2\overset{O}{\overset{\|}{C}}OC_2H_5$

(e) $CH_3\underset{\underset{C_2H_5}{|}}{\underset{OH}{\underset{|}{CH}}}CHCH_2OH$ $\xleftarrow[(2)\ H_3O^+]{(1)\ LiAlH_4}$ $CH_3\overset{O}{\overset{\|}{C}}\underset{\underset{C_2H_5}{|}}{CH}\overset{O}{\overset{\|}{C}}OC_2H_5$ $\xleftarrow[(2)CH_3CH_2Br]{(1)\ NaOC_2H_5}$

$CH_3\overset{O}{\overset{\|}{C}}CH_2\overset{O}{\overset{\|}{C}}OC_2H_5$

(f) $CH_3\underset{\underset{OH}{|}}{CH}CH_2\underset{\underset{OH}{|}}{CH}C_6H_5$ $\xleftarrow{NaBH_4}$ $CH_3\overset{O}{\overset{\|}{C}}CH_2\overset{O}{\overset{\|}{C}}C_6H_5$ $\longleftarrow$ compare Problem 19.13

19.28 (a) $CH_3CH_2\underset{\underset{CH_3}{|}}{C}HCO_2H$ $\xleftarrow[-CO_2]{heat}$ $\underset{\underset{CH_3}{|}}{\overset{CH_3CH_2}{C}}\overset{CO_2H}{\underset{CO_2H}{}}$ $\xleftarrow[(2)\,H_3O^+]{(1)\,OH,\,H_2O,\,heat}$

$\underset{\underset{CH_3}{|}}{\overset{CH_3CH_2}{C}}\overset{CO_2C_2H_5}{\underset{CO_2C_2H_5}{}}$ $\xleftarrow[(2)\,CH_3I]{(1)\,KOC(CH_3)_3}$ $CH_3CH_2-\underset{\underset{CO_2C_2H_5}{|}}{C}H\overset{CO_2C_2H_5}{}$

$\xleftarrow[(2)\,CH_3CH_2Br]{(1)\,NaOC_2H_5}$ $H_2C\overset{CO_2C_2H_5}{\underset{CO_2C_2H_5}{}}$

(b) $CH_3\underset{\underset{CH_3}{|}}{C}HCH_2CH_2CH_2OH$ $\xleftarrow[(2)\,H_3O^+]{(1)\,LiAlH_4}$ $CH_3\underset{\underset{CH_3}{|}}{C}HCH_2CH_2CO_2H$

[from Problem 19.16(c)]

(c) $CH_3CH_2\underset{\underset{CH_2OH}{|}}{C}HCH_2OH$ $\xleftarrow[(2)\,H_3O^+]{(1)\,LiAlH_4}$ $CH_3CH_2-\underset{\underset{CO_2C_2H_5}{|}}{C}H\overset{CO_2C_2H_5}{}$

[from (a) above]

(d) $HOCH_2CH_2CH_2CH_2OH$ $\xleftarrow[(2)\,H_3O^+]{(1)\,LiAlH_4}$ $HO_2CCH_2CH_2CO_2H$ $\xleftarrow[-CO_2]{heat}$

$\underset{\underset{HO_2C}{}}{\overset{HO_2C}{}}CHCH_2CO_2H$ $\xleftarrow[heat]{HCl}$ $\underset{\underset{C_2H_5O_2C}{}}{\overset{C_2H_5O_2C}{}}CHCH_2CO_2C_2H_5$

$\xleftarrow{}$ $\underset{\underset{C_2H_5O_2C}{}}{\overset{C_2H_5O_2C}{}}CH_2$ + $NaOC_2H_5$ + $BrCH_2CO_2C_2H_5$

19.29 The following reaction took place:

$\underset{O}{\overset{O}{\underset{\|}{CH_3C}}}\underset{O}{\overset{O}{\underset{\|}{CH_2C}}}OC_2H_5$ + $BrCH_2CH_2CH_2Br$ $\xrightarrow{NaOC_2H_5}$ $BrCH_2CH_2CH_2\underset{\underset{O}{\overset{\|}{COC_2H_5}}}{C}H\overset{\overset{CH_3}{|}}{\underset{}{C=O}}$

19.30   (a) $BrCH_2CH_2Br$ + $\underset{\overset{|}{CO_2C_2H_5}}{\overset{CO_2C_2H_5}{CH_2}}$ + $NaOC_2H_5$ ⟶

D
Racemic form

E
Meso compound

(c) $BrCH_2CH_2CH_2CH_2Br$ $\xrightarrow{NaCH(CO_2C_2H_5)_2}$ $BrCH_2CH_2CH_2CH_2CH\overset{\displaystyle CO_2C_2H_5}{\underset{\displaystyle CO_2C_2H_5}{|}}$

$\xrightarrow{KOC(CH_3)_3}$ [cyclopentane with $CO_2C_2H_5$, $CO_2C_2H_5$] $\xrightarrow[\substack{(2)\ H_3O^+ \\ (3)\ heat,\ -CO_2}]{(1)\ OH^-,\ H_2O}$ [cyclopentane with $-CO_2H$]

19.31 (a) $CH_2(CO_2C_2H_5)_2$ + $^-OC_2H_5$ $\rightleftharpoons$ $\ddot{:}CH(CO_2C_2H_5)_2$ + C_2H_5OH

$C_6H_5CH=\!CH-\overset{O}{\overset{||}{C}}OC_2H_5$ + $\ddot{:}CH(CO_2C_2H_5)_2$ $\rightleftharpoons$ $C_6H_5CHCH\!=\!\overset{O}{\overset{||}{C}}OC_2H_5$
$\underset{\displaystyle CH(CO_2C_2H_5)_2}{|}$

$\xrightarrow{+H^+}$ $C_6H_5CHCH_2\text{-}\overset{O}{\overset{||}{C}}OC_2H_5$
$\underset{\displaystyle CH(CO_2C_2H_5)_2}{|}$

(b) $CH_3\overset{..}{N}H_2$ + $CH_2\!=\!CH-\overset{O}{\overset{||}{C}}OCH_3$ $\rightleftharpoons$ $CH_3-\overset{\displaystyle H}{\underset{\displaystyle H}{N^+}}-CH_2-CH\!=\!\overset{O}{\overset{||}{C}}OCH_3$ $\rightleftharpoons$

$CH_3\underset{\displaystyle H}{\overset{\displaystyle |}{N}}-CH_2-CH_2-\overset{O}{\overset{||}{C}}OCH_3$ $\xrightarrow[\displaystyle]{CH_2\!=\!CH-\overset{O}{\overset{||}{C}}OCH_3}$ $CH_3N(CH_2CH_2CO_2CH_3)_2$

$\xrightarrow{base}$ $CH_3-N\overset{\displaystyle CH_2-\overset{-}{C}H\overset{\displaystyle CO_2CH_3}{\underset{\displaystyle}{}}}{\underset{\displaystyle CH_2-CH_2}{\overset{\displaystyle |}{}}}\overset{\displaystyle}{\underset{\displaystyle C}{\overset{\displaystyle O-CH_3}{}}}$ $\overset{\text{Dieckmann}}{\underset{\substack{\text{condensation} \\ \text{(several steps)}}}{\xrightarrow{\hspace{1cm}}}}$ CH_3-N[piperidine ring with CO_2CH_3 and $=O$]

The Michael reaction is reversible, and the reaction just given is an example of a reverse Michael reaction.

19.32 Two reactions take place. The first is a normal Knoevenagel condensation,

Then the α, β-unsaturated diketone reacts with a second mole of the active methylene compound in a Michael addition.

19.37 (a) $ClCH_2CO_2C_2H_5$ + $C_2H_5O^-$ ⇌ $Cl-\overset{..}{\overset{-}{C}}HCO_2C_2H_5$ + C_2H_5OH

(b) Decarboxylation of the epoxy acid gives an enol anion which, on protonation, gives an aldehyde.

(c)

β-Ionone

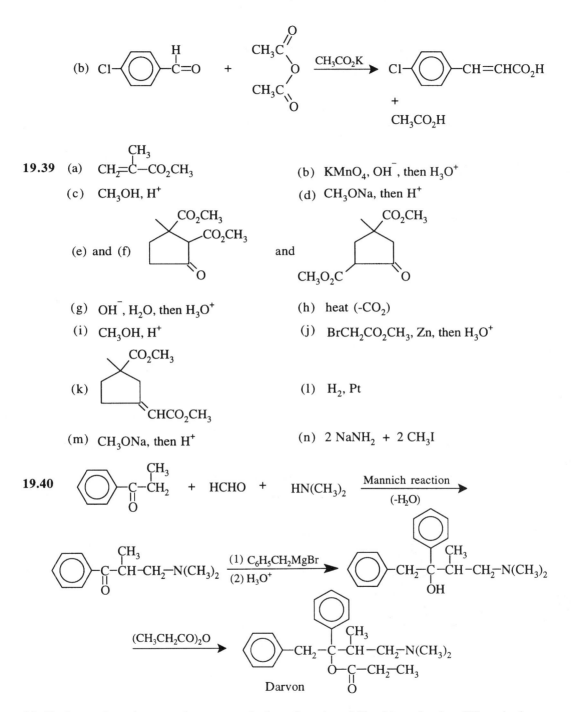

19.39 (a) $CH_2{=}\overset{\underset{\textstyle CH_3}{|}}{C}{-}CO_2CH_3$

(b) $KMnO_4$, OH^-, then H_3O^+

(c) CH_3OH, H^+

(d) CH_3ONa, then H^+

(e) and (f)

and

(g) OH^-, H_2O, then H_3O^+

(h) heat ($-CO_2$)

(i) CH_3OH, H^+

(j) $BrCH_2CO_2CH_3$, Zn, then H_3O^+

(k)

(l) H_2, Pt

(m) CH_3ONa, then H^+

(n) $2\ NaNH_2$ + $2\ CH_3I$

19.40

Darvon

19.41 In a polar solvent, such as water, the keto form is stabilized by solvation. When the interaction with the solvent becomes minimal, the enol form achieves stability by internal hydrogen bonding.

19.42 Intramolecular cyclization (which would give a product of formula $C_6H_8O_3$) is not favored because of ring strain. The formula of the product actually obtained suggests a 1:1 intermolecular reaction:

19.43 A gamma hydrogen is abstracted by base (as is an alpha hydrogen in the usual Claisen) to give a resonance-stabilized species:

Ethyl crotonate differs from ethyl acetate by $-CH=CH-$, a vinyl group. The transmission of the stabilizing effect of the $-COOC_2H_5$ group is an example of the **principle of vinylogy.**

19.44 The synthesis actually involves two sequential Claisen condensations, with ethyl acetate serving as the source of the carbanionic species.

*19.45

*19.46

QUIZ

19.1 Which hydrogen atoms in the following ester are most acidic?

$$\overset{a}{CH_3}-\overset{b}{CH_2}-\overset{O}{\overset{\|}{C}}-\overset{c}{CH_2}-\overset{O}{\overset{\|}{C}}-O\overset{d}{CH_2}-\overset{e}{CH_3}$$

(a) a (b) b (c) c (d) d (e) e

19.2 What would be the product of the following reaction?

$$CH_3CH_2\overset{O}{\overset{\|}{C}}OEt \xrightarrow[\text{(2) H}^+]{\text{(1) NaOC}_2\text{H}_5} \quad ?$$

(a) $CH_3CH_2\overset{O}{\overset{\|}{C}}CH_2CH_2\overset{O}{\overset{\|}{C}}OEt$

(b) $CH_3CH_2\overset{O}{\overset{\|}{C}}CH_2\overset{O}{\overset{\|}{C}}CH_3$

(c) $CH_3CH_2\overset{O}{\overset{\|}{C}}CH_2\overset{O}{\overset{\|}{C}}OEt$

(d) $CH_3\overset{O}{\overset{\|}{C}}CH_2CH_2\overset{O}{\overset{\|}{C}}OEt$

(e) $CH_3CH_2\overset{O}{\overset{\|}{C}}\overset{O}{\overset{\|}{\underset{\underset{CH_3}{|}}{C}}}HCOEt$

19.3 What starting materials could be used in a crossed Claisen condensation to prepare the following compound?

$$EtO-\overset{O}{\overset{\|}{C}}-\overset{O}{\overset{\|}{C}}-\underset{\underset{CH_3}{|}}{CH}-\overset{O}{\overset{\|}{C}}OEt$$

(a) CH_3CO_2Et and $EtO-\overset{O}{\overset{\|}{C}}-\overset{O}{\overset{\|}{C}}-CH_2CH_3$

(d) $EtO_2C\underset{\underset{CH_3}{|}}{CH}CO_2Et$ and HCO_2Et

(b) $CH_3CH_2CO_2Et$ and $EtO_2C—CO_2Et$

(e) More than one of the above

(c) $CH_3CH_2CO_2Et$ and HCO_2Et

19.4 Supply the missing reagents, intermediates, and products.

(a) $CH_3CH_2CH_2\overset{O}{\overset{\|}{C}}OEt$ + $EtO\overset{O}{\overset{\|}{C}}OEt$ $\xrightarrow[\text{(2) H}^+\text{(-EtOH)}]{\text{(1) NaOC}_2\text{H}_5}$

(d)

SOLUTIONS TO PROBLEMS

C.1 (a)

C.2 (a) (b) (c) (d) (e)

C.3 $CH_2=CHCH_2Br$ + ...

C.4 $CH_2=CHCH_2OH$ ...

C.5 (a) $\overset{\overset{\displaystyle O}{\displaystyle \|}}{\text{ClCH}_2\text{CH}_2\text{C}}(\text{CH}_2)_4\text{CO}_2\text{C}_2\text{H}_5$ (This step is the Friedel-Crafts alkylation of an alkene.)

(b) SOCl_2

(c) $2\ \text{C}_6\text{H}_5\text{CH}_2\text{SH}$ and KOH

(d) H_3O^+

(e)

C.6 $\text{H}_2\ddot{\text{S}}:\ +\ \text{H}_2\text{C}{-}\text{CH}_2 \longrightarrow \ddot{\text{H}}\ddot{\text{S}}{-}\text{CH}_2\text{CH}_2\text{OH} \xrightarrow{\ \ \overset{\displaystyle O}{\text{H}_2\text{C}{-}\text{CH}_2}\ \ }$

$$\underset{(\text{C}_4\text{H}_{10}\text{SO}_2)}{\text{HOCH}_2\text{CH}_2\text{SCH}_2\text{CH}_2\text{OH}} \xrightarrow[\text{ZnCl}_2]{\text{HCl}} \underset{\text{Mustard gas}}{\text{ClCH}_2\text{CH}_2\text{SCH}_2\text{CH}_2\text{Cl}}$$

SPECIAL TOPIC
Thiol Esters and Lipid Biosynthesis

SOLUTIONS TO PROBLEM

D.1

Farnesol

Bisabolene

20 AMINES

PREPARATION AND REACTIONS OF AMINES

A. Preparation

1. Preparation via nucleophilic substitution reactions.

2. Preparation through reduction of nitro compounds.

3. Preparation via reductive amination.

$$R \overset{R}{\underset{R'}{\diagdown}} C=O \xrightarrow[\text{[H]}]{NH_3} R-\overset{R'}{\underset{|}{CH}}-NH_2$$

Aldehyde or ketone

$$R \overset{R}{\underset{R'}{\diagdown}} C=O \xrightarrow[\text{[H]}]{R''NH_2} R-\overset{R'}{\underset{|}{CH}}-NHR''$$

$$R \overset{R}{\underset{R'}{\diagdown}} C=O \xrightarrow[\text{[H]}]{R''R'''NH} R-\overset{R'}{\underset{|}{CH}}-NR''R'''$$

4. Preparation of amines through reduction of amides, oximes, and nitriles.

$$RCH_2Br \xrightarrow{CN^-} RCH_2CN \xrightarrow[\text{Ni}]{H_2} RCH_2CH_2NH_2$$

$$R-NH_2 + R'\overset{O}{\underset{\|}{C}}Cl \longrightarrow R-NH\overset{O}{\underset{\|}{C}}R' \xrightarrow[\text{(2) H}_2\text{O}]{\text{(1) LiAlH}_4} RNHCH_2R'$$

$$RNHR' + R''\overset{O}{\underset{\|}{C}}Cl \longrightarrow R''-\overset{O}{\underset{\|}{C}}NRR' \xrightarrow[\text{(2) H}_2\text{O}]{\text{(1) LiAlH}_4} R''CH_2NRR'$$

5. Preparation through the Hofmann rearrangement of amides.

$$\overset{O}{\underset{\|}{RCOH}} \xrightarrow{SOCl_2} \overset{O}{\underset{\|}{RCCl}} \xrightarrow{NH_3} \overset{O}{\underset{\|}{RCNH_2}} \xrightarrow[\text{(NaOBr)}]{\text{Br}_2/\text{NaOH}} RNH_2 + CO_2^{2-}$$

B. Reaction of Amines

1. As a base or a nucleophile

   As a base

$$-\overset{}{\underset{}{N}}: + H\overset{+}{\underset{\underset{H}{|}}{O}}-H \rightleftharpoons -\overset{|}{\underset{|}{N}}\overset{+}{-}H + H_2O$$

   As a nucleophile in alkylation

$$-\overset{}{\underset{}{N}}: + RCH_2-X \longrightarrow -\overset{|}{\underset{|}{N}}\overset{+}{-}CH_2R + X^-$$

   As a nucleophile in acylation

$$\underset{H}{\overset{}{\underset{}{N}}}: + R-\overset{O}{\underset{\|}{C}}-Cl \xrightarrow{(-HCl)} -\overset{|}{\underset{\cdot\cdot}{N}}-\overset{O}{\underset{\|}{C}}-R$$

2. With nitrous acid

$$R-NH_2 \xrightarrow[\text{HX}]{\text{HONO}} R-N_2^+X^- \xrightarrow{-N_2} R^+ \longrightarrow \begin{array}{l}\text{alkenes,}\\\text{alcohols,}\\\text{and so on}\end{array}$$

1° aliphatic   (unstable)

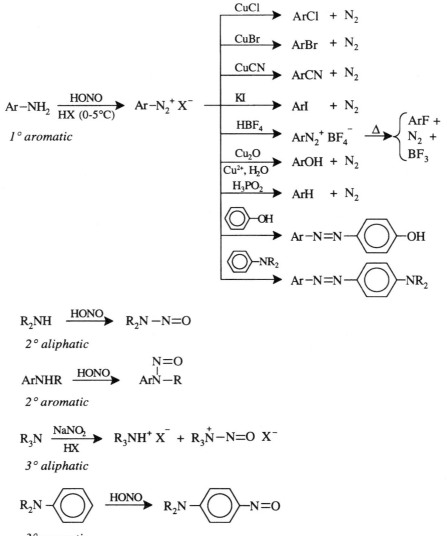

3. With sulfonyl chlorides

$$R-NH_2 + ArSO_2Cl \xrightarrow[(-HCl)]{} RNHSO_2Ar \underset{H^+}{\overset{OH^-}{\rightleftarrows}} [RNSO_2Ar]^- + H_2O$$
  1° amine

$$R_2NH + ArSO_2Cl \xrightarrow[(-HCl)]{} R_2NSO_2Ar$$
  2° amine

4. The Hofmann elimination

SOLUTIONS TO PROBLEMS

20.1 Dissolve both compounds in diethyl ether and extract with aqueous HCl. This procedure gives an ether layer that contains cyclohexane and an aqueous layer that contains hexylaminium chloride. Cyclohexane may then be recovered from the ether layer by distillation. Hexylamine may be recovered from the aqueous layer by adding aqueous NaOH (to convert hexylaminium chloride to the hexylamine) and then by ether extraction and distillation.

20.2 We begin by dissolving the mixture in a water-immiscible organic solvent such as CH_2Cl_2 or diethyl ether. Then, extractions with aqueous acids and bases allow us to separate the components. [We separate 4-methylphenol (*p*-cresol) from benzoic acid by taking advantage of benzoic acid's solubility in the more weakly basic aqueous $NaHCO_3$, whereas *p*-cresol requires the more strongly basic, aqueous NaOH.]

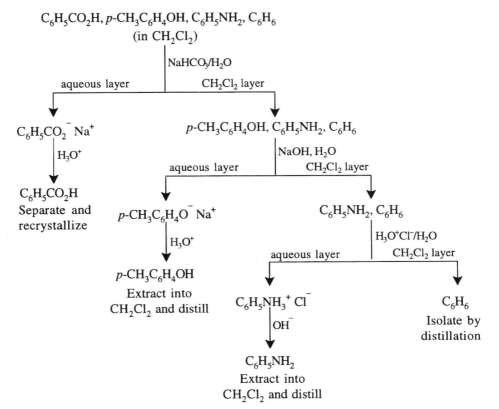

20.3   (a)   Neglecting Kekulé forms of the ring, we can write the following resonance structures for the phthalimide anion.

(b)   Phthalimide is more acidic than benzamide because its anion is stabilized by resonance to a greater extent than the anion of benzamide. (Benzamide has only one carbonyl group attached to the nitrogen atom, and thus fewer resonance contributors are possible.)

(c)

Then,

20.4

20.5 (a) $CH_3(CH_2)_3CHO + NH_3 \xrightarrow[Ni]{H_2} CH_3(CH_2)_3CH_2NH_2$

(b) $C_6H_5CH_2\overset{O}{\underset{\|}{C}}CH_3 + NH_3 \xrightarrow[Ni]{H_2} C_6H_5CH_2\underset{NH_2}{CHCH_3}$

(c) $CH_3(CH_2)_4CHO + C_6H_5NH_2 \xrightarrow[CH_3OH]{LiBH_3CN} CH_3(CH_2)_4CH_2NHC_6H_5$

(d) $C_6H_5CHO + (CH_3)_2NH \xrightarrow[CH_3OH]{LiBH_3CN} C_6H_5CH_2N(CH_3)_2$

20.6 The reaction of a secondary halide with ammonia would inevitably be accompanied by considerable elimination, thus decreasing the yield.

$$R\overset{R'}{\underset{|}{C}H}-X + NH_3 \text{ (excess)} \begin{cases} \xrightarrow{\text{substitution}} R\overset{R'}{\underset{|}{C}H}NH_2 \\ \xrightarrow{\text{elimination}} \text{alkene} \end{cases}$$

20.7 (a) $C_6H_5CO_2H \xrightarrow{SOCl_2} C_6H_5COCl \xrightarrow{CH_3CH_2NH_2}$

$C_6H_5CONHCH_2CH_3 \xrightarrow{LiAlH_4} C_6H_5CH_2NHCH_2CH_3$

(b) $CH_3CH_2CH_2CH_2CH_2Br \xrightarrow{NaCN} CH_3CH_2CH_2CH_2CH_2CN$

$\xrightarrow{LiAlH_4} CH_3CH_2CH_2CH_2CH_2CH_2NH_2$

(c) $CH_3CH_2CO_2H \xrightarrow{SOCl_2} CH_3CH_2COCl \xrightarrow{(CH_3CH_2CH_2)_2NH}$

$CH_3CH_2CON(CH_2CH_2CH_3)_2 \xrightarrow{LiAlH_4} N(CH_2CH_2CH_3)_3$

(d) $CH_3\overset{O}{\underset{\|}{C}}CH_2CH_3 \xrightarrow{NH_2OH} CH_3\overset{NOH}{\underset{\|}{C}}CH_2CH_3 \xrightarrow{Na/ethanol} CH_3\overset{NH_2}{\underset{|}{C}H}CH_2CH_3$

20.8 (a) CH$_3$O—⬡ $\xrightarrow[\text{H}_2\text{SO}_4]{\text{HNO}_3}$ CH$_3$O—⬡—NO$_2$ $\xrightarrow[\text{HCl}]{\text{Fe}}$ CH$_3$O—⬡—NH$_2$

(b) CH$_3$O—⬡ $\xrightarrow[\text{AlCl}_3]{\text{CH}_3\text{COCl}}$ CH$_3$O—⬡—$\overset{\displaystyle O}{\overset{\|}{\text{C}}}CH_3$ $\xrightarrow[\text{Ni}]{\text{NH}_3, \text{H}_2}$

CH$_3$O—⬡—$\underset{\text{NH}_2}{\overset{}{\text{CHCH}_3}}$

(c) ⬡—CH$_3$ $\xrightarrow{\underset{hv}{\text{Cl}_2}}$ ⬡—CH$_2$Cl $\xrightarrow{\text{(CH}_3)_3\text{N}}$ ⬡—CH$_2\overset{+}{\text{N}}$(CH$_3$)$_3$Cl$^-$
(excess)

(d) O$_2$N—⬡—CH$_3$ $\xrightarrow[\text{(2) H}^+]{\text{(1) KMnO}_4, \text{OH}^-}$ O$_2$N—⬡—CO$_2$H $\xrightarrow{\text{SOCl}_2}$

O$_2$N—⬡—$\overset{\displaystyle O}{\overset{\|}{\text{C}}}$—Cl $\xrightarrow{\text{NH}_3}$ O$_2$N—⬡—$\overset{\displaystyle O}{\overset{\|}{\text{C}}}NH_2$ $\xrightarrow[\text{OH}]{\text{Br}_2}$ O$_2$N—⬡—NH$_2$

(e) ⬡—CH$_3$ + NBS $\xrightarrow{\text{ROOR}}$ ⬡—CH$_2$Br $\xrightarrow{\text{KCN}}$

⬡—CH$_2$CN $\xrightarrow[\text{(2)H}_2\text{O}]{\text{(1) LiAlH}_4, \text{Et}_2\text{O}}$ ⬡—CH$_2$CH$_2$NH$_2$

20.9 An amine acting as a base.

CH$_3$CH$_2$N̈H$_2$ + H$_3$O$^+$ ⇌ CH$_3$CH$_2$NH$_3^+$ + H$_2$O

An amine acting as a nucleophile in an alkylation reaction.

(CH$_3$CH$_2$)$_3$N: + CH$_3$—I ⟶ (CH$_3$CH$_2$)$_3\overset{+}{\text{N}}$CH$_3$ I$^-$

An amine acting as a nucleophile in an acylation reaction.

(CH$_3$)$_2$N̈H + CH$_3$C$\overset{\displaystyle O}{\underset{\displaystyle Cl}{}}$ ⟶ (CH$_3$)$_2$N$\overset{\displaystyle O}{\overset{\|}{\text{C}}}CH_3$ + (CH$_3$)$_2\overset{+}{\text{N}}$H$_2$ Cl$^-$
(excess)

An amino group acting as an activating group and as an ortho-para director in electrophilic aromatic substitution.

20.10 (a, b)

$$^-O-N=O + H_3O^+ \rightleftharpoons HO-N=O + H_2O$$

$$HO-N=O + H_3O^+ \rightleftharpoons \overset{+}{HO}-N=O + H_2O \\ | \\ H$$

$$\overset{+}{HO}-N=O \rightleftharpoons H_2O + \overset{+}{N}=O \\ | \\ H$$

(c) The $\overset{+}{N}O$ ion is a weak electrophile. For it to react with an aromatic ring, the ring must have a powerful activating group such as –OH or $-NR_2$.

20.11 (a)

fuming HNO_3 / H_2SO_4, heat → H_2S / NH_3, C_2H_5OH →

(b) [as in (a)]

$HCl/NaNO_2$ / H_2O (0-5°C) → $CuCl$ → (1) Fe, HCl, heat (2) OH^- →

(c) (by nitration of benzene)

Br_2 / $FeBr_3$ → (1) Fe, HCl, heat (2) OH^- →

(d) [as in (c)]

(e) [from part(d)]   (plus a trace of ortho)

20.12 *p*-Toluidine

20.13 (a) Toluene $\xrightarrow[H_2SO_4]{HNO_3}$ *p*-Nitrotoluene (+ *o*-nitrotoluene)

20.14

20.15

Orange II

20.16

20.17

A

B

C

20.18 (1)   That **A** reacts with benzenesulfonyl chloride in aqueous KOH to give a clear solut which on acidification yields a precipitate, shows that **A** is a primary amine.

(2)   That diazotization of **A** followed by treatment with 2-naphthol gives an intensely colc precipitate shows that **A** is a primary aromatic amine; that is, **A** is a substituted aniline.

(3)   Consideration of the molecular formula of **A** leads us to conclude that **A** is a methyl line (i.e., a toluidine).

But is **A** 2-methylaniline, 3-methylaniline, or 4-methylaniline?

(4)   This question is answered by the IR data. A single absorption peak in the 680–840 cr region at 815 cm^{-1} is indicative of a para substituted benzene. Thus, **A** is 4-methylani (*p*-toluidine).

A NH$_2$

20.19 First convert the sulfonamide to its anion, then alkylate the anion with an alkyl halide, then remove the $-SO_2C_6H_5$ group by hydrolysis. For example,

20.20 (a)

Aniline

Sulfathiazole

(b)

Succinylsulfathiazole

20.21 (a) $C_6H_5CH_2NHCH_3$ (b) $(CH_3CH)_3N$ with CH_3

(c)

(d)

(e)

(f)

(g)

(h)

(i)

(j)

(k) CH₃–N⁺–H Cl⁻

(l)

(m) H₂NCH₂CH₂CH₂OH

(n) (CH₃CH₂CH₂)₄N⁺ Cl⁻

(o)

(p) CH₃––N⟨CH₃ CH₃⟩

(q) CH₃O––NH₂

(r) (CH₃)₄N⁺ OH⁻

(s)

(t)

20.22 (a) Propylamine

(b) *N*-Methylaniline

(c) Isopropyltrimethylammonium iodide

(d) 2-Methylaniline (*o*-toluidine)

(e) 2-Methoxyaniline (or *o*-methoxyaniline)

(f) Pyrazole

(g) 2-Aminopyrimidine

(h) Benzylaminium chloride

(i) *N,N*-Dipropylaniline

(j) Benzenesulfonamide

(k) Methylaminium acetate

(l) 3-Amino-1-propanol

(m) Purine

(n) *N*-Methylpyrrole

20.23 (a) C≡N + LiAlH₄ →(2)H₂O→ CH₂NH₂

(b) C–NH₂ + LiAlH₄ →(2)H₂O→ CH₂NH₂

(c), (d), (e), (f), (g) reactions

20.24 (a), (b), (c)

Hofmann rearrangement

20.25 (a)

(b) $CH_3(CH_2)_2CH_2Br \xrightarrow{NaCN} CH_3(CH_2)_3CN \xrightarrow[(2) H_2O]{(1) LiAlH_4, Et_2O} CH_3(CH_2)_3CH_2NH_2$

[from part(a)]

(c) $CH_3(CH_2)_2CH_2OH \xrightarrow[(2) H^+]{(1) KMnO_4, OH^-} CH_3CH_2CH_2CO_2H$

$\xrightarrow[(2) NH_3]{(1) SOCl_2} CH_3CH_2CH_2CONH_2 \xrightarrow[OH^-]{Br_2} CH_3CH_2CH_2NH_2$

(d) $CH_3(CH_2)_2CH_2OH \xrightarrow[CH_2Cl_2]{PCC} CH_3CH_2CH_2CHO \xrightarrow[H_2, Ni]{CH_3NH_2}$

$CH_3CH_2CH_2CH_2NHCH_3$

20.26 (a)

(b)

(c)

20.27 (a) $CH_3CH_2CH_2NH_2$ $\xrightarrow[\text{NaNO}_2/\text{HCl}]{\text{HONO}}$ $[CH_3CH_2CH_2N_2^+]$ $\xrightarrow{-N_2}$

(b) $(CH_3CH_2CH_2)_2NH$ $\xrightarrow[\text{NaNO}_2/\text{HCl}]{\text{HONO}}$ $(CH_3CH_2CH_2)_2N-N=O$

(c)

(d)

(e) $CH_3CH_2CH_2$—⟨⟩—NH_2 $\xrightarrow[\text{NaNO}_2/\text{HCl}]{\text{HONO, 0-5°C}}$ $CH_3CH_2CH_2$—⟨⟩—N_2^+ Cl^-

20.28 (a) $CH_3CH_2CH_2NH_2$ + $C_6H_5SO_2Cl$ $\xrightarrow[\text{H}_2\text{O}]{\text{KOH}}$ $CH_3CH_2CH_2\overset{-}{N}SO_2C_6H_5$
K^+
Clear solution

$\xrightarrow{\text{H}_3\text{O}^+}$ $CH_3CH_2CH_2NHSO_2C_6H_5$
Precipitate

(b) $(CH_3CH_2CH_2)_2NH$ + $C_6H_5SO_2Cl$ $\xrightarrow[\text{H}_2\text{O}]{\text{KOH}}$ $(CH_3CH_2CH_2)_2NSO_2C_6H_5$
Precipitate

$\xrightarrow{\text{H}_3\text{O}^+}$ No reaction
(precipitate remains)

(c)

+ $C_6H_5SO_2Cl$ $\xrightarrow[\text{H}_2\text{O}]{\text{KOH}}$

Precipitate

$\xrightarrow{\text{H}_3\text{O}^+}$ No reaction
(precipitate remains)

(d)

+ $C_6H_5SO_2Cl$ $\xrightarrow[\text{H}_2\text{O}]{\text{KOH}}$ No reaction
(3° amine is insoluble)

$\xrightarrow{\text{H}_3\text{O}^+}$ ⟨⟩—$\overset{+}{N}H(CH_2CH_2CH_3)_2$

3° Amine dissolves

(e)

Clear solution

$\xrightarrow{H_3O^+}$ C_3H_7—⟨benzene⟩—$NHSO_2C_6H_5$

Precipitate

20.29 (a)

(b) ⟨piperidine⟩N–H + $C_6H_5SO_2Cl$ $\xrightarrow[H_2O]{KOH}$ ⟨piperidine⟩N–$SO_2C_6H_5$

20.30 (a) $2\ CH_3CH_2NH_2$ + C_6H_5COCl $\longrightarrow$ $CH_3CH_2NHCOC_6H_5$ +

$CH_3CH_2NH_3^+\ Cl^-$

(b) $2\ CH_3NH_2$ + $(CH_3\overset{O}{\overset{||}{C}})_2O$ $\longrightarrow$ $CH_3NH\overset{O}{\overset{||}{C}}CH_3$ + $CH_3\overset{+}{N}H_3\ CH_3\overset{O}{\overset{||}{C}}O^-$

(c)

(d) [product of (c)] $\xrightarrow{heat}$ + H_2O + CH_3NH_2

(e) ⟨pyrrolidine⟩ + ⟨phthalic anhydride⟩ $\longrightarrow$ ⟨benzene with $\overset{O}{\overset{||}{C}}$–N–pyrrolidine and $\overset{O}{\overset{||}{C}}OH$⟩

(f)

(g) 2

(h)

(i)

(j)

20.31 (a)

Separated

(b) [from part (a)]

(1) Fe, HCl
(2) OH⁻

HONO
0–5°C

KI

(c) [from part (a)]

(1) Fe, HCl, heat
(2) OH⁻

(1) H₂SO₄/NaNO₂
(2) Cu₂O, Cu²⁺, H₂O

(d) [by reduction of *m*-dinitrobenzene, cf. Problem 20.11(a)]

HCl/NaNO₂

CuCl

(e) [from part (d)]

CuCN

(f) [from Problem 20.11(a)]

(1) H₂SO₄/NaNO₂
(2) KI

(1) Fe, HCl, heat
(2) OH⁻

(1) HONO
(2) Cu₂O, Cu²⁺, H₂O

(g) [from Problem 20.11(a)]

(1) HBr/NaNO₂
(2) CuBr

(1) Fe, HCl, heat
(2) OH⁻

(1) H₂SO₄/NaNO₂
(2) CuCN

(h)

[from Problem 20.11(e)]

(i)

[from part (h)]

(j)

[from part (h)]

(k)

[from part (j)]

(l)

[from part (h)]

(q) H₃C—⟨◯⟩—NH₂ [from part (c)]

$\xrightarrow{\text{H}_2\text{SO}_4/\text{NaNO}_2}$ H₃C—⟨◯⟩—N₂⁺ X⁻

⟨◯⟩—OH, pH 8-10

H₃C—⟨◯⟩—N=N—⟨◯⟩—OH

(r) H₃C—⟨◯⟩—N₂⁺ X⁻ [from part (q)]

$\xrightarrow[\text{pH 8-10}]{\text{H}_3\text{C}—⟨◯⟩—\text{OH} \; [\text{from part (c)}]}$ H₃C—⟨◯⟩—N=N—⟨◯⟩(HO, CH₃)

20.32 (a) Benzylamine dissolves in dilute HCl at room temperature,

$$C_6H_5CH_2NH_2 \;+\; H_3O^+ \;+\; Cl^- \xrightarrow{25°C} C_6H_5CH_2\overset{+}{N}H_3 \; Cl^-$$

benzamide does not dissolve:

$$C_6H_5CONH_2 \;+\; H_3O^+ \;+\; Cl^- \xrightarrow{25°C} \text{No reaction}$$

(b) Allylamine reacts with (and decolorizes) bromine in carbon tetrachloride instantly,

$$CH_2{=}CHCH_2NH_2 \;+\; Br_2 \xrightarrow{CCl_4} \underset{\underset{Br}{|} \; \underset{Br}{|}}{CH_2CHCH_2NH_2}$$

propylamine does not:

$$CH_3CH_2CH_2NH_2 \;+\; Br_2 \xrightarrow{CCl_4} \text{No reaction if the mixture is not heated or irradiated}$$

(c) The Hinsberg test:

H₃C—⟨◯⟩—NH₂ + C₆H₅SO₂Cl $\xrightarrow[\text{H}_2\text{O}]{\text{KOH}}$ H₃C—⟨◯⟩—N̄SO₂C₆H₅ (K⁺) $\xrightarrow{\text{H}_3\text{O}^+}$

Soluble

H₃C—⟨◯⟩—NHSO₂C₆H₅

Precipitate

(d)   The Hinsberg test:

(e)   Pyridine dissolves in dilute HCl,

benzene does not:

(f)   Aniline reacts with nitrous acid at 0-5°C to give a stable diazonium salt that couples with 2-naphthol, yielding an intensely colored azo compound.

Cyclohexylamine reacts with nitrous acid at 0-5°C to yield a highly unstable diazonium salt—one that decomposes so rapidly that the addition of 2-naphthol gives no azo compound.

(g)   The Hinsberg test:

$$(C_2H_5)_3N \ + \ C_6H_5SO_2Cl \ \xrightarrow[\text{H}_2\text{O}]{\text{KOH}} \ \text{No reaction} \ \xrightarrow{\text{H}_3\text{O}^+} \ (C_2H_5)_3\overset{+}{N}H$$
$$\qquad\qquad\qquad\qquad\qquad\qquad\qquad\qquad\qquad\qquad \text{Soluble}$$

$$(C_2H_5)_2NH \ + \ C_6H_5SO_2Cl \ \xrightarrow[\text{H}_2\text{O}]{\text{KOH}} \ (C_2H_5)_2NSO_2C_6H_5 \ \xrightarrow{\text{H}_3\text{O}^+} \ \begin{array}{l}\text{Precipitate}\\ \text{remains}\end{array}$$
$$\qquad\qquad\qquad\qquad\qquad\qquad\qquad \text{Precipitate}$$

(h)   Tripropylaminium chloride reacts with aqueous NaOH to give a water insoluble tertiary amine.

$$(CH_3CH_2CH_2)_3NH^+ \ Cl^- \ \xrightarrow[\text{H}_2\text{O}]{\text{NaOH}} \ (CH_3CH_2CH_2)_3N$$
$$\quad\ \text{Water soluble} \qquad\qquad\qquad\qquad\quad \text{Water insoluble}$$

Tetrapropylammonium chloride does not react with aqueous NaOH (at room temperature), and the tetrapropylammonium ion remains in solution.

$$(CH_3CH_2CH_2)_4N^+ \ Cl^- \ \xrightarrow[\text{H}_2\text{O}]{\text{NaOH}} \ (CH_3CH_2CH_2)_4N^+ \ [Cl^- \text{ or } OH^-]$$
$$\quad\ \text{Water soluble} \qquad\qquad\qquad\qquad\qquad \text{Water soluble}$$

(i)   Tetrapropylammonium chloride dissolves in water to give a neutral solution. Tetrapropylammonium hydroxide dissolves in water to give a strongly basic solution.

20.33   Follow the procedure outlined in the answer to Problem 20.2. Toluene will show the same solubility behavior as benzene.

20.34

20.35   (a) $HOCH_2(CH_2)_8CH_2OH \ \xrightarrow{2\ PBr_3} \ BrCH_2(CH_2)_8CH_2Br \ \xrightarrow{2(CH_3)_3N}$

$$(CH_3)_3\overset{+}{N}CH_2(CH_2)_8CH_2\overset{+}{N}(CH_3)_3 \ 2\ Br^-$$

(b) $HO_2CCH_2CH_2CO_2H \ + \ 2 \ BrCH_2CH_2OH \ \xrightarrow[(-2\ \text{H}_2\text{O})]{\text{H}^+}$

$$BrCH_2CH_2O_2CCH_2CH_2CO_2CH_2CH_2Br \ \xrightarrow{2(CH_3)_3N}$$

$$(CH_3)_3\overset{+}{N}CH_2CH_2O_2CCH_2CH_2CO_2CH_2CH_2\overset{+}{N}(CH_3)_3 \ 2\ Br^-$$

20.36

20.37 The results of the Hinsberg test indicate that compound **W** is a tertiary amine. The ^{1}H NMR provides evidence for the following:

(1) Two different C_6H_5–groups (one absorbing at δ 7.2 and one at δ 6.7).
(2) A CH_3CH_2–group (the quartet at δ 3.5 and the triplet at δ 1.2).
(3) An unsplit –CH_2–group (the singlet at δ 4.5).

There is only one reasonable way to pull all of this together.

Thus **W** is *N*-benzyl-*N*-ethylaniline.

20.38 Compound **X** is benzyl bromide, $C_6H_5CH_2Br$. This is the only structure consistent with the ^{1}H NMR and IR data. (The monosubstituted benzene ring is strongly indicated by the (5H), δ 7.3 ^{1}H NMR absorption and is confirmed by the peaks at 690 and 770 cm^{-1} in the IR spectrum.)

Compound **Y**, therefore, must be phenylacetonitrile, $(C_6H_5CH_2CN)$ and **Z** must be 2-phenylethylamine, $C_6H_5CH_2CH_2NH_2$.

Interpretations of the IR and ¹H NMR spectra of **Z** are as follows.

(*a*)   singlet δ 1.0
(*b*)   triplet δ 2.7
(*c*)   triplet δ 2.9
(*d*)   multiplet δ 7.25

20.39

20.40

20.41 That **A** contains nitrogen and is soluble in dilute HCl suggests that **A** is an amine. The two IR absorption bands in the 3300–3500-cm^{-1} region suggest that **A** is a primary amine. The ^{13}C spectrum shows only two signals in the upfield aliphatic region. There are four signals downfield in the aromatic region. The information from the DEPT spectra suggests an ethyl group *or two equivalent ethyl groups*. Assuming the latter, and assuming that **A** is a primary amine, we can conclude from the molecular formula that **A** is 2,6-diethylaniline. The assignments are

(a) δ 12.9
(b) δ 24.2
(c) δ 118.1
(d) δ 125.9
(e) δ 127.4
(f) δ 141.5

(An equally plausible answer would be that **A** is 3,5-diethylaniline.)

20.42 That **B** dissolves in dilute HCl suggests that **B** is an amine. That the IR spectrum of **B** lacks bands in the 3300–3500-cm^{-1} region suggests that **B** is a tertiary amine. The upfield signals in the ^{13}C spectrum and the DEPT information suggest two equivalent ethyl groups (as was also true of **A** in the preceding problem). The DEPT information for the downfield peaks (in the aromatic region) is consistent with a monosubstituted benzene ring. Putting all of these observations together with the molecular formula leads us to conclude that **B** is *N,N*-diethylaniline. The assignments are

(b) (a)
N(CH$_2$CH$_3$)$_2$
(f)
(c)
(e)
(d)

(a) δ 12.5
(b) δ 44.2
(c) δ 112.0
(d) δ 115.5
(e) δ 128.1
(f) δ 147.8

20.43 That **C** gives a positive Tollens' test indicates the presence of an aldehyde group; the solubility of **C** in aqueous HCl suggests that **C** is also an amine. The absence of bands in the 3300–3500-cm^{-1} region of the IR spectrum of **C** suggests that **C** is a tertiary amine. The signal at δ 189.7 in the ^{13}C spectrum can be assigned to the aldehyde group. The signal at δ 39.7 is the only one in the aliphatic region and is consistent with a methyl group or with two equivalent methyl groups. The remaining signals are in the aromatic region. If we assume that **C**

has a benzene ring containing a $-\overset{\overset{\displaystyle O}{\displaystyle \|}}{C}H$ group and a $-N(CH_3)_2$ group, then the aromatic signals and their DEPT spectra are consistent with **C** being *p*-(*N,N*-dimethylamino)benzaldehyde. The assignments are

(a) δ 39.7
(b) δ 110.8
(c) δ 124.9
(d) δ 131.6
(e) δ 154.1
(f) δ 189.7

20.44

20.45

20.46

Diethylpropion

20.47 Carry out the Hofmann reaction using a mixture of ^{15}N labeled benzamide, $C_6H_5CO*NH_2$, and *p*-chlorobenzamide. If the process is intramolecular, only labeled aniline, $C_6H_5*NH_2$, and *p*-chloroaniline, will be produced.

   If the process is one in which the migrating moiety truly separates from the remainder of the molecule, then, in addition to the two products mentioned above, there should be produced both the unlabeled aniline and labeled *p*-chloroaniline, *p*-ClC$_6$H$_4$*NH$_2$.

Note: When this experiment is actually carried out, analysis of the reaction mixture by mass spectrometry shows that the process is intramolecular.

20.48

*20.52   An abbreviated version of one of several possible routes:

"triacetoneamine"

QUIZ

20.1   Which of the following would be soluble in dilute aqueous HCl?

(a) $C_6H_5NH_2$

(b) $C_6H_5CH_2NH_2$

(c) $C_6H_5\overset{\overset{\displaystyle O}{\|}}{C}NH_2$

(d)   Two of the above

(e)   All of the above

20.2   Which would yield propylamine?

(a) $CH_3CH_2Br \xrightarrow[\text{(2) LiAlH}_4]{\text{(1) NaCN}}$

(b) $CH_3CH_2\overset{\overset{\displaystyle O}{\|}}{C}H \xrightarrow[\text{H}_2/\text{Ni}]{\text{NH}_3}$

(c) $CH_3CH_2CH_2\overset{\overset{\displaystyle O}{\|}}{C}NH_2 \xrightarrow[\text{OH}^-]{\text{Br}_2}$

(d)   Two of the above

(e)   All of the above

20.3 Select the reagent from the list below that could be the basis of a simple chemical test that would distinguish between each of the following:

(a) C$_6$H$_5$$\overset{\overset{\displaystyle O}{\|}}{C}NH_2$ and   C$_6$H$_5$CH$_2$NH$_2$

(b) C$_6$H$_5$CH$_2$NH$_2$ and   C$_6$H$_5$NHCH$_3$

(c)

1. Cold dilute NaHCO$_3$
2. Cold dilute HCl
3. NaNO$_2$, HCl, 5°C, then 2-naphthol
4. C$_6$H$_5$SO$_2$Cl, OH$^-$, then HCl
5. Cold dilute NaOH

20.4 Complete the following syntheses:

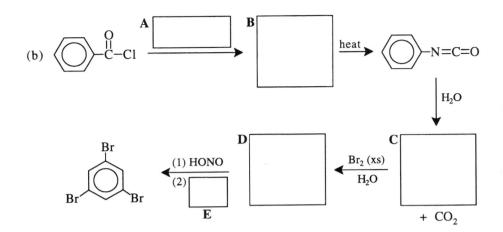

20.5 Select the stronger base from each pair:

SPECIAL TOPIC
Reactions and Synthesis of Heterocyclic Amines

SOLUTIONS TO PROBLEMS

E.1 (a), (b), (c), (d), (e)

E.2 (a) The cyclopentadienyl anion.

(b) The pyrrole anion is a resonance hybrid of the following structures:

The imidazole anion is a hybrid of these:

E.3 A mechanism involving a "pyridyne" intermediate would involve a net loss (of 50%) of the deuterium label.

Since in the actual experiment there was no loss of deuterium, this mechanism was disallowed.

The mechanism given in Section E.4 would not be expected to result in a loss of deuterium; thus, it is consistent with the labeling experiment.

E.4   When pyridine undergoes nucleophilic substitution, the leaving group is a hydride ion—an ion that is a strong base and, consequently, a poor leaving group. With 2-halopyridines, on the other hand, the leaving groups are halide ions—ions that are weak bases and thus good leaving groups.

E.5   If we write the reactants in the following way, we can better see how the reaction occurs.

E.6 (a)

SPECIAL TOPIC
Alkaloids

SOLUTIONS TO PROBLEMS

F.1   (a)   The first step is similar to a crossed Claisen condensation (see Section 19.2A):

(b)   This step involves hydrolysis of an amide (lactam) and can be carried out with either acid or base. Here we use acid.

(c)   This step is the decarboxylation of a substituted malonic acid; it requires only the application of heat and takes place during the acid hydrolysis of step (b).

(d)   This is the reduction of a ketone to a secondary alcohol. A variety of reducing agents can be used, sodium borohydride, for example.

(e)   Here we convert the secondary alcohol to an alkyl bromide with hydrogen bromide; this reagent also gives a hydrobromide salt of the aliphatic amine.

(f)   Treating the salt with base produces the secondary amine; it then acts as a nucleophile and attacks the carbon atom bearing the bromine. This reaction leads to the formation of a five-membered ring and (±) nicotine.

F.2   (a)   The stereocenter adjacent to the ester carbonyl group is racemized by base (probably through the formation of an anion that can undergo inversion of configuration; cf. Section 17.3).

(b)

F.3   (a)

Tropine   (±) Tropic acid

(b)   Tropine is a meso compound; it has a plane of symmetry that passes through the $\rangle$CHOH group, the $\rangle$NCH$_3$ group, and between the two –CH$_2$ – groups of the five-membered ring.

(c)

ψ-Tropine

F.4

Tropine $C_8H_{13}N$ $C_9H_{16}NI$

$C_9H_{15}N$ $C_{10}H_{18}NI$

F.5   One possible sequence of steps is the following:

Tropinone

F.6

$C_{20}H_{25}NO_5$

Dihydropapaverine

Papaverine

F.7 A Diels-Alder reaction was carried out using 1,3-butadiene as the diene component.

F.8 Acetic anhydride acetylates both –OH groups.

Heroin

F.9 (a) A Mannich reaction (see Section 19.10).

(b) CH_2O + $HN(CH_3)_2$ $\rightleftharpoons$ ($-H_2O$, $+H^+$) $CH_2\overset{+}{=}N(CH_3)_2$

Gramine

21

PHENOLS AND ARYL HALIDES: NUCLEOPHILIC AROMATIC SUBSTITUTION

SOLUTIONS TO PROBLEMS

21.1 An electron-releasing group (i.e., $-CH_3$) changes the charge distribution in the molecule so as to make the hydroxyl oxygen less positive, causing the proton to be held more strongly; it also destabilizes the phenoxide anion by intensifying its negative charge. These effects make the substituted phenol less acidic than phenol itself.

Electron-releasing $-CH_3$ destabilizes the anion more than the acid—pK_a is larger than for phenol.

21.2 An electron-withdrawing group such as chlorine changes the charge distribution in the molecule so as to make the hydroxyl oxygen more positive, causing the proton to be held less strongly; it also can stabilize the phenoxide ion by dispersing its negative charge. These effects make the substituted phenol more acidic than phenol itself.

Electron-withdrawing chlorine stabilizes the anion by dispersing the negative charge. pK_a is smaller than for phenol.

Nitro groups are very powerful electron-withdrawing groups by their inductive and resonance effects. Resonance structures (**B–D**) below place a positive charge on the hydroxyl oxygen. This effect makes the hydroxyl oxygen dramatically more positive, causing the proton to be held much less strongly. These contributions explain why 2,4,6-trinitrophenol (picric acid) is so exceptionally acidic.

21.3   Dissolve the mixture in a solvent such as CH_2Cl_2 (one that is immiscible with water). Using a separatory funnel, extract this solution with an aqueous solution of sodium bicarbonate. This extraction will remove the benzoic acid from the CH_2Cl_2 solution and transfer it (as a sodium benzoate) to the aqueous bicarbonate solution. Acidification of this aqueous extract will cause benzoic acid to precipitate; it can then be separated by filtration and purified by recrystallization.

   The CH_2Cl_2 solution can now be extracted with an aqueous solution of sodium hydroxide. This will remove the 4-methylphenol (as its sodium salt). Acidification of the aqueous extract will cause the formation of 4-methylphenol as a water-insoluble layer. The 4-methylphenol can then be extracted into ether, the ether removed, and the 4-methylphenol purified by distillation.

   The CH_2Cl_2 solution will now contain only toluene (and CH_2Cl_2). These can be separated easily by fractional distillation.

21.4   (a)   The para-sulfonated phenol, because it is the major product at the higher temperature—when the reaction is under equilibrium control.

   (b)   For ortho sulfonation, because it is the major reaction pathway at the lower temperature—when the reaction is under rate control.

21.5 If the mechanism involved dissociation into an allyl cation and a phenoxide ion, then re-combination would lead to two products: one in which the labeled carbon atom is bonded to the ring and one in which an unlabeled carbon atom is bonded to the ring.

The fact that all of the product has the labeled carbon atom bonded to the ring eliminates this mechanism from consideration.

21.9   (a)

21.10   That *o*-chlorotoluene leads to the formation of two products (*o*-cresol and *m*-cresol) when submitted to the conditions used in the Dow process suggests that an elimination-addition mechanism takes place.

Inasmuch as chlorobenzene and *o*-chlorotoluene should have similar reactivities under these conditions, it is reasonable to assume that chlorobenzene reacts by an elimination-addition mechanism in the Dow process.

21.11   2-Bromo-1,3-dimethylbenzene, because it has no α-hydrogen, cannot undergo an elimination. Its lack of reactivity toward sodium amide in liquid ammonia suggests that those compounds (e.g., bromobenzene) that do react do so by a mechanism that begins with an elimination.

21.12

(c) (major)

(d)

(e)

(f)

(g)

(h)

(i) Same as (h)

(j)

(k)

(l) Same as (k)

(m)

21.14 (a) 4-Chlorophenol will dissolve in aqueous NaOH; 4-chloro-1-methylbenzene will not.

(b) 4-Methylbenzoic acid will dissolve in aqueous $NaHCO_3$; 4-methylphenol will not.

(c) Phenyl vinyl ether will react with bromine in carbon tetrachloride by addition (thus decolorizing the solution); ethyl phenyl ether will not.

(d) 2,4,6-Trinitrophenol, because it is so exceptionally acidic ($pK_a = 0.38$), will dissolve in aqueous $NaHCO_3$; 4-methylphenol ($pK_a = 10.17$) will not.

(e) 4-Ethylphenol will dissolve in aqueous NaOH; ethyl phenyl ether will not.

21.15 Both *o*- and *m*-toluidine are formed in the following way:

Both *m*- and *p*-toluidine are formed from another benzyne-type intermediate.

21.16 (a)   4-Fluorophenol because a fluorine substituent is more electron withdrawing than a methyl group.

(b)   4-Nitrophenol because a nitro group is more electron withdrawing than a methyl group.

(c)   4-Nitrophenol because the nitro group can exert an electron-withdrawing effect through a resonance effect.

Whereas in 3-nitrophenol this resonance effect is not possible.

(d)   4-Methylphenol because it is a phenol, not an alcohol.

(e)   4-Fluorophenol because fluorine is more electronegative than bromine.

21.17 (a)

(b)   Because the phenolic radical is highly stabilized by resonance (see the following structures), it is relatively unreactive.

21.18

Dibenzo-18-crown-6

21.19 **X** is a phenol because it dissolves in aqueous NaOH but not in aqueous NaHCO$_3$. It gives
a dibromo derivative and must therefore be substituted in the ortho or para position. The
broad IR peak at 3250 cm^{-1} also suggests a phenol. The peak at 830 cm^{-1} indicates para
substitution. The ^{1}H NMR singlet at δ 1.3 (9H) suggests nine equivalent methyl hydrogen atoms, which must be a *tert*-butyl group. The structure of **X** is

21.20 (a)

BHA

(b)

BHT

Notice that both reactions are Friedel-Crafts alkylations.

21.21

2,4-D

21.22 The broad IR peak at 3200–3600 cm^{-1} suggests a hydroxyl group. The two ^{1}H NMR peaks at δ 1.67 and δ 1.74 are not a doublet because their separation is not equal to other splittings; therefore, these peaks are singlets. Reaction with Br$_2$/CCl$_4$ suggests an alkene. If we put these bits of information together, we conclude that **Z** is 3-methyl-2-buten-1-ol.

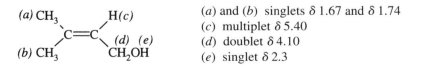

(a) and (b) singlets δ 1.67 and δ 1.74
(c) multiplet δ 5.40
(d) doublet δ 4.10
(e) singlet δ 2.3

21.23

Toliprolol

21.24 The proximity of the two -OH groups results in the two naphthalene nuclei being non-coplanar. As a result, the two enantiomeric forms are nonequivalent and can be separated by a resolution technique.

21.25 One can draw a resonance structure for the 2,6-di-*tert*-butylphenoxide ion which shows the carbon para to the oxygen as a nucleophilic center; that is, the species is an ambident nucleophile.

Given the steric hindrance about the oxygen, the nucleophilic character of the para carbon is dominant and an S$_N$Ar reaction occurs at that position to produce this biphenyl derivative:

21.26 The phenoxide ion has nucleophilic character both at the oxygen and at the carbon para to it; it is ambient.

The benzyne produced by the elimination phase of the reaction can undergo addition then by attack by either nucleophilic center, reaction at oxygen producing 1 and reaction at carbon producing **2**.

21.27 H^- is expected to be a very poor leaving group because of the strong basicity of the hydride ion. In this case the reaction is favored not only by the activating nitro groups but also by the oxidant ferricyanide ion, which converts H^-, as it forms, to H_2.

QUIZ

21.1   Which of the following would be the strongest acid?

(a) O_2N—⬡—OH   (b) H_3C—⬡—OH   (c) ⬡—OH

(d) H_3CH_2C—⬡—OH   (e) ⬡—OH

21.2   What products would you expect from the following reaction?

(a) [OCH₃, NH₂ structure] alone   (b) [OCH₃, NH₂ structure] alone   (c) [OCH₃, NH₂ structure] alone

(d)   More than one of the above   (e)   All of the above

21.3   Which of the reagents listed here would serve as the basis for a simple chemical test to distinguish between

H_3C—⬡—OH and $(CH_3)_3CCH_2OH$?

(a) $Ag(NH_3)_2OH$ (d)   Cold concd. H_2SO_4

(b) $NaOH/H_2O$ (e)   None of the above

(c)   Dilute HCl

21.4   Indicate the correct product, if any, of the following reaction.

H_3C—⬡—OH + HBr ⟶ ?

(a) H_3C—⬡—Br   (b) H_3C—⬡—OH (with Br)   (c) H_3C—⬡—OH (with Br)

(d) H_3C—⬡—OH (with Br, Br)   (e)   There is no reaction.

21.5   Complete the following synthesis:

21.6   Give the products:

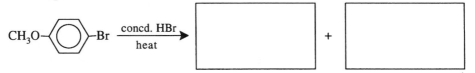

21.7   Select the stronger acid.

ANSWERS TO
SECOND REVIEW PROBLEM SET

The problems review concepts from Chapters 13-21.

1. Increasing acidity ➔

(a) $CH_3\overset{O}{\overset{\|}{C}}CH_3$ < CH_3CH_2OH < $CH_3O\overset{O}{\overset{\|}{C}}CH_2\overset{O}{\overset{\|}{C}}OCH_3$ < $CH_3\overset{O}{\overset{\|}{C}}OH$

(b) [cyclohexyl]$-C\equiv CH$ < [cyclohexyl]$-OH$ < [phenyl]$-OH$ < [cyclohexyl]$-\overset{O}{\overset{\|}{C}}OH$

(c) $(CH_3)_3C-$[phenyl]$-\overset{O}{\overset{\|}{C}}OH$ < [phenyl]$-\overset{O}{\overset{\|}{C}}OH$ < $(CH_3)_3\overset{+}{N}-$[phenyl]$-\overset{O}{\overset{\|}{C}}OH$

(d) $CH_3CH_2\overset{O}{\overset{\|}{C}}OH$ < $CH_3CHCl\overset{O}{\overset{\|}{C}}OH$ < $CH_3CCl_2\overset{O}{\overset{\|}{C}}OH$

(e) [phenyl]$-NH_2$ < [phenyl]$-\overset{O}{\overset{\|}{C}}NH_2$ < [phthalimide ring with two C=O and NH]

2. Increasing basicity ➔

(a) $CH_3\overset{O}{\overset{\|}{C}}NH_2$ < NH_3 < $CH_3CH_2NH_2$

(b) [phenyl]$-NH_2$ < H_3C-[phenyl]$-NH_2$ < [cyclohexyl]$-NH_2$

(c) O_2N-[phenyl]$-NH_2$ < [phenyl]$-NH_2$ < H_3C-[phenyl]$-NH_2$

(d) $CH_3CH_2CH_3$ < CH_3OCH_3 < CH_3NHCH_3

450

5.

2-Methyl-1,3-butadiene   +   Diethyl fumarate   →   Diels-Alder reaction   **A** + enantiomer

(1) LiAlH₄ / (2) H₃O⁺   →   **B** + enantiomer

PBr₃   →   **C** + enantiomer

Zn / H⁺   →   **D** + enantiomer

6. (a) **A** is $CH_2=CHC(CH_3)(OH)C\equiv CH$ **C** is $BrMgOCH_2CH=C(CH_3)C\equiv CMgBr$

(b) **A** is an allylic alcohol and thus forms a carbocation readily. **B** is a conjugated enyne and is therefore more stable than **A**.

Vitamin A acetate

Bisphenol A

Procaine

Ethinamate

11. (a) $C_6H_5\overset{\overset{O}{\|}}{C}H$ $\xrightarrow[\text{(2) } H_3O^+]{\text{(1) } C_6H_5MgBr}$ $\underset{\underset{C_6H_5}{|}}{C_6H_5CHOH}$ $\xrightarrow{PBr_3}$ $\underset{\underset{C_6H_5}{|}}{C_6H_5CHBr}$
                                             A   **B**

$\xrightarrow[\text{-HBr}]{(CH_3)_2NCH_2CH_2OH}$ $\underset{\underset{C_6H_5}{|}}{C_6H_5CHOCH_2CH_2N(CH_3)_2}$

Diphenhydramine

(b) The last step probably takes place by an S_N1 mechanism. Diphenylmethyl bromide, **B**, ionizes readily because it forms the resonance-stabilized benzylic carbocation,

$$\underset{\underset{C_6H_5}{|}}{C_6H_5\overset{+}{C}H}$$

12. (a) For this synthesis we need to prepare the benzylic halide, $Br-\langle\bigcirc\rangle-\underset{\underset{C_6H_5}{|}}{CHBr}$, and then allow it to react with $(CH_3)_2NCH_2CH_2OH$ as in Problem 11.

This benzylic halide can be made as follows:

(b) For this synthesis we can prepare the requisite benzylic halide in two ways:

$\underset{\overset{|}{\underset{CH}{\overset{\|}{O}}}}{\overset{CH_3}{\bigcirc}}$ $\xrightarrow[\text{(2) } H_3O^+]{\text{(1) } C_6H_5MgBr}$ $\underset{\underset{C_6H_5}{|}}{\overset{CH_3}{\bigcirc}-CHOH}$ $\xrightarrow{PBr_3}$ $\underset{\underset{C_6H_5}{|}}{\overset{CH_3}{\bigcirc}-CHBr}$

or

$C_6H_5\overset{\overset{O}{\|}}{C}H$ $\xrightarrow[\text{(2) } H_3O^+]{(1) \overset{CH_3}{\bigcirc}-MgBr}$ $\underset{\underset{C_6H_5}{|}}{\overset{CH_3}{\bigcirc}-CHOH}$ $\xrightarrow{PBr_3}$ $\underset{\underset{C_6H_5}{|}}{\overset{CH_3}{\bigcirc}-CHBr}$

We shall then allow the benzylic halide to react with $(CH_3)_2NCH_2CH_2OH$ as in Problem 11.

13. $\underset{\underset{Br}{|}}{CH_3CHCO_2C_2H_5}$ $\xrightarrow{CH_2(CO_2C_2H_5)_2, \text{ EtO}^-}$ $\underset{\underset{\underset{\underset{CO_2C_2H_5}{}}{C_2H_5O_2C}}{\overset{|}{CH}}}{\overset{C_2H_5O_2C}{\underset{|}{\overset{\diagdown}{C}H}\diagup}\overset{CH_3}{\diagup}}$ $\xrightarrow[\text{(Michael addition)}]{CH_2=CHCN, \text{ EtO}^-}$

                                                     A

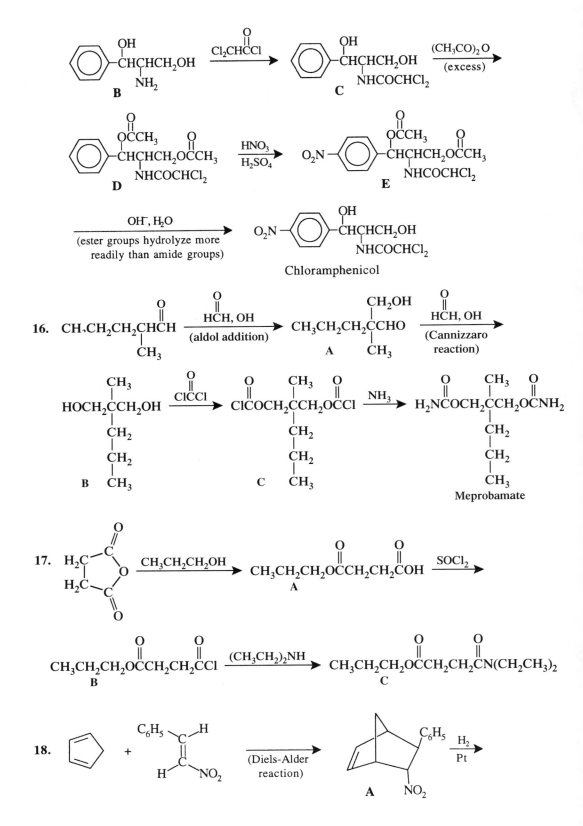

Chloramphenicol

16.

Meprobamate

17.

18.

Fencamfamine

19.

Infrared band in 3200-3550-cm^{-1} region

Infrared band in 1650-1730-cm^{-1} region

Notice that the second step involves the oxidation of a secondary alcohol in the presence of a tertiary alcohol. This selectivity is possible because tertiary alcohols do not undergo oxidation readily (Section 11.8)

20. Working backward, we notice that methyl *trans*-4-isopropylcyclohexanecarboxylate has both large groups equatorial and is, therefore, more stable than the corresponding cis isomer. This stability of the trans isomer means that, if we were to synthesize the cis isomer or a mixture of both the cis and trans isomers, we could obtain the desired trans isomer by a base-catalyzed isomerization (epimerization):

CO_2CH_3 ← CO_2CH_3

$CH(CH_3)_2$ $CH(CH_3)_2$

(more stable   (cis isomer or mixture
trans isomer)   of cis and trans isomers)

We could synthesize a mixture of the desired isomers from phenol in the following way:

21. That **Y** gives a green opaque solution when treated with CrO_3 in aqueous H_2SO_4 indicates that **Y** is a primary or secondary alcohol. That **Y** gives a negative iodoform test indicates that **Y** does not contain the grouping $-\underset{\underset{OH}{|}}{C}HCH_3$. The ^{13}C spectrum of **Y** contains only four signals indicating that some of the carbons in **Y** are equivalent. The information from DEPT spectra helps us conclude that **Y** is 2-ethyl-1-butanol.

$$\underset{(CH_3CH_2)_2CHCH_2OH}{\overset{(a)\ (b)\quad (c)\ (d)}{}}$$

(a) δ 11.1
(b) δ 23.0
(c) δ 43.6
(d) δ 64.6

Notice that the most downfield signal is a CH_2 group. This indicates that this carbon atom bears the –OH group and that **Y** is a primary alcohol. The most upfield signals indicate the presence of the ethyl groups.

22. That **Z** decolorizes bromine in CCl_4 indicates that **Z** is an alkene. We are told that **Z** is the more stable isomer of a pair of stereoisomers. This fact suggests that **Z** is a *trans* alkene. That the ^{13}C spectrum contains only three signals, even though **Z** contains eight carbon atoms, indicates that **Z** is highly symmetric. The information from DEPT spectra indicates that the upfield signals of the alkyl groups arise from equivalent isopropyl groups. We conclude, therefore, that **Z** is *trans*-2,5-dimethyl-3-hexene.

(a) δ 22.8
(b) δ 31.0
(c) δ 134.5

23.

SPECIAL TOPIC
Electrocyclic and Cycloaddition Reactions

SOLUTIONS TO PROBLEMS

G.1   According to the Woodward-Hoffmann rule for electrocyclic reactions of $4n$ π-electron systems (Section G.2A), the photochemical cyclization of *cis, trans*-2,4-hexadiene should proceed with *disrotatory motion.* Thus, it should yield *trans*-3,4-dimethylcyclobutene:

  cis, trans-2,4-Hexadiene   *trans*-3,4-Dimethylcyclobutene

G.2   (a)

ψ_2 of a hexadiene
(Section G.2A)

(b)   This is a thermal electrocyclic reaction of a $4n$ π-electron system; it should, *and does,* proceed with conrotatory motion.

G.3

  trans, trans-2,4-Hexadiene   *cis*-3,4-Dimethylcyclobutene

  cis, trans-2,4-Hexadiene

Here we find that two consecutive electrocyclic reactions (the first photochemical, the second thermal), provide a stereospecific synthesis of *cis,trans*-2,4-hexadiene from *trans, trans*-2,4-hexadiene.

G.4   (a)   This is a photochemical electrocyclic reaction of an eight π-electron system—a 4n π-electron system where n = 2. It should, therefore, proceed with disrotatory motion.

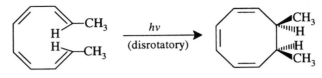

cis-7, 8-Dimethyl-1, 3, 5-cyclooctatriene

(b)   This is a thermal electrocyclic reaction of the eight π-electron system. It should proceed with conrotatory motion.

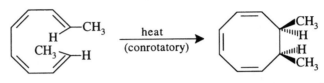

cis-7, 8-Dimethyl-1, 3, 5-cyclooctatriene

G.5   (a)   This is conrotatory motion, and since this is a 4n π-electron system (where n = 1) it should occur under the influence of heat.

(b)   This is conrotatory motion, and since this is also a 4n π-electron system (where n = 2) it should occur under the influence of heat.

+ enantiomer

(c)   This is disrotatory motion. This, too, is a 4n π-electron system (where n = 1); thus it should occur under the influence of light.

G.6   (a)   This is a $(4n + 2)$ π-electron system (where $n = 1$); a thermal reaction should take place with disrotatory motion:

(b)   This is also a $(4n + 2)$ π-electron system; a photochemical reaction should take place with conrotatory motion.

G.7   Here we need a conrotatory ring opening of *trans*-5,6-dimethyl-1,3-cyclohexadiene (to produce *trans,cis,trans*-2,4,6-octatriene); then we need a disrotatory cyclization to produce *cis*-5,6-dimethyl-1,3-cyclohexadiene.

trans-5,6-Dimethyl-1,3-
cyclohexadiene

trans, cis, trans-2,4,6-
Octatriene

cis-5,6-Dimethyl-1,3-
cyclohexadiene

Since both reactions involve $(4n + 2)$ π-electron systems, we apply light to accomplish the first step and heat to accomplish the second. It would also be possible to use heat to produce *trans,cis,cis*-2,4,6-octatriene and then use light to produce the desired product.

G.8   The first electrocyclic reaction is a thermal, conrotatory ring opening of a $4n$ π-electron system. The second electrocyclic reaction is a thermal, disrotatory ring closure of a $(4n + 2)$ π-electron system.

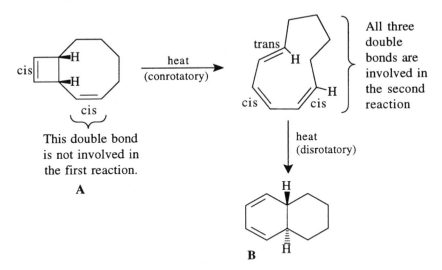

All three double bonds are involved in the second reaction

This double bond is not involved in the first reaction.

A

B

G.9   (a)   There are two possible products that can result from a concerted cycloaddition. They are formed when *cis*-2-butene molecules come together in the following ways:

(b)   There are two possible products that can be obtained from *trans*-2-butene as well.

G.10   This is an intramolecular [2 + 2] cycloaddition.

G.11   (a)

(b)

+

Enantiomer

SOLUTIONS TO PROBLEMS

H.1

H.2 An elimination reaction.

H.3 (a)

(b)

H.4 An S_N2 reaction:

SOLUTIONS TO PROBLEMS

I.1

Cyclobutadiene iron
tricarbonyl

$$d^n = \begin{matrix} \text{Total number of} \\ \text{valence electrons} \\ \text{(both } s \text{ and } d \text{ electrons)} \\ \text{of elemental iron} \end{matrix} \quad - \quad \begin{matrix} \text{oxidation state} \\ \text{of the metal} \\ \text{in the complex} \end{matrix}$$

$d^n = 8 - 0 = 8$

$$\begin{matrix} \text{Total number} \\ \text{of valence electrons} \\ \text{of iron in the} \\ \text{complex} \end{matrix} = \begin{matrix} \text{electrons} \\ d^n + \text{donated by} \\ \text{ligands} \end{matrix}$$

$$= \quad 8 + 3(CO) + \text{cyclobutadiene}$$

$$= \quad 8 + 3(2) + 4 = 18$$

Cyclopentadienylmanganese
tricarbonyl

$d^n = 7 - 1 = 6$

$$\begin{matrix} \text{Total number of} \\ \text{valence electrons} \\ \text{of Mn in complex} \end{matrix} = \quad 6 + 3(CO) + Cp$$

$$= \quad 6 + 3(2) + 6 = 18$$

468

Benzene chromium
tricarbonyl

$d^n = 6 - 0 = 6$

Total number of valence electrons of Cr in complex	=	$6 + 3(CO) + benzene$
	=	$6 + 3(2) + 6 = 18$

I.2   A syn addition of D_2 to the *trans* alkene would produce the following racemic form.

I.3 $(Ph_3P)_3RhCl$ + CH_3Li $\xrightarrow{\text{ligand exchange}}$ $(Ph_3P)_3RhCH_3$ + LiCl

(16 electrons)   (16 electrons)

Rh^I Rh^I

Oxidative addition

$(Ph_3P)_3Rh(CH_3)I$

(18 electrons)

Rh^{III}

$\xleftarrow{\text{reductive elimination}}$

+ $(Ph_3P)_3RhI$

(16 electrons)

Rh^I

I.4 $(Ph_3P)_2Rh(CO)Cl$ + CH_3Li $\xrightarrow{(a)}$ $(Ph_3P)_2Rh(CO)(CH_3)$ + $LiCl$

1   **2**

(16 electrons)   (16 electrons)

Rh^I Rh^I

(b) | $C_6H_5\overset{O}{\underset{\|}{C}}Cl$

$(Ph_3P)_2Rh(CO)Cl$ + $C_6H_5\overset{O}{\underset{\|}{C}}CH_3$ $\xleftarrow{(c)}$ $(Ph_3P)_2Rh(CO)(COC_6H_5)(CH_3)Cl$

3

(16 electrons)   (18 electrons)

Rh^I Rh^{III}

(a)   Is a ligand exchange

(b)   Is an oxidative addition

(c)   Is a reductive elimination

I.5   1. $(Ph_3P)_3Rh(CO)H$ $\xrightarrow{-Ph_3P}$ $(Ph_3P)_2Rh(CO)H$ **Ligand dissociation**

(18 electrons, Rh^I)   (16 electrons, Rh^I)

2. $(Ph_3P)_2Rh(CO)H$ $\xrightarrow{CH_3O\overset{O}{\underset{\|}{C}}C\equiv C\overset{O}{\underset{\|}{C}}OCH_3}$ **Ligand association**

(16 electrons, Rh^I)

(18 electrons, Rh^I)

3.   (18 electrons, Rh^I) $\longrightarrow$ (16 electrons, Rh^I)   **Insertion**

22 CARBOHYDRATES

SUMMARY OF SOME REACTIONS OF MONOSACCHARIDES

SOLUTIONS TO PROBLEMS

22.1 (a) Two, CHO
          *CHOH
          *CHOH
          CH₂OH

(b) Two, CH₂OH
          C=O
         *CHOH
         *CHOH
         CH₂OH

(c) There would be four stereoisomers (two sets of enantiomers) with each general structure: $2^2 = 4$.

22.2

22.3 (a)

D-(+)-Glucose   2-Hydroxybenzyl alcohol

(b)

Salicin

22.4 Dissolve D-glucose in ethanol and then bubble in gaseous HCl.

22.5 Since glycosides are acetals, they undergo hydrolysis in aqueous acid to form cyclic hemi-acetals that then undergo mutarotation.

22.6 α-D-Glucopyranose will give a positive test with Benedict's or Tollens' solution because it is a cyclic hemiacetal. Methyl α-D-glucopyranoside, because it is a cyclic acetal, will not.

22.7 (a) Yes

(b)

D-Mannaric acid

(c) Yes

(d)

(e) No

(f)

D-Threose

D-Tartaric acid

(g) The aldaric acid obtained from D-erythrose is *meso*-tartaric acid; the aldaric acid obtained from D-threose is D-tartaric acid.

22.8   and

22.9 One way of predicting the products from a periodate oxidation is to place an –OH group on each carbon atom at the point where C–C bond cleavage has occurred:

Then if we recall (Section 16.7A) that *gem*-diols are usually unstable and lose water to produce carbonyl compounds, we get the following results:

Let us apply this procedure to several examples here while we remember that for every C–C bond that is broken 1 mol of HIO_4 is consumed.

(h)

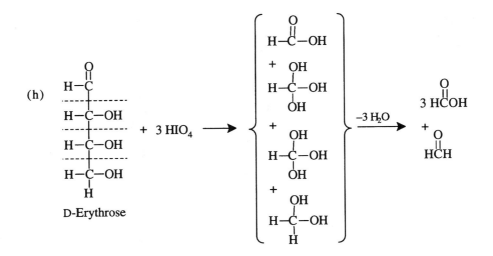

D-Erythrose

22.10 Oxidation of an aldohexose and a ketohexose would each require 5 mol of HIO_4 but would give different results.

22.11 (a) Yes, D-glucitol would be optically active; only those alditols whose molecules possess a plane of symmetry would be optically inactive.

(b)

(b) This experiment shows that D-glucose and D-fructose have the same configurations at C3, C4, and C5.

22.13 (a)

(b) L-Glyceraldehyde

22.14 (a)

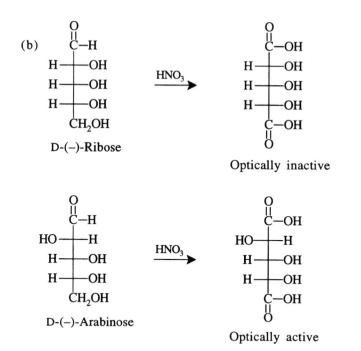

(b)

D-(–)-Ribose → HNO₃ → Optically inactive

D-(–)-Arabinose → HNO₃ → Optically active

22.15 A Kiliani-Fischer synthesis starting with D-(−)-threose would yield **I** and **II**.

I	**II**
D-(+)-Xylose	D-(−)-Lyxose

I must be D-(+)-xylose because, when oxidized by nitric acid, it yields an optically inactive aldaric acid:

I $\xrightarrow{\text{HNO}_3}$

Optically inactive

II must be D-(−)-lyxose because, when oxidized by nitric acid, it yields an optically active aldaric acid:

Optically active

22.16

L-(+)-Ribose   L-(+)-Arabinose   L-(−)-Xylose   L-(+)-Lyxose

22.17 Since D-(+)-galactose yields an optically inactive aldaric acid, it must have either structure **III** or structure **IV**.

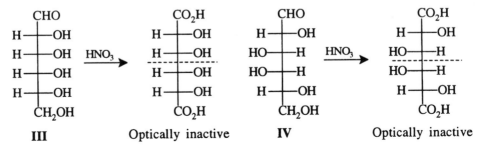

III   Optically inactive   **IV**   Optically inactive

A Ruff degradation beginning with **III** would yield D-(−)-ribose

D-(−)-Ribose

A Ruff degradation beginning with **IV** would yield D-(−)-lyxose: thus, D-(+)-galactose must have structure **IV**.

CHO
HO——H
HO——H
H——OH
CH₂OH

D-(−)-Lyxose

22.18 D-(+)-glucose, as shown here.

The other γ-lactone of D-glucaric acid

D-(+)-Glucose

22.19

D-Galactose

D-Galacturonic acid

22.20 (a)

CHO
CHOH
CHOH
CHOH
CH₂OH

(b)

CH₂OH
C=O
CHOH
CHOH
CHOH
CH₂OH

(c)

CHO
(CHOH)ₙ
HO——C——H
CH₂OH

or

CH₂OH
C=O
(CHOH)ₙ
HO——C——H
CH₂OH

(d) (e) (f)

(g) (h)

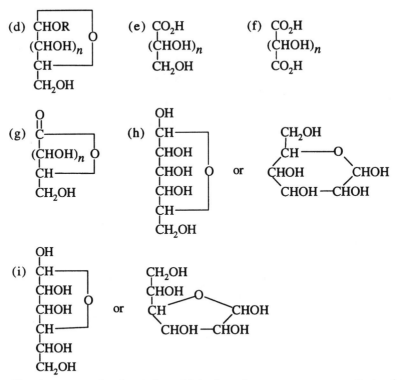

(i)

(j) Any sugar that has a free aldehyde or ketone group or one that exists as a cyclic hemiacetal. The following are examples:

(k) (l)

(m) Any two aldoses that differ only in configuration at C2. (See also Section 22.8 for a broader definition.) D-Erythrose and D-threose are examples.

D-Erythrose   D-Threose

(n) Cyclic sugars that differ only in the configuration of C1. The following are examples:

(o)

(p) Maltose is an example:

(q) Amylose is an example:

(r) Any sugar in which all potential carbonyl groups are present as acetals (i.e., as glycosides). Sucrose (Section 22.12A) is an example of a nonreducing disaccharide; the methyl D-glucopyranosides (Section 22.4) are examples of nonreducing monosaccharides.

22.21 (a)   (b)

(c)

22.22 (a)   and

(b)

A methyl ribofuranoside would consume only 1 mol of HIO_4; a methyl ribopyranoside would consume 2 mol of HIO_4 and would also produce 1 mol of formic acid.

22.23 One anomer of D-mannose is dextrorotatory ($[\alpha]_D^{25} = +29.3°$); the other is levorotatory ($[\alpha]_D^{25} = -17.0°$).

22.24 The microorganism selectively oxidizes the –CHOH group of D-glucitol that corresponds to C5 of D-glucose.

D-Glucose   D-Glucitol   L-Sorbose

22.25 L-Gulose and L-idose would yield the same phenylosazone as L-sorbose.

22.26

22.27 **A** is D-altrose, **B** is D-talose, **C** is D-galactose.

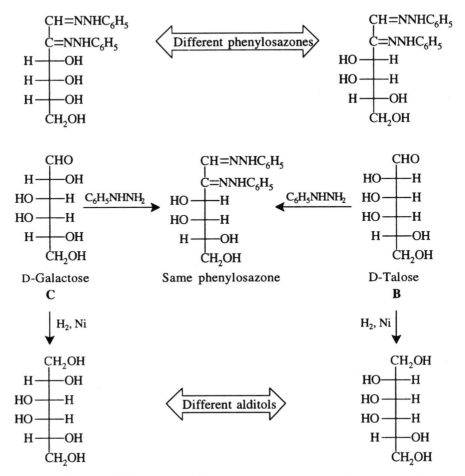

D-Galactose
C

Same phenylosazone

D-Talose
B

(*Note*: If we had designated D-talose as **A**, and D-altrose as **B**, then **C** is D-allose).

22.28

D-Xylose

D-Xylitol

22.29

D-Glucose

D-Mannose

22.30 The conformation of D-idopyranose with four equatorial –OH groups and an axial –CH$_2$OH group is more stable than the one with four axial –OH groups and an equatorial –CH$_2$OH group.

More stable

4 Equatorial –OH groups
1 Axial –CH$_2$OH group

Less stable

4 Axial –OH groups
1 Equatorial –CH$_2$OH group

22.31 (a)   The anhydro sugar is formed when the axial –CH$_2$OH group reacts with C1 to form a cyclic acetal.

β-D-Altropyranose

Anhydro sugar

Because the anhydro sugar is an acetal (i.e., an internal glycoside), it is a nonreducing sugar.

Methylation followed by acid hydrolysis converts the anhydro sugar to 2,3,4-tri-*O*-methyl-D-altrose:

(b)   Formation of an anhydro sugar requires that the monosaccharide adopt a chair conformation with the –CH$_2$OH group axial. With β-D-altropyranose this requires that two –OH groups be axial as well. With β-D-glucopyranose, however, it requires that all four –OH groups become axial and thus that the molecule adopt a very unstable conformation:

22.32 1.   The molecular formula and the results of acid hydrolysis show that lactose is a disaccharide composed of D-glucose and D-galactose. The fact that lactose is hydrolyzed by a β-*galactosidase* indicates that galactose is present as a glycoside and that the glycosidic linkage is beta to the galactose ring.

2.   That lactose is a reducing sugar, forms a phenylosazone, and undergoes mutarotation indicates that one ring (presumably that of D-glucose) is present as a hemiacetal and thus is capable of existing to a limited extent as an aldehyde.

3.   This experiment confirms that the D-glucose unit is present as a cyclic hemiacetal and that the D-galactose unit is present as a cyclic glycoside.

4.   That 2,3,4,6-tetra-*O*-methyl-D-galactose is obtained in this experiment indicates (by virtue of the free –OH at C5) that the galactose ring of lactose is present as a pyranoside. That the methylated gluconic acid obtained from this experiment has a free –OH group at C4 indicates that the C4 oxygen atom of the glucose unit is connected in a glycosidic linkage to the galactose unit.

  Now only the size of the glucose ring remains in question, and the answer to this is provided by experiment 5.

5.   That methylation of lactose and subsequent hydrolysis gives 2,3,6-tri-*O*-methyl-D-glucose—that it gives a methylated glucose derivative with a free –OH at C4 and C5—demonstrates that the glucose ring is present as a pyranose. (We know already that the oxygen at C4 is connected in a glycosidic linkage to the galactose unit; thus, a free –OH at C5 indicates that the C5 oxygen atom is a part of the hemiacetal group of the glucose unit and that the ring is six membered.)

22.33

6-*O*-(α-D-Galactopyranosyl)-D-glucopyranose

  We arrive at this conclusion from the data given:

1.   That melibiose is a reducing sugar and that it undergoes mutarotation and forms a phenylosazone indicate that one monosaccharide is present as a cyclic hemiacetal.

2.   That acid hydrolysis gives D-galatose and D-glucose indicates that melibiose is a disaccharide composed of one D-galactose unit and one D-glucose unit. That melibiose is hydrolyzed by an α-galactosidase suggests that melibiose is an α-D-galactosyl-D-glucose.

3.   Oxidation of melibiose to melibionic acid and subsequent hydrolysis to give D-galactose and D-gluconic acid confirms that the glucose unit is present as a cyclic hemiacetal and that the galactose unit is present as a glycoside. (Had the reverse been true, this experiment would have yielded D-glucose and D-galactonic acid.)

  Methylation and hydrolysis of melibionic acid produces 2,3,4,6-tetra-*O*-methyl-D-galactose and 2,3,4,5-tetra-*O*-methyl-D-gluconic acid. Formation of the first product—a galactose derivative with a free –OH at C5—demonstrates that the galactose ring is six membered; formation of the second product—a gluconic acid derivative with a free –OH at C6—demonstrates that the oxygen at C6 of the glucose unit is joined in a glycosidic linkage to the galactose unit.

4.   That methylation and hydrolysis of melibiose gives a glucose derivative (2,3,4-tri-*O*-methyl-D-glucose) with free –OH groups at C5 and C6 shows that the glucose ring is also six membered. Melibiose is, therefore, 6-*O*-(α-D-galactopyranosyl-D-glucopyranose.

22.34 Trehalose has the following structure:

α-D-Glucopyranosyl-α-D-glucopyranoside

We arrive at this structure in the following way:

1. Acid hydrolysis shows that trehalose is a disaccharide consisting only of D-glucose units.

2. Hydrolysis by α-glucosidases and not by β-glucosidases shows that the glycosidic linkages are alpha.

3. That trehalose is a nonreducing sugar, that it does not form a phenylosazone, and that it does not react with bromine water indicate that no hemiacetal groups are present. This means that C1 of one glucose unit and C1 of the other must be joined in a glycosidic linkage. Fact 2 (just cited) indicates that this linkage is alpha to each ring.

4. That methylation of trehalose followed by hydrolysis yields only 2,3,4,6-tetra-*O*-methyl-D-glucose demonstrates that both rings are six membered.

22.35 (a) Tollens' reagent or Benedict's reagent will give a positive test with D-glucose but will give no reaction with D-glucitol.

(b) D-Glucaric acid will give an acidic aqueous solution that can be detected with blue litmus paper. D-Glucitol will give a neutral aqueous solution.

(c) D-Glucose will be oxidized by bromine water and the red brown color of bromine will disappear. D-Fructose will not be oxidized by bromine water since it does not contain an aldehyde group.

(d) Nitric acid oxidation will produce an *optically active* aldaric acid from D-glucose but an *optically inactive* aldaric acid will result from D-galactose.

(e) Maltose is a reducing sugar and will give a positive test with Tollens' or Benedict's solution. Sucrose is a nonreducing sugar and will not react.

(f) Maltose will give a positive Tollens' or Benedict's test; maltonic acid will not.

(g) 2,3,4,6-Tetra-*O*-methyl-β-D-glucopyranose will give a positive test with Tollens' or Benedict's solution; methyl β-D-glucopyranoside will not.

(h) Periodic acid will react with methyl α-D-ribofuranoside because it has hydroxyl groups on adjacent carbons. Methyl 2-deoxy-α-D-ribofuranoside will not react.

22.36 That the Schardinger dextrins are nonreducing shows that they have no free aldehyde or hemiacetal groups. This lack of reaction strongly suggests the presence of a *cyclic* structure. That methylation and subsequent hydrolysis yields only 2,3,6-tri-*O*-methyl-D-glucose indicates that the glycosidic linkages all involve C1 of one glucose unit and C4 of the next. That α-glucosidases cause hydrolysis of the glycosidic linkages indicates that they are α-glycosidic linkages. Thus, we are led to the following general structure.

$n = 3, 4, \text{ or } 5$

Note: Schardinger dextrins are extremely interesting compounds. They are able to form complexes with a wide variety of compounds by incorporating these compounds in the cavity in the middle of the cyclic dextrin structure. Complex formation takes place, however, only when the cyclic dextrin and the guest molecule are the right size. Anthracene molecules, for example, will fit into the cavity of a cyclic dextrin with eight glucose units but will not fit into one with seven. For more information about these fascinating compounds, see Bergeron, R. J., "Cycloamyloses," *J. Chem. Educ.,* **1977,** *54,* 204–207.

22.37 Isomaltose has the following structure:

6-*O*-(α-D-Glucopyranosyl)-D-glucopyranose

(1)   The acid and enzymic hydrolysis experiments tell us that isomaltose has two glucose units linked by an α linkage.

(2)   That isomaltose is a reducing sugar indicates that one glucose unit is present as a cyclic hemiacetal.

(3)   Methylation of isomaltonic acid followed by hydrolysis gives us information about the size of the nonreducing pyranoside ring and about its point of attachment to the reducing ring. The formation of the first product (2,3,4,6-tetra-*O*-methyl-D-glucose)—a compound with an –OH at C5—tells us that the nonreducing ring is present as a pyranoside. The formation of 2,3,4,5-tetra-*O*-methyl-D-gluconic acid—a compound with an –OH at C6—shows that the nonreducing ring is linked to C6 of the reducing ring.

(4)   Methylation of maltose itself tells the size of the reducing ring. That 2,3,4-tri-*O*-methyl-D-glucose is formed shows that the reducing ring is also six membered; we know this because of the free –OH at C5.

22.38   Stachyose has the following structure:

Raffinose has the following structure:

The enzymic hydrolyses (as just indicated) give the basic structure of stachyose and raffinose. The only remaining question is the ring size of the first galactose unit of stachyose. That methylation of stachyose and subsequent hydrolysis yields 2,3,4,6-tetra-*O*-methyl-D-galactose establishes that this ring is a pyranoside.

22.39 Arbutin has the following structure:

or

p-Hydroxyphenyl-β-D-glucopyranoside

Compounds **X**, **Y**, and **Z** are hydroquinone, *p*-methoxyphenol, and *p*-dimethoxybenzene, respectively.

X

Hydroquinone

(a) Singlet δ 7.9 [2H]
(b) Singlet δ 6.8 [4H]

Y

p-Methoxyphenol

(a) Singlet δ 4.8 [1H]
(b) Multiplet δ 6.8 [4H]
(c) Singlet δ 3.9 [3H]

Z

p-Dimethoxybenzene

(a) Singlet δ 3.75 [6H]
(b) Singlet δ 6.8 [4H]

The reactions that take place are the following:

D-Glucose   Hydroquinone

2,3,4,6-Tetra-*O*-methyl-D-glucose   *p*-Methoxyphenol

p-Methoxyphenol

Z

p-Dimethoxybenzene

22.40 Aldotetrose **B** must be D-threose because the alditol derived from it (D-threitol) is optically active (the alditol from D-erythrose, the other possible D-aldotetrose, would be meso). Due to rotational symmetry, however, the alditol from **B** (D-threitol) would produce only two ^{13}C NMR signals. Compounds **A-F** are thus in the family of aldoses stemming from D-threose. Since reduction of aldopentose **A** produces an optically inactive alditol, **A** must be D-xylose. The two diastereomeric aldohexoses **C** and **D** produced from A by a Kiliani-Fischer synthesis must therefore be D-idose and D-gulose, respectively. **E** and **F** are the alditols derived from **C** and **D**, respectively. Alditol **E** would produce only three ^{13}C NMR signals due to rotational symmetry while **F** would produce six signals.

22.41 There are four closely spaced upfield alkyl signals in the ^{13}C NMR spectrum (δ26.5, δ25.6, δ24.9, δ24.2), corresponding to the four methyls of the two acetonide protecting groups. (The compound is, therefore, the 1,2,5,6-bis-acetonide of mannofuranose, below.)

22.42 The final product is the acetonide of glyceraldehyde (below); two molar equivalents are formed from each molar equivalent of the 1,2,5,6-bis-acetonide of mannitol.

*22.43 The β-anomer can hydrogen bond intramolecularly, as in:

In contrast, the α-anomer can only hydrogen bond intermolecularly.

*22.44

*22.45   (a)   The proton at Cl. (b) Because of the single neighboring hydrogen (at C2).
(c and d).

V   VI

QUIZ

22.1   Supply the appropriate structural formula or complete the partial formula for each of
the following:

(a)	(b)	(c)	(d)
	CHO $-$C$-$ $-$C$-$ $-$C$-$ $-$C$-$ CH_2OH	CHO $-$C$-$ $-$C$-$ CH_2OH	
A ketotetrose	A D-sugar	An L-sugar	An aldose

CHO
H——OH
H——OH
HO——H
H——OH
CH_2OH

D-Gulose

(e)	(f)	(g)	(h)
		The compound that gives the same osazone as D-gulose	The compound that gives the same aldaric acid as D-gulose
α-D-Gulo-pyranose	β-D-Gulo-pyranose		

22.2   Which of the following monosaccharides yields an optically inactive alditol
on $NaBH_4$ reduction?

A   B   C   D

Answer: ☐

22.3 Give the structural formula of the monosaccharide that you could use as starting material in the Kiliani-Fischer synthesis of the following compound:

22.4 The D-aldopentose, (a), is oxidized to an aldaric acid, (b), which is optically active. Compound (a) undergoes a Ruff degradation to form an aldotetrose, (c), which undergoes oxidation to an optically inactive aldaric acid, (d). Supply the reagents for these transformations and the structural formulas of (a), (b), (c), and (d).

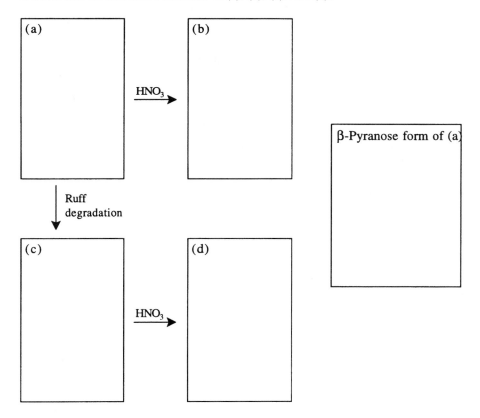

22.5 Give the structural formula of the β-pyranose form of (a) in the space just given.

22.6 Complete the following skeletal formulas and statements by filling in the blanks and circling the words that make the statements true.

The Haworth and conformational formulas of the β-cyclic hemiacetal

of

CHO
HO——H
HO——H
H——OH
H——OH
CH₂OH

D-**Mannose**

are

(a)   and (b)

This cyclic hemiacetal is   (c)   reducing, nonreducing;   on reaction with Br_2/H_2O

it gives an optically   (d)   active, inactive   (e)   aldaric, aldonic   acid. On reaction

with dilute HNO_3 it gives an optically   (f)   active, inactive   (g)   aldaric, aldonic

acid. Reaction of the cyclic hemiacetal with   (h)   converts it into an

optically   (i)   active, inactive   alditol.

22.7 Outline chemical tests that would allow you to distinguish between:

(a)   Glucose

CHO
H——OH
HO——H
H——OH
H——OH
CH₂OH

and galactose

CHO
H——OH
HO——H
HO——H
H——OH
CH₂OH

(b)   Glucose and fructose

22.8   Hydrolysis of (+)-sucrose (ordinary table sugar) yields

(a)   D-glucose

(b)   D-mannose

(c)   D-fructose

(d)   D-galactose

(e)   More than one of the above.

22.9   Select the reagent needed to perform the following transformation:

(a) CH$_3$OH, KOH   (b) (CH$_3$C)$_2$O   (c) (CH$_3$)$_2$SO$_4$, OH$^-$

(d) CH$_3$OH, HCl   (e) CH$_3$OCH$_3$, HCl

SOLUTIONS TO PROBLEMS

23.1 (a) There are two sets of enantiomers, giving a total of four stereoisomers:

erythro   *threo*

(b)

(±)-*threo*-9,10-Dibromohexadecanoic acids

Formation of a bromonium ion at the other face of palmitoleic acid gives a result such that the *threo* enantiomers are the only products formed (obtained as a racemic modification).

The designations *erythro* and *threo* come from the names of the sugars called *erythrose* and *threose* (Section 22.9A).

23.2

Zingiberene
(a sesquiterpene)

β-Selinene
(a sesquiterpene)

Caryophyllene
(a sesquiterpene)

Squalene
(a triterpene)

23.3 (a) Myrcene

$\dfrac{(1)\ O_3}{(2)\ Zn,\ HOAc}$

(b) Limonene

$\dfrac{(1)\ O_3}{(2)\ Zn,\ HOAc}$

(c) α-Farnesene
(see Section 23.3)

$\dfrac{(1)\ O_3}{(2)\ Zn,\ HOAc}$

(d) Geraniol (see Section 23.3) $\xrightarrow[\text{(2) Zn, HOAc}]{\text{(1) O}_3}$ $CH_3\overset{O}{\underset{\|}{C}}CH_3$ + $H\overset{O}{\underset{\|}{C}}CH_2CH_2\overset{O}{\underset{\|}{C}}CH_3$

+ $H\overset{O}{\underset{\|}{C}}CH_2OH$

(e) Squalene (see Section 23.3) $\xrightarrow[\text{(2) Zn, HOAc}]{\text{(1) O}_3}$ $2\ CH_3\overset{O}{\underset{\|}{C}}CH_3$ + $H\overset{O}{\underset{\|}{C}}CH_2CH_2\overset{O}{\underset{\|}{C}}H$

+ $4\ CH_3\overset{O}{\underset{\|}{C}}CH_2CH_2\overset{O}{\underset{\|}{C}}H$

23.4 (a)

+ CO_2

(+ further oxidation products)

(b)

(c)

(+ rearranged products)

(d)

23.5 Br$_2$ in CCl$_4$ or KMnO$_4$ in H$_2$O at room temperature. Either reagent would give a positive result with geraniol and a negative result with menthol.

23.6

5α-Series

5β-Series

23.7 (a)

3α-Hydroxy-5α-androstan-17-one
(androsterone)

(b)

17α-Ethynyl-17β-hydroxy-5(10)-estren-3-one
(norethynodrel)

23.8

Absolute configuration of cholesterol
(5-cholesten-3β-ol)

23.9 Estrone and estradiol are *phenols* and thus are soluble in aqueous sodium hydroxide. Extraction with aqueous sodium hydroxide separates the estrogens from the androgens.

23.10 (a)

Cholesterol

$$\xrightarrow[\text{CCl}_4]{\text{Br}_2}$$

5α,6β-Dibromocholestan-3β-ol

(b)

5α,6α-Epoxycholestan-3β-ol
(prepared by epoxidation
of cholesterol; cf. Section 23.4G)

Cholestan-3β,5α,6β-triol

(c)

5α–Cholestan-3β-ol
(prepared by hydrogenation
of cholesterol; cf. Section 23.4G)

5α–Cholestan-3-one

(d)

Cholesterol

6α-Deuterio-5α-cholestan-3β-ol

(e)

5α,6α-Epoxycholestan-3β-ol

6β-Bromocholestan-3β,5α-diol

23.11 (a) $\begin{array}{l}CH_2OH\\CHOH\\CH_2OH\end{array}$ + $R\overset{\displaystyle O}{\overset{\|}{C}}OH$ + $R'\overset{\displaystyle O}{\overset{\|}{C}}OH$ + H_3PO_4 + $HOCH_2CH_2\overset{+}{N}(CH_3)_3$ X^-

(b) $\begin{array}{l}CH_2OH\\CHOH\\CH_2OH\end{array}$ + $R\overset{\displaystyle O}{\overset{\|}{C}}OH$ + $R'\overset{\displaystyle O}{\overset{\|}{C}}OH$ + H_3PO_4 + $HOCH_2CH_2NH_2$

(c) $\begin{array}{l}CH_2OH\\CHOH\\CH_2OH\end{array}$ + $CH_3(CH_2)_nCH_2\overset{\displaystyle O}{\overset{\|}{C}}H$ + $R'\overset{\displaystyle O}{\overset{\|}{C}}OH$ + H_3PO_4

$\qquad$ + $HOCH_2CH_2\overset{+}{N}(CH_3)_3$ X^-

23.12 (a) $CH_3(CH_2)_{16}CO_2H$ + C_2H_5OH $\underset{}{\overset{H^+}{\rightleftharpoons}}$ $CH_3(CH_2)_{16}CO_2C_2H_5$ + H_3O

$CH_3(CH_2)_{16}CO_2H$ $\xrightarrow{SOCl_2}$ $CH_3(CH_2)_{16}COCl$ $\xrightarrow{C_2H_5OH}$ $CH_3(CH_2)_{16}CO_2C_2H_5$

(b) $CH_3(CH_2)_{16}COCl$ $\xrightarrow{(CH_3)_3COH}$ $CH_3(CH_2)_{16}CO_2C(CH_3)_3$

(c) $CH_3(CH_2)_{16}COCl$ $\xrightarrow{NH_3}$ $CH_3(CH_2)_{16}CONH_2$

(d) $CH_3(CH_2)_{16}COCl$ $\xrightarrow{(CH_3)_2NH}$ $CH_3(CH_2)_{16}CON(CH_3)_2$

(e) $CH_3(CH_2)_{16}CONH_2$ $\xrightarrow{LiAlH_4}$ $CH_3(CH_2)_{16}CH_2NH_2$

(f) $CH_3(CH_2)_{16}CONH_2$ $\xrightarrow[OH^-]{Br_2}$ $CH_3(CH_2)_{15}CH_2NH_2$

(g) $CH_3(CH_2)_{16}COCl$ $\xrightarrow{LiAlH[OC(CH_3)_3]_3}$ $CH_3(CH_2)_{16}CHO$

(h) $CH_3(CH_2)_{16}CO_2C_2H_5$ $\xrightarrow{\underset{Ni}{H_2}}$ $CH_3(CH_2)_{16}CH_2OH$

$\qquad\qquad\qquad$ $CH_3(CH_2)_{16}COCl$

$\qquad\qquad\qquad\qquad$ $CH_3(CH_2)_{16}CO_2CH_2(CH_2)_{16}CH_3$

(i) $CH_3(CH_2)_{16}CO_2H$ $\xrightarrow[(2)\ H_2O]{(1)\ LiAlH_4}$ $CH_3(CH_2)_{16}CH_2OH$

$\qquad\quad$ $CH_3(CH_2)_{16}CO_2C_2H_5$ $\xrightarrow[Ni]{H_2}$ $CH_3(CH_2)_{16}CH_2OH$

(j) $CH_3(CH_2)_{16}COCl$ + $(CH_3)_2CuLi$ $\longrightarrow$ $CH_3(CH_2)_{16}COCH_3$

(k) $CH_3(CH_2)_{16}CH_2OH \xrightarrow{PBr_3} CH_3(CH_2)_{16}CH_2Br$

(l) $CH_3(CH_2)_{16}CH_2Br \xrightarrow[\text{(2)}H^+, H_2O, \text{heat}]{\text{(1) NaCN}} CH_3(CH_2)_{16}CH_2CO_2H$

23.13 (a) $CH_3(CH_2)_{11}CH_2CO_2H \xrightarrow{Br_2, P} CH_3(CH_2)_{11}\underset{Br}{CHCO_2H}$

(b) $CH_3(CH_2)_{11}\underset{Br}{CHCO_2H} \xrightarrow[\text{(2) }H_3O^+]{\text{(1) }OH^-, \text{heat}} CH_3(CH_2)_{11}\underset{OH}{CHCO_2H}$

(c) $CH_3(CH_2)_{11}\underset{Br}{CHCO_2H} \xrightarrow[\text{(2) }H^+]{\text{(1) NaCN}} CH_3(CH_2)_{11}\underset{CN}{CHCO_2H}$

(d) $CH_3(CH_2)_{11}\underset{Br}{CHCO_2H} \xrightarrow[\text{(2) }H^+]{\text{(1) }NH_3 \text{ (excess)}} CH_3(CH_2)_{11}\underset{NH_2}{CHCO_2H}$

or $CH_3(CH_2)_{11}\underset{NH_3^+}{CHCO_2^-}$

23.14 (a) $CH_3(CH_2)_5CH=CH(CH_2)_7CO_2H \xrightarrow{Br_2} CH_3(CH_2)_5CHBrCHBr(CH_2)_7CO_2H$

(b) $CH_3(CH_2)_5CH=CH(CH_2)_7CO_2H \xrightarrow[Ni]{H_2} CH_3(CH_2)_{14}CO_2H$

(c) $CH_3(CH_2)_5CH=CH(CH_2)_7CO_2H \xrightarrow{KMnO_4}$
$CH_3(CH_2)_5CHOHCHOH(CH_2)_7CO_2H$

(d) $CH_3(CH_2)_5CH=CH(CH_2)_7CO_2H \xrightarrow{HCl} CH_3(CH_2)_5CH_2CHCl(CH_2)_7CO_2H$
$+$
$CH_3(CH_2)_5CHClCH_2(CH_2)_7CO_2H$

23.15 Elaidic acid is *trans*-9-octadecenoic acid:

It is formed by the isomerization of oleic acid.

23.16 (a)

(b) Infrared spectroscopy

(c) A peak in the 675–730-cm^{-1} region would indicate that the double bond is cis; a peak in the 960–975-cm^{-1} region would indicate that it is trans.

23.17 A reverse Diels-Alder reaction takes place.

23.18

CH$_3$ CH$_2$

α-Phellandrene   β-Phellandrene

Note: On permanganate oxidation, the =CH$_2$ group of β-phellandrene is converted to CO$_2$ and thus is not detected in the reaction.

23.19   CH$_3$(CH$_2$)$_5$C≡CH + NaNH$_2$ $\xrightarrow{\text{liq. NH}_3}$ CH$_3$(CH$_2$)$_5$C≡CNa
                                                                                    A

$\xrightarrow{\text{ICH}_2(\text{CH}_2)_7\text{CH}_2\text{Cl}}$ CH$_3$(CH$_2$)$_5$C≡CCH$_2$(CH$_2$)$_7$CH$_2$Cl $\xrightarrow{\text{NaCN}}$
                                   B

CH$_3$(CH$_2$)$_5$C≡CCH$_2$(CH$_2$)$_7$CH$_2$CN $\xrightarrow[\text{H}_2\text{O}]{\text{KOH}}$ CH$_3$(CH$_2$)$_5$C≡CCH$_2$(CH$_2$)$_7$CH$_2$CO$_2$K
                     C   **D**

$\xrightarrow{\text{H}_3\text{O}^+}$ CH$_3$(CH$_2$)$_5$C≡CCH$_2$(CH$_2$)$_7$CH$_2$CO$_2$H $\xrightarrow[\text{Pd-BaSO}_4]{\text{H}_2}$
                     E

$$\underset{\text{Vaccenic acid}}{\overset{\text{CH}_3(\text{CH}_2)_5}{\underset{\text{H}}{}}\!\!C\!=\!C\!\!\overset{\text{CH}_2(\text{CH}_2)_7\text{CH}_2\text{CO}_2\text{H}}{\underset{\text{H}}{}}}$$

23.20 $FCH_2(CH_2)_6CH_2Br$ + $HC\equiv CNa$ $\longrightarrow$ $FCH_2(CH_2)_6CH_2C\equiv CH$

F

$\xrightarrow[\text{(2) I(CH}_2)_7\text{Cl}]{\text{(1) NaNH}_2}$ $FCH_2(CH_2)_6CH_2C\equiv C(CH_2)_7Cl$ $\xrightarrow{\text{NaCN}}$

G

$FCH_2(CH_2)_6CH_2C\equiv C(CH_2)_7CN$ $\xrightarrow[\text{(2) H}^+]{\text{(1) KOH}}$ $FCH_2(CH_2)_6CH_2C\equiv C(CH_2)_7CO_2H$

H   I

$\xrightarrow[\text{Ni}_2\text{B (P-2)}]{\text{H}_2}$

5α-Cholest-2-ene   **A**

B

Here we find that epoxidation takes place at the less hindered α face (cf. Section 23.4G). Ring opening by HBr takes place in an anti fashion to give a product with diaxial substituents.

23.22 (a) $CH_2{=}CH{-}CH{=}CH_2$

(b) OH^- (Removal of the α hydrogen atom allows isomerization to the more stable compound with a trans ring junction.)

(c) $LiAlH_4$

(d) H_3O^+ and heat. (Hydrolysis of the enol ether is followed by dehydration of one alcohol group.)

(e) $HCO_2C_2H_5$, C_2H_5ONa

(f) OsO_4, then $NaHSO_3$

(g) $CH_3\overset{\overset{\textstyle O}{\|}}{C}CH_3$, H^+

(h) H_2, Pd catalyst

(i) H_3O^+, H_2O

(j) HIO_4

(k) Base and heat. (This reaction is an aldol condensation.)

(l) and (m) Na_2CrO_4, CH_3CO_2H to oxidize the aldehyde to an acid, followed by esterification.

(n) H_2 and Pt. (Hydrogen addition takes place from the less hindered α face of the molecule.)

(o), (p), and (q) $NaBH_4$ to reduce the keto group; OH^-, H_2O to hydrolyze the ester; and acetic anhydride to esterify the OH at the 3-position.

(r) and (s) $SOCl_2$ to make the acid chloride, followed by treatment with $(CH_3)_2Cd$.

(t)
$$\underset{\underset{\text{(t)} \;\; CH_3CHCH_2CH_2CH_2MgBr,\text{ followed by } H_3O^+.}{|}}{\overset{CH_3}{}}$$

(u), (v), and (w) Acetic acid and heat to dehydrate the tertiary alcohol; followed by acetic anhydride to acetylate the secondary alcohol; followed by H_2, Pt to hydrogenate the double bond.

23.23 (a) $CH_3(CH_2)_4\overset{\overset{O}{\|}}{C}H$ (b) C_4H_9Li (c) [structure]

(d) [structure: $NC(CH_2)_6$ ... H / H ... NO_2]   (e) Michael addition using a basic catalyst.

23.24 First: an elimination takes place,

$$R_3\overset{+}{N}CH_2CH_2\overset{\overset{O}{\|}}{C}CH_2CH_3 + NH_2^- \longrightarrow CH_2{=}CH\overset{\overset{O}{\|}}{C}CH_2CH_3 + R_3N + NH_3$$

Then a conjugate addition occurs, followed by an aldol addition:

*23.25

*23.26 The reaction is an intramolecular transesterification.

QUIZ

23.1 Write an appropriate formula in each box.

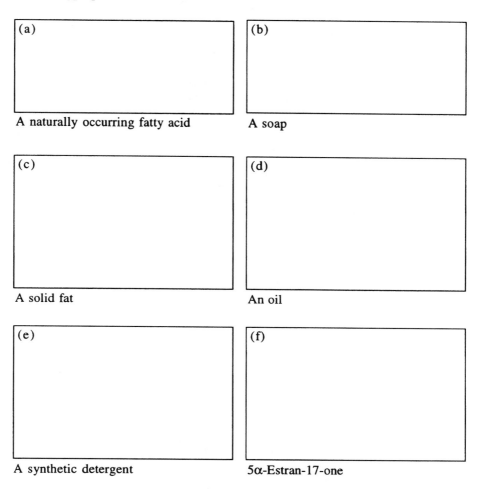

(a)

A naturally occurring fatty acid

(b)

A soap

(c)

A solid fat

(d)

An oil

(e)

A synthetic detergent

(f)

5α-Estran-17-one

23.2 Give a reagent that would distinguish between each of the following:

(a) Pregnane and 20-pregnanone

(b)   Stearic acid and oleic acid

(c) 17α-Ethynyl-1,3,5(10)-estratriene-3,17β-diol (ethynylestradiol) and 1,3,5(10)-estratriene-3,17β-diol (estradiol)

23.3   What product would be obtained by catalytic hydrogenation of 4-androstrene?

23.4   Supply the missing compounds:

23.5 The following compound is a:

(a) Monoterpene   (b) Sesquiterpene   (c) Diterpene

(d) Triterpene   (e) Tetraterpene

23.6 Mark off the isoprene units in the previous compound.

23.7 Which is a systematic name for the steroid shown here?

(a) 5α-Androstan-3α-ol

(b) 5β-Androstan-3β-ol

(c) 5α-Pregnan-3α-ol

(d) 5β-Pregnan-3β-ol

(e) 5α-Estran-3α-ol

24 AMINO ACIDS AND PROTEINS

SOLUTIONS TO PROBLEMS

24.1 (a) $HO_2CCH_2CH_2\underset{\underset{NH_3^+}{|}}{C}HCO_2H$ (b) $^-O_2CCH_2CH_2\underset{\underset{NH_2}{|}}{C}HCO_2^-$

(c) $HO_2CCH_2CH_2\underset{\underset{NH_3^+}{|}}{C}HCO_2^-$ predominates at the isoelectric point rather than

$^-O_2CCH_2CH_2\underset{\underset{NH_3^+}{|}}{C}HCO_2H$ because of the acid-strengthening inductive effect of the α-aminium group.

(d)   Since glutamic acid is a dicarboxylic acid, acid must be added (i.e., the pH must be made lower) to suppress the ionization of the second carboxyl group and thus achieve the isoelectric point. Glutamine, with only one carboxyl group, is similar to glycine or phenylalanine and has its isoelectric point at a higher pH.

24.2   The conjugate acid is highly stabilized by resonance.

24.3 (a)

516

24.5   Because of the presence of an electron-withdrawing 2,4-dinitrophenyl group, the labeled amino acid is relatively nonbasic and is, therefore, insoluble in dilute aqueous acid. The other amino acids (those that are not labeled) dissolve in dilute aqueous acid.

24.8 (a) Two structures are possible with the sequence Glu•Cys•Gly. Glutamic acid may be linked to cysteine through its α-carboxyl group,

$$HO_2CCH_2CH_2\underset{\underset{NH_3^+}{|}}{C}HCO-\underset{\underset{CH_2SH}{|}}{N}HCHCO-NHCH_2CO_2^-$$

or through its γ-carboxyl group,

$$H_3\overset{+}{N}\underset{\underset{CO_2^-}{|}}{C}HCH_2CH_2CO-\underset{\underset{CH_2SH}{|}}{N}HCHCO-NHCH_2CO_2^-$$

(b) This result shows that the second structure is correct, that in glutathione the γ-carboxyl group is linked to cysteine.

24.9   We look for points of overlap to determine the amino acid sequence in each case.

(a)   Ser • Thr
                         Thr • Hyp
    Pro • Ser
    ————————————————————
    Pro • Ser • Thr • Hyp

(b)   Ala • Cys
                         Cys • Arg
                                      Arg • Val
    Leu • Ala
    ————————————————————
    Leu • Ala • Cys • Arg • Val

24.10   Sodium in liquid ammonia brings about reductive cleavage of the disulfide linkage of oxytocin to two thiol groups; then air oxidizes the two thiol groups back to a disulfide linkage:

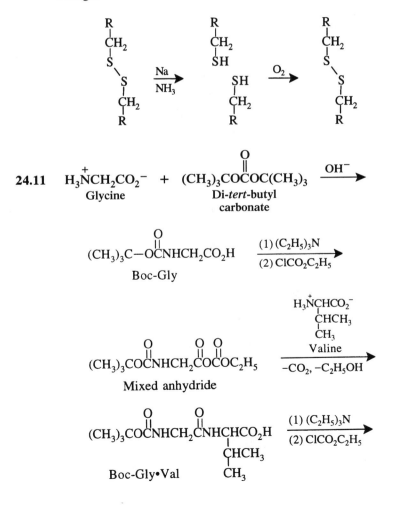

24.11 $H_3\overset{+}{N}CH_2CO_2^-$ +   (CH₃)₃COCOC(CH₃)₃ $\xrightarrow{OH^-}$

Glycine   Di-*tert*-butyl
                            carbonate

(CH₃)₃C—OCNHCH₂CO₂H $\xrightarrow[(2)\ ClCO_2C_2H_5]{(1)\ (C_2H_5)_3N}$

Boc-Gly

(CH₃)₃COCNHCH₂COCOC₂H₅ $\xrightarrow[-CO_2,\ -C_2H_5OH]{\substack{H_3\overset{+}{N}CHCO_2^- \\ CHCH_3 \\ CH_3 \\ Valine}}$

Mixed anhydride

(CH₃)₃COCNHCH₂CNHCHCO₂H $\xrightarrow[(2)\ ClCO_2C_2H_5]{(1)\ (C_2H_5)_3N}$
                                                     CHCH₃

Boc-Gly•Val   CH₃

(b) $3\ C_6H_5CH_2O\overset{O}{\overset{\|}{C}}Cl\ +\ H_2N\overset{NH}{\overset{\|}{C}}NHCH_2CH_2CH_2\underset{NH_2}{\overset{|}{C}}HCO_2^-\ \xrightarrow[25°C]{OH^-}$

3 $C_6H_5CH_2Br$ + 3 CO_2 + $H_3\overset{NH}{\overset{\|}{N}}CNHCH_2CH_2CH_2\underset{NH_2}{\overset{|}{C}}HCONH\underset{CH_3}{\overset{|}{C}}HCO_2^-$

Arg•Ala

24.13 The weakness of the benzyl-oxygen bond allows these groups to be removed by catalytic hydrogenolysis.

24.14 (a) An electrophilic substitution reaction:

(b) The linkage between the resin and the polypeptide is a benzylic ester. It is cleaved by HBr in CF_3CO_2H at room temperature because the carbocation that is formed initially is the relatively stable, benzylic cation.

24.15

$$\text{\Large \bigcirc}-CH_2Cl \ + \ HO\overset{\overset{O}{\|}}{C}\overset{}{\underset{\underset{CH_3}{|}}{C}}HNH\overset{\overset{O}{\|}}{C}OC(CH_3)_3$$

1. Add Boc-Ala.

↓ base

$$\text{\Large \bigcirc}-CH_2O\overset{\overset{O}{\|}}{C}\overset{}{\underset{\underset{CH_3}{|}}{C}}HNH\overset{\overset{O}{\|}}{C}OC(CH_3)_3$$

2. Purify by washing.

↓ CF_3CO_2H, CH_2Cl_2

3. Remove protecting group.

$$\text{\Large \bigcirc}-CH_2O\overset{\overset{O}{\|}}{C}\overset{}{\underset{\underset{CH_3}{|}}{C}}HNH_2$$

4. Purify by washing.

↓ $HO\overset{\overset{O}{\|}}{C}\overset{}{\underset{\underset{CH_2C_6H_5}{|}}{C}}HNH\overset{\overset{O}{\|}}{C}OC(CH_3)_3$
and
dicyclohexylcarbodiimide

5. Add Boc-Phe.

$$\text{\Large \bigcirc}-CH_2O\overset{\overset{O}{\|}}{C}\overset{}{\underset{\underset{CH_3}{|}}{C}}HNH\overset{\overset{O}{\|}}{C}\overset{}{\underset{\underset{\underset{C_6H_5}{|}}{CH_2}}{|}}CHNH\overset{\overset{O}{\|}}{C}OC(CH_3)_3$$

6. Purify by washing.

↓ CF_3CO_2H, CH_2Cl_2

7. Remove protecting group.

$$\text{\Large \bigcirc}-CH_2O\overset{\overset{O}{\|}}{C}\overset{}{\underset{\underset{CH_3}{|}}{C}}HNH\overset{\overset{O}{\|}}{C}\overset{}{\underset{\underset{\underset{C_6H_5}{|}}{CH_2}}{|}}CHNH_2$$

8. Purify by washing.

↓ $HO\overset{\overset{O}{\|}}{C}\overset{}{\underset{\underset{\underset{O=COC(CH_3)_3}{NH}}{|}}{C}}HCH_2CH_2CH_2CH_2NH\overset{\overset{O}{\|}}{C}OC(CH_3)_3$
and
dicyclohexylcarbodiimide

9. Add protected Lys.

10. Purify by washing.

CF_3CO_2H, CH_2Cl_2

11. Remove protecting groups.

12. Purify by washing.

HBr, CF_3CO_2H

13. Detach tripeptide.

14. Isolate product.

Lys•Phe•Ala $+ NH_3$

24.16 (a) Isoleucine, threonine, hydroxyproline, and cystine.

(b)

(With cystine, both stereocenters are α-carbon atoms; thus, according to the problem, both must have the L-configuration, and no isomers of this type can be written.)

(c)   Diastereomers

24.17 (a)   Alanine

$$CH_3\underset{\underset{NH_3^+}{|}}{C}HCO_2^- \;+\; HONO \;\longrightarrow\; CH_3\underset{\underset{OH}{|}}{C}HCO_2H \;+\; N_2$$

(b)   Proline and hydroxyproline. All of the other amino acids have at least one primary amino group.

(c)   (d)

(e) $CH_3\underset{\underset{\underset{\underset{C_6H_5}{|}}{C=O}}{\underset{NH}{|}}}{C}HCO_2^-$

24.18 (a)

(−)-Serine

A
$(C_4H_{10}ClNO_3)$

B
$(C_4H_9Cl_2NO_2)$

C
$(C_3H_6ClNO_2)$

L-(+)-Alanine

(b) **B** $\xrightarrow{OH^-}$ $\xrightarrow{NaSH}$

E
(C₄H₉NO₂S)

$\xrightarrow[\text{(2) OH}^-]{\text{(1) H}_3\text{O}^+,\ \text{H}_2\text{O, heat}}$

$\underset{\underset{\displaystyle CH_2SH}{|}}{\overset{\overset{\displaystyle CO_2^-}{|}}{H_3\overset{+}{N}-\!\!\!\!\!\!\!-H}}$

L-(+)-Cysteine

(c) $\underset{\underset{\underset{\displaystyle O}{\|}}{\underset{\displaystyle CH_2CNH_2}{|}}}{\overset{\overset{\displaystyle CO_2^-}{|}}{H_3\overset{+}{N}-\!\!\!\!\!\!\!-H}}$ $\xrightarrow[\text{Hofmann rearrangement}]{\text{NaOBr, OH}^-}$ $\underset{\underset{\displaystyle CH_2NH_2}{|}}{\overset{\overset{\displaystyle CO_2^-}{|}}{H_2N-\!\!\!\!\!\!\!-H}}$ $\xleftarrow{NH_3}$ $\underset{\underset{\displaystyle CH_2Cl}{|}}{\overset{\overset{\displaystyle CO_2^-}{|}}{H_3\overset{+}{N}-\!\!\!\!\!\!\!-H}}$

L-(−)-Asparagine   **F**   **C**
                    (C₃H₇N₂O₂)   [from part (a)]

24.19 (a)

$\underset{\displaystyle \textbf{G}}{\underset{\underset{\displaystyle CO_2C_2H_5}{|}}{\overset{\overset{\displaystyle CO_2C_2H_5}{|}}{CH_3\overset{\overset{\displaystyle O}{\|}}{C}NH-C-CH_2CH_2C\equiv N}}}$ $\xrightarrow[\substack{\text{reflux 6 h} \\ (66\%\ \text{yield})}]{\text{concd. HCl}}$ $\underset{\underset{\displaystyle NH_3^+}{|}}{HO_2CCH_2CH_2CHCO_2^-}$

DL-Glutamic acid

$+\ CH_3CO_2H\ +\ 2\ C_2H_5OH\ +\ NH_4^+\ +\ CO_2$

(b) $\underset{\underset{\displaystyle CO_2C_2H_5}{|}}{\overset{\overset{\displaystyle CO_2C_2H_5}{|}}{CH_3\overset{\overset{\displaystyle O}{\|}}{C}NH-C-CH_2CH_2C\equiv N}}$ $\xrightarrow[\substack{68°C,\ 1000\ \text{psi} \\ (90\%\ \text{yield})}]{H_2,\ Ni}$ $\left[\underset{\underset{\displaystyle CO_2C_2H_5}{|}}{\overset{\overset{\displaystyle CO_2C_2H_5}{|}}{CH_3\overset{\overset{\displaystyle O}{\|}}{C}NH-C-CH_2CH_2CH_2NH_2}}\right]$

DL-Ornithine hydrochloride

$+\ CH_3CO_2H\ +\ C_2H_5OH\ +\ CO_2$

24.20 We look for points of overlap:

```
                                  Phe • Ser
                     Pro • Gly • Phe
              Pro • Pro                      Ser • Pro • Phe
   Arg • Pro                                        Phe • Arg
   ─────────────────────────────────────────────────────────
   Arg • Pro • Pro • Gly • Phe • Ser • Pro • Phe • Arg
```

Bradykinin

24.21 1. This experiment shows that valine is the N-terminal amino acid and that valine is attached to leucine. (Lysine labeled at the ϵ-amino group is to be expected if lysine is not the N-terminal amino acid and if it is linked in the polypeptide through its α-amino group.)

2. This experiment shows that alanine is the C-terminal amino acid and that it is linked to glutamic acid.

  At this point, then, we have the following information about the structure of the heptapeptide.

Val • Leu (Ala, Lys, Phe) Glu • Ala

The sequence here is
unknown.

3. (a) This experiment shows that the dipeptide, **A,** is

Leu • Lys

(b) The carboxypeptidase reaction shows that the C-terminal amino acid of the tripeptide, **B,** is glutamic acid; the DNP labeling experiment shows that the N-terminal amino acid is phenylalanine. Thus, the tripeptide **B** is

Phe • Ala • Glu

Putting these pieces together in the only way possible, we arrive at the following amino acid sequence for the heptapeptide.

```
Val • Leu
      Leu • Lys
                  Phe • Ala • Glu
                        Glu • Ala
      ─────────────────────────────────
Val • Leu • Lys • Phe • Ala • Glu • Ala
```

24.22 At pH 2-3 the γ-carboxyl groups of polyglutamic acid are uncharged. (They are present as $-CO_2H$ groups.) At pH 5 the γ-carboxyl groups ionize and become negatively charged. (They become γ-CO_2^- groups.) The repulsive forces between these negatively charged groups cause an unwinding of the α helix and the formation of a random coil.

24.23 The observation that the ^{1}H NMR spectrum taken at room temperature shows two different signals for the methyl groups suggests that they are in different environments. This would be true if rotation about the carbon-nitrogen bond was not taking place.

We assign the δ 2.80 signal to the methyl group that is on the same side as the electronegative oxygen atom.

The fact that the methyl signals appear as doublets (and that the formyl proton signal is a multiplet) indicates that long-range coupling is taking place between the methyl protons and the formyl proton.

That the two doublets are not simply the result of spin-spin coupling is indicated by the observation that the distance that separates one doublet from the other changes when the applied magnetic field strength is lowered. [*Remember!* The magnitude of a chemical shift is proportional to the strength of the applied magnetic field, while the magnitude of a coupling constant is not.]

That raising the temperature (to 111°C) causes the doublets to coalesce into a single signal indicates that at higher temperatures the molecules have enough energy to surmount the energy barrier of the carbon-nitrogen bond. Above 111°C, rotation is taking place so rapidly that the spectrometer is unable to discriminate between the two methyl groups.

*24.24

*24.25

(By pyridine displacement of the sulfonyloxy group of the initially formed ethyl ester.)

*24.26 (a)   A large difference in the solubilities of the (*S,S*) and (*S,R*) diastereomers.

(b)   Le Chatelier's Principle, which says that an equilibrium system will try to produce more of any species that is removed (here by crystallization).

(c)   Hydrogenolysis of benzyl groups, in both instances (one is preceded by simple hydrogenation of the ketone carbonyl group).

QUIZ

24.1   Write the structural formula of the principal ionic species present in aqueous solutions at pH 2, 7, and 12 of isoleucine (2-amino-3-methylpentanoic acid).

At pH = 2	At pH = 7	At pH = 12
(a)	(b)	(c)

24.2   A hexapeptide gave the following products:

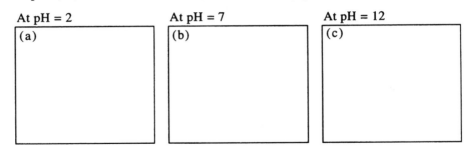

= Proline (Pro)

Hexapeptide $\xrightarrow{3\,N\,\text{HCl, 100°C}}$ 2 Gly, 1 Leu, 1 Phe, 1 Pro, 1 Tyr

Hexapeptide $\xrightarrow{1\,N\,\text{HCl, 80°C}}$ Phe•Gly•Tyr + Gly•Phe•Gly + Pro•Leu•Gly + Leu•Gly•Phe

The structure of the hexapeptide (using abbreviations such as Gly•Leu•etc.) is

25 NUCLEIC ACIDS AND PROTEIN SYNTHESIS

SOLUTIONS TO PROBLEMS

25.1 Adenine:

Guanine:

Cytosine:

Thymine (R=CH₃) or Uracil (R=H):

25.2 (a) The nucleosides have an *N*-glycosidic linkage that (like an *O*-glycosidic linkage) is rapidly hydrolyzed by aqueous acid but is one that is stable in aqueous base.

25.3   The reaction appears to take place through an S_N2 mechanism. Attack occurs preferentially at the primary 5′-carbon atom rather than at the secondary 3′-carbon atom.

25.4

25.5 (a) The isopropylidene group is part of a cyclic acetal and is thus susceptible to hydrolysis by mild acid.

(b) It can be installed by treating the nucleoside with acetone and a trace of acid and by simultaneously removing the water that is produced.

25.6 (a) Cordycepin is

(3'-Deoxyadenosine)

(b)

2'-Deoxyadenosine

25.7   (a) 6×10^9 base pairs $\times \dfrac{34 \text{ Å}}{10 \text{ base pairs}} \times \dfrac{10^{-10} \text{ m}}{\text{Å}} \cong 2 \text{ m}$

(b) $6 \times 10^{-12} \dfrac{\text{g}}{\text{ovum}} \times 3 \times 10^9 \text{ ova} = 1.8 \times 10^{-2} \text{ g}$

25.8   (a)

Lactim form   Thymine
of guanine

(b)   Thymine would pair with adenine and thus adenine would be introduced into the complementary strand where guanine should occur.

25.9   (a)   A diazonium salt and a heterocyclic analog of a phenol.

Hypoxanthine
nucleotide

(b)

Hypoxanthine   Cytosine

(c) Original double strand

First replication

Second replication

Errors

No errors in
daughter strands

25.10

Uracil
(in mRNA)

Adenine
(in DNA)

25.11 (a) UGG ┊ GGG ┊ UUU ┊ UAC ┊ AGC *m*RNA

       (b) Tyr ┊ Gly ┊ Phe ┊ Tyr ┊ Ser Amino acids

       (c) ACC ┊ CCC ┊ AAA ┊ AUG ┊ UCG Anticodons

25.12   Arg • Ile • Cys • Tyr • Val Amino acids

       (a) AGA ┊ AUA ┊ UGC ┊ UGG ┊ GUA ┊ *m*RNA

       (b) TCT ┊ TAT ┊ ACG ┊ ACC ┊ CAT ┊ DNA

       (c) UCU ┊ UAU ┊ ACG ┊ ACC ┊ CAU ┊ Anticodons

25.13 A change from C–T–T to C–A–T or a change from C–T–C to C–A–C.

A

APPENDIX
Empirical and Molecular Formulas

In Section 1.2B, we discussed briefly the pioneering work of Berzelius, Dumas, Liebig, and Cannizzaro in devising methods for determining the formulas of organic compounds. Although the experimental procedures for these analyses have been refined, the basic methods for determining the elemental composition of an organic compound today are not substantially different from those used in the nineteenth century. A carefully weighed quantity of the compound to be analyzed is oxidized completely to carbon dioxide and water. The weights of carbon dioxide and water are carefully measured and used to find the percentages of carbon and hydrogen in the compound. The percentage of nitrogen is usually determined by measuring the volume of nitrogen (N_2) produced in a separate procedure.

Special techniques for determining the percentage composition of other elements typically found in organic compounds have also been developed, but the direct determination of the percentage of oxygen is difficult. However, if the percentage composition of all the other elements is known, then the percentage of oxygen can be determined by difference. The following examples will illustrate how these calculations can be carried out.

EXAMPLE A

A new organic compound is found to have the following elemental analysis.

Carbon	67.95%
Hydrogen	5.69
Nitrogen	26.20
Total:	99.84%

Since the total of these percentages is very close to 100% (within experimental error), we can assume that no other element is present. For the purpose of our calculation it is convenient to assume that we have a 100-g sample. If we did, it would contain the following:

67.95 g of carbon
5.69 g of hydrogen
26.20 g of nitrogen

In other words, we use percentages *by weight* to give us the ratios *by weight* of the elements in the substance. To write a formula for the substance, however, we need *ratios by moles*.

We now divide each of these weight-ratio numbers by the atomic weight of the particular element and obtain the number of moles of each element, respectively, in 100 g of the compound. This operation gives us the ratios *by moles* of the elements in the substance:

$$C \quad \frac{67.95 \text{ g}}{12.01 \text{ g mol}^{-1}} \quad = \quad 5.66 \text{ mol}$$

$$H \quad \frac{5.69 \text{ g}}{1.008 \text{ g mol}^{-1}} \quad = \quad 5.64 \text{ mol}$$

$$N \quad \frac{26.20 \text{ g}}{14.01 \text{ g mol}^{-1}} \quad = \quad 1.87 \text{ mol}$$

One possible formula for the compound, therefore, is $C_{5.66}H_{5.64}N_{1.87}$.

By convention, however, we use *whole* numbers in formulas. Therefore, we convert these fractional numbers of moles to whole numbers by dividing each by 1.87, the smallest number.

$$C \quad \frac{5.66}{1.87} \quad = \quad 3.03 \text{ which is } \sim 3$$

$$H \quad \frac{5.64}{1.87} \quad = \quad 3.02 \text{ which is } \sim 3$$

$$N \quad \frac{1.87}{1.87} \quad = \quad 1.00$$

Thus, within experimental error, the ratios by moles are 3 C to 3 H to 1 N, and C_3H_3N is the *empirical formula*. By empirical formula, we mean the formula in which the subscripts are the smallest integers that give the ratio of atoms in the compound. In contrast, a *molecular* formula discloses the complete composition of one molecule. The molecular formula of this particular compound could be C_3H_3N or some whole number multiple of C_3H_3N; that is, $C_6H_6N_2$, $C_9H_9N_3$, $C_{12}H_{12}N_4$, and so on. If, in a separate determination, we find that the molecular weight of the compound is 108 ± 3, we can be certain that the *molecular formula* of the compound is $C_6H_6N_2$.

FORMULA	MOLECULAR WEIGHT
C_3H_3N	53.06
$C_6H_6N_2$	106.13 (which is within the range 108 ± 3)
$C_9H_9N_3$	159.19
$C_{12}H_{12}N_4$	212.26

The most accurate method for determining molecular weights is by mass spectrometry. A variety of other methods based on freezing point depression, boiling point elevation, osmotic pressure, and vapor density can also be used to determine molecular weights.

EXAMPLE B

Histidine, an amino acid isolated from protein, has the following elemental analysis:

Carbon	46.38%
Hydrogen	5.90
Nitrogen	27.01
Total:	79.29
Difference	20.71 (assumed to be oxygen)
	100.00%

Since no elements, other than carbon, hydrogen, and nitrogen, are found to be present in histidine, the difference is assumed to be oxygen. Again, we assume a 100-g sample and divide the weight of each element by its gram-atomic weight. This gives us the ratio of moles (A).

$$
\begin{array}{ccccccc}
 & (A) & & (B) & & & (C) \\
C & \dfrac{46.38}{12.01} = 3.86 & \dfrac{3.86}{1.29} = 2.99 & \times & 2 & = & 5.98 \sim 6 \text{ carbon atoms} \\[2mm]
H & \dfrac{5.90}{1.008} = 5.85 & \dfrac{5.85}{1.29} = 4.53 & \times & 2 & = & 9.06 \sim 9 \text{ hydrogen atoms} \\[2mm]
N & \dfrac{27.01}{14.01} = 1.93 & \dfrac{1.93}{1.29} = 1.50 & \times & 2 & = & 3.00 = 3 \text{ nitrogen atoms} \\[2mm]
O & \dfrac{20.71}{16.00} = 1.29 & \dfrac{1.29}{1.29} = 1.00 & \times & 2 & = & 2.00 = 2 \text{ oxygen atoms}
\end{array}
$$

Dividing each of the moles (A) by the smallest of them does not give a set of numbers (B) that is close to a set of whole numbers. Multiplying each of the numbers in column (B) by 2 does, however, as seen in column (C). The empirical formula of histidine is, therefore, $C_6H_9N_3O_2$.

In a separate determination, the molecular weight of histidine was found to be 158 ± 5. The empirical formula weight of $C_6H_9N_3O_2$ (155.15) is within this range; thus, the molecular formula for histidine is the same as the empirical formula.'

PROBLEMS

A.1 What is the empirical formula of each of the following compounds?

(a) Hydrazine, N_2H_4 (d) Nicotine, $C_{10}H_{14}N_2$
(b) Benzene, C_6H_6 (e) Cyclodecane, $C_{10}H_{20}$
(c) Dioxane, $C_4H_8O_2$ (f) Acetylene, C_2H_2

A.2 The empirical formulas and molecular weights of several compounds are given next. In each case, calculate the molecular formula for the compound.

EMPIRICAL FORMULA	MOLECULAR WEIGHT
(a) CH_2O	179 ± 5
(b) CHN	80 ± 5
(c) CCl_2	410 ± 10

A.3   The widely used antibiotic, penicillin G, gave the following elemental analysis: C, 57.45%; H, 5.40%; N, 8.45%; S, 9.61%. The molecular weight of penicillin G is 330 ± 10. Assume that no other elements except oxygen are present and calculate the empirical and molecular formulas for penicillin G.

ADDITIONAL PROBLEMS

A.4   Calculate the percentage composition of each of the following compounds.
(a) $C_6H_{12}O_6$
(b) $CH_3CH_2NO_2$
(c) $CH_3CH_2CBr_3$

A.5   An organometallic compound called *ferrocene* contains 30.02% iron. What is the minimum molecular weight of ferrocene?

A.6   A gaseous compound gave the following analysis: C, 40.04%; H, 6.69%. At standard temperature and pressure, 1.00 g of the gas occupied a volume of 746 mL. What is the molecular formula of the compound?

A.7   A gaseous hydrocarbon has a density of 1.251 g L^{-1} at standard temperature and pressure. When subjected to complete combustion, a 1.000-L sample of the hydrocarbon gave 3.926 g of carbon dioxide and 1.608 g of water. What is the molecular formula for the hydrocarbon?

A.8   Nicotinamide, a vitamin that prevents the occurrence of pellagra, gave the following analysis: C, 59.10%; H, 4.92%; N, 22.91%. The molecular weight of nicotinamide was shown in a separate determination to be 120 ± 5. What is the molecular formula for nicotinamide?

A.9   The antibiotic chloramphenicol gave the following analysis: C, 40.88%; H, 3.74%; Cl, 21.95%; N, 8.67%. The molecular weight was found to be 300 ± 30. What is the molecular formula for chloramphenicol?

SOLUTIONS TO PROBLEMS OF APPENDIX A

A.1 (a) NH_2 (b) CH (c) C_2H_4O (d) C_5H_7N (e) CH_2 (f) CH

A.2

EMPIRICAL FORMULA	EMPIRICAL FORMULA WEIGHT	$\left(\dfrac{\text{MOLECULAR WEIGHT}}{\text{EMP. FORM. WT.}}\right)$	MOLECULAR FORMULA
(a) CH_2O	30	$\dfrac{179}{30} \cong 6$	$C_6H_{12}O_6$
(b) CHN	27	$\dfrac{80}{27} \cong 3$	$C_3H_3N_3$
(c) CCl_2	83	$\dfrac{410}{83} \cong 5$	C_5Cl_{10}

A.3 If we assume that we have a 100-g sample, the amounts of the elements are

	WEIGHT	Moles (A)	B
C	57.45	$\dfrac{57.45}{12.01} = 4.78$	$\dfrac{4.78}{0.300} = 15.9 = 16$
H	5.40	$\dfrac{5.40}{1.008} = 5.36$	$\dfrac{5.36}{0.300} = 17.9 = 18$
N	8.45	$\dfrac{8.45}{14.01} = 0.603$	$\dfrac{0.603}{0.300} = 2.01 = 2$
S	9.61	$\dfrac{9.61}{32.06} = 0.300$	$\dfrac{0.300}{0.300} = 1.00 = 1$
O*	$\dfrac{19.09}{100.00}$	$\dfrac{19.09}{16.00} = 1.19$	$\dfrac{1.19}{0.300} = 3.97 = 4$

(* by difference from 100)

The empirical formula is thus $C_{16}H_{18}N_2SO_4$. The empirical formula weight (334.4) is within the range given for the molecular weight (330 ± 10). Thus, the molecular formula for penicillin G is the same as the empirical formula.

A.4 (a) To calculate the percentage composition from the molecular formula, first determine the weight of each element in 1 mol of the compound. For $C_6H_{12}O_6$,

$$C_6 = 6 \times 12.01 = 72.06 \qquad \dfrac{72.06}{180.2} = 0.400 = 40.0\%$$

$$H_{12} = 12 \times 1.008 = 12.10 \qquad \dfrac{12.10}{180.2} = 0.0671 = 6.7\%$$

$$O_6 = 6 \times 16.00 = 96.00 \qquad \dfrac{96.00}{180.2} = 0.533 = 53.3\%$$

MW $\overline{\qquad\qquad\qquad\qquad 180.16}$

(MW = molecular weight)

Then determine the percentage of each element using the formula

$$\text{Percentage of A} = \frac{\text{Weight of A}}{\text{Molecular Weight}} \times 100$$

(b) $C_2 = 2 \times 12.01 = 24.02$ $\dfrac{24.02}{75.07} = 0.320 = 32.0\%$

$H_5 = 5 \times 1.008 = 5.04$ $\dfrac{5.04}{75.07} = 0.067 = 6.7\%$

$N = 1 \times 14.01 = 14.01$ $\dfrac{14.01}{75.07} = 0.187 = 18.7\%$

$O_2 = 2 \times 16.00 = \underline{32.00}$ $\dfrac{32.00}{75.07} = 0.426 = 42.6\%$

Total = 75.07

(c) $C_3 = 3 \times 12.01 = 36.03$ $\dfrac{36.03}{280.77} = 0.128 = 12.8\%$

$H_5 = 5 \times 1.008 = 5.04$ $\dfrac{5.04}{280.77} = 0.018 = 1.8\%$

$Br_3 = 3 \times 79.90 = \underline{239.70}$ $\dfrac{239.70}{280.77} = 0.854 = 85.4\%$

Total = 280.77

A.5 If the compound contains iron, each molecule must contain at least one atom of iron, and 1 mol of the compound must contain at least 55.85 g of iron. Therefore,

$$\text{MW of ferrocene} = 55.85\ \frac{\text{g of Fe}}{\text{mol}} \times \frac{1.000\ \text{g}}{0.3002\ \text{g of Fe}}$$

$$= 186.0\ \frac{\text{g}}{\text{mol}}$$

A.6 First, we must determine the empirical formula. Assuming that the difference between the percentages given and 100% is due to oxygen, we calculate:

C   40.04 $\dfrac{40.04}{12.01} = 3.33$ $\dfrac{3.33}{3.33} = 1$

H   6.69 $\dfrac{6.69}{1.008} = 6.64$ $\dfrac{6.64}{3.33} \cong 2$

O $\dfrac{53.27}{100.00}$ $\dfrac{53.27}{16.00} = 3.33$ $\dfrac{3.33}{3.33} = 1$

The empirical formula is thus CH_2O.

To determine the molecular formula, we must first determine the molecular weight. At standard temperature and pressure, the volume of 1 mol of an ideal gas is 22.4 L. Assuming ideal behavior,

$$\frac{1.00 \text{ g}}{0.746 \text{ L}} = \frac{MW}{22.4 \text{ L}} \quad \text{where MW = molecular weight}$$

$$MW = \frac{(1.00)(22.4)}{0.746} = 30.0 \text{ g}$$

The empirical formula weight (30.0) equals the molecular weight; thus, the molecular formula is the same as the empirical formula.

A.7   As in Problem A.6, the molecular weight is found by the equation

$$\frac{1.251 \text{ g}}{1.00 \text{ L}} = \frac{MW}{22.4 \text{ L}}$$
$$MW = (1.251)(22.4)$$
$$MW = 28.02$$

To determine the empirical formula, we must determine the amount of carbon in 3.926 g of carbon dioxide, and the amount of hydrogen in 1.608 g of water.

$$C \quad \left(3.926 \text{ g } CO_2\right) \left(\frac{12.01 \text{ g C}}{44.01 \text{ g } CO_2}\right) = 1.071 \text{ g carbon}$$

$$H \quad \left(1.608 \text{ g } H_2O\right) \left(\frac{2.016 \text{ g H}}{18.016 \text{ g } H_2O}\right) = \frac{0.180 \text{ g hydrogen}}{1.251 \text{ g sample}}$$

The weight of C and H in a 1.251-g sample is 1.251 g. Therefore, there are no other elements present.

To determine the empirical formula, we proceed as in Problem A.6 except that the sample size is 1.251 g instead of 100 g.

$$C \quad \frac{1.071}{12.01} = 0.0892 \qquad \frac{0.0892}{0.0892} = 1$$

$$H \quad \frac{0.180}{1.008} = 0.179 \qquad \frac{0.179}{0.0892} = 2$$

The empirical formula is thus CH_2. The empirical formula weight (14) is one-half the molecular weight. Thus, the molecular formula is C_2H_4.

A.8   Use the procedure of Problem A.3.

$$C \quad 59.10 \qquad \frac{59.10}{12.01} = 4.92 \qquad \frac{4.92}{0.817} = 6.02 \cong 6$$

$$H \quad 4.92 \quad \frac{4.92}{1.008} = 4.88 \quad \frac{4.88}{0.817} = 5.97 \cong 6$$

$$N \quad 22.91 \quad \frac{22.91}{14.01} = 1.64 \quad \frac{1.64}{0.817} = 2$$

$$O \quad \frac{13.07}{100.00} \quad \frac{13.07}{16.00} = 0.817 \quad \frac{0.817}{0.817} = 1$$

The empirical formula is thus $C_6H_6N_2O$. The empirical formula weight is 122.13, which is equal to the molecular weight within experimental error. The molecular formula is thus the same as the empirical formula.

A.9

$$C \quad 40.88 \quad \frac{40.88}{12.01} = 3.40 \quad \frac{3.40}{0.619} = 5.5 \quad 5.5 \times 2 = 11$$

$$H \quad 3.74 \quad \frac{3.74}{1.008} = 3.71 \quad \frac{3.71}{0.619} = 6 \quad 6 \times 2 = 12$$

$$Cl \quad 21.95 \quad \frac{21.95}{35.45} = 0.619 \quad \frac{0.619}{0.619} = 1 \quad 1 \times 2 = 2$$

$$N \quad 8.67 \quad \frac{8.67}{14.01} = 0.619 \quad \frac{0.619}{0.619} = 1 \quad 1 \times 2 = 2$$

$$O \quad \frac{24.76}{100.00} \quad \frac{24.76}{16.00} = 1.55 \quad \frac{1.55}{0.619} = 2.5 \quad 2.5 \times 2 = 5$$

The empirical formula is thus $C_{11}H_{12}Cl_2N_2O_5$. The empirical formula weight (323) is equal to the molecular weight; therefore, the molecular formula is the same as the empirical formula.

B APPENDIX
Answers to Quizzes

CHAPTER 1

1.1 (d), **1.2** (d), **1.3** (e), **1.4** (d), **1.5** (c) **1.6**

1.9 (a) sp^2 (b) sp^3 (c) 0 (d) trigonal planar (e) 0

1.10 (a) +1 (b) 0 (c) −1

CHAPTER 2

2.1 (e) **2.2** (a) **2.3** (e)

(d) H—C(=O)—O—CH$_3$ (e) Cl$_2$C=CH$_2$ (as drawn) (f) CH$_3$OCH$_2$CH$_3$

(g) CH$_3$CH$_2$—N(CH$_3$)$_2$

2.5 ☐ or ▷—

2.6 (a) CH$_3$CH$_2$CH$_2$OH (b) ⬡N—H (c) CH$_3$CH$_2$COH (C=O)

(d) CH$_3$OCH$_2$CH$_2$OH (e) CH$_3$CH$_2$CNHCH$_3$ (C=O)

2.7 (a) Isopropyl phenyl ether

(b) Ethylmethylphenylamine

(c) Isopropylamine

CHAPTER 3

3.1 (a) **3.2** (c) **3.3** (b) **3.4** (e) **3.5** (b)

3.6 (b)

3.7 H$_2$SO$_4$ + NaF ⟶ NaHSO$_4$ + HF

3.8 (CH$_3$)$_2$NH

3.9 (a) CH$_3$CH$_2$Li (b) D$_2$O

3.10 (a) CH$_3$Li (b) CH$_3$CHCH$_2$OH (with CH$_3$) (c) CH$_3$CHCH$_2$OLi (with CH$_3$)

CHAPTER 4

4.1 (c) **4.2** (c) **4.3** (b) **4.4** (a) **4.5** (b)

4.6 (a) **4.7** (a)

4.8 (a) (b) or

(c)

4.9 (a) H$_2$, Pt or H$_2$, Ni

(b) Zn, H$^+$

4.10 CH$_3$—CH⟨CH$_3$⟩ [cyclohexane with CH$_3$]

CHAPTER 5

5.1 (a)   **5.2** (b)   **5.3** (b)   **5.4** (e)   **5.5** (b)

5.6

H$_3$C—C(H)—CH$_2$CH$_3$, CH$_2$CH$_2$CH$_3$ and   H$_3$C—C(H)—CH$_2$CH$_3$, CH(H$_3$C)(CH$_3$)

5.7 H$_3$C—C(H)—CH$_2$CH$_3$, CH=CH$_2$ **5.8** CH$_3$CH=CHCHCH$_2$CH$_3$, CH$_3$

CHAPTER 6

6.1 (b)   **6.2** (b)   **6.3** (a)

6.4 CH$_3$CH$_2$CH$_2$CH$_2$Br > CH$_3$CHCH$_2$Br(CH$_3$) > CH$_3$CH$_2$CHBr(CH$_3$) > CH$_3$C(CH$_3$)(CH$_3$)—Br

6.5 A = Br—C(''''H)(CH$_2$CH$_3$)(CH$_3$)

B = CH$_3$CH$_2$CH=CH$_2$

C = H—C(CH$_3$)(CH$_2$OCH$_3$)(CH$_2$CH$_3$)

D = H$_3$C— [cyclohexane ring with CN]

6.6 (b)

6.7 A = CH$_3$CH$_2$—C(CH$_2$Br)(H)—CH$_3$

B = CH$_3$CH$_2$—C(CH$_2$C≡CH)(H)—CH$_3$ $\xrightarrow[\text{Pt}]{\text{H}_2}$

C = CH$_3$CH$_2$—C(CH$_2$CH$_2$CH$_3$)(H)—CH$_3$

CHAPTER 7

7.1 (c)   **7.2** (d)   **7.3** (a)

7.4 (a) Li, C$_2$H$_5$NH$_2$, −78°C, then NH$_4$Cl

(b) H$_2$/Ni$_2$B(P–2)   or   H$_2$/Pd/CaCO$_3$ (Lindlar's catalyst)

(c) H$_2$/Ni   or   H$_2$/Pt using at least 2 molar equivalents of H$_2$

(d) C$_2$H$_5$ONa/C$_2$H$_5$OH

(e) (CH$_3$)$_3$COK/(CH$_3$)$_3$COH

(f) Zn/CH$_3$CO$_2$H   or   NaI/acetone

7.5 CH$_3$C(CH$_3$)=CHCH$_3$ > [H$_3$C, H / C=C / H, CH$_2$CH$_3$] > [H$_3$C, CH$_2$CH$_3$ / C=C / H, H] >

CH$_2$=CHCH$_2$CH$_2$CH$_3$

7.6 (a) CH$_3$CHCH$_2$Br (with Br on CH)   (b) CH$_3$C≡CNa   (c) CH$_3$C≡CH

(d) CH$_3$C≡CNa   (e) CH$_3$CH$_2$Br

CHAPTER 8

8.1   (e)   8.2   (c)   8.3   (e)   8.4   (a)   8.5   (d)   8.6   (c)

8.7   (e)   8.8   (b)

CHAPTER 9

9.1   (a) CH₃ĊCH₃   (b) BrCH₂ĊCH₂Br   (c) CH₂=CCH₂ĊCH₃

(d) ⬡—CH₂ĊCH₃   (e) CH₃CH₂C≡CCH₂NO₂

9.2   (c)   9.3   (a)

CHAPTER 10

10.1   (d)   10.2   (b)   10.3   (c)   10.4   (b)   10.5   (c)

10.6   CH₃CH₂ĊCH₃ (with CH₃ above)

10.7   Six

10.8   (d)

CHAPTER 11

11.1   (d)   11.2   (a)   11.3   (e)

11.4   A = C₆H₅CH₂ONa   B = C₆H₅CH₂OCH₂CH₂OH
       C = C₆H₅CH₂OCH₂CH₂OSO₂CH₃   D = C₆H₅CH₂OCH₂CH₂OCH₂CH₃

CHAPTER 12

12.1   (b)   12.2   (a)

12.3   A = CH₃C≡CLi   or   CH₃C≡CMgBr

       B = NaH

       C = CH₃I

12.4   **A =** $(CH_3)_3CCH_2CH_2OH$

B = PCC/CH_2Cl_2

C = $(CH_3)_3CCH_2\overset{O}{\overset{||}{C}}H$

D = C_6H_5MgBr

12.5   **A =** $CH_3\underset{CH_3}{\underset{|}{C}H}CH_2\overset{O}{\overset{||}{C}}OR$ or $CH_3\underset{CH_3}{\underset{|}{C}H}CH_2\overset{O}{\overset{||}{C}}CH_3$

CHAPTER 13

13.1 (d)   **13.2** (c)   **13.3** (c)   **13.4** (c)   **13.5** (b)

CHAPTER 14

14.1 (e)   **14.2** (a)   **14.3** (b)   **14.4** (b)

14.5 $\langle\bigcirc\rangle$—CH_2Cl **14.6** Azulene

CHAPTER 15

15.1 (a)   **15.2** (a)   **15.3** (b)

15.4 (a)   **A =** SO_3/H_2SO_4

B =

C = H_2O, H_2SO_4, heat   **D =**

(b)   **A =** $SOCl_2$ or PCl_5 **B =** $\bigcirc$ + $AlCl_3$

C = $Zn(Hg)$, HCl, reflux   **D =** $Br_2/FeBr_3$

CHAPTER 16

16.1 (d) **16.2** (b) **16.3** (b)

16.4 (a) A = ⟨benzene ring⟩—CH₂Br   **B** = NaCN

C (1) DIBAL-H, hexane, -78°C, (2) H₂O

(b) **A** = PCC, CH₂Cl₂   **B** = HOCH₂CH₂OH, H⁺ **C** = H₂O, H₃O⁺

(c) **A** = (C₆H₅)₃P   **B** = CH₃CH₂CH₂—P⁺(C₆H₅)₃ Br⁻   **C** = ⟨cyclopentanone⟩

(d) **A** = (CH₃)₂CuLi   **B** = HCN   **C** = (1) LiAlH₄
                                                        (2) H₂O

CHAPTER 17

17.1 (a)
$$\underset{\underset{CH_3}{|}}{CH_3CH_2CHCHCH}\overset{O}{\overset{||}{}}$$

(b)
$$\underset{\underset{CH_3}{|}}{CH_3CH_2CHCHCH_2OH}$$

(c) CH₃CH₂CH =CCH
    with O double bond and CH₃

(d) LiAlH₄   (e) H₂, Ni   (f) CH₃OH (excess), H⁺

(g) CH₃CH₂CH₂CHCH(OCH₃)₂   (h) CH₃CH₂CH₂CHCH
        CH₃   CH₃   (with =O)

(i) (1) CH₃CHBrCO₂CH₂CH₃, Zn   (2) H₃O⁺

17.2 (a) ⟨phenyl⟩—CCH₃ (C=O) (b) ⟨phenyl⟩—C—CH=CH⟨phenyl⟩ (c) HCN

17.3 (a) ⟨bicyclic ketone with OH and CH₃⟩ (b) ⟨bicyclic enone with CH₃⟩ (c) (CH₃)₂CuLi

(d) ⟨bicyclic ketone with '''CH₃ and CH₃⟩ (e) Zn(Hg)/HCl

17.4 (e)   **17.5** (a)

CHAPTER 18

18.1 (b)   **18.2** (d)   **18.3** (d)

18.4 **A** = 3-Chlorobutanoic acid

B = Methyl 4-nitrobenzoate

C = *N*-Methylaniline

18.5 (a) **A** = (1) KMnO$_4$, OH⁻, heat   **B** = SOCl$_2$ or PCl$_5$
(2) H$_3$O⁺

C = (structure)   **D** = (structure)

E = (structure)   **F** = (structure)

(b) **A** = (structure)   **B** = CH$_3$—¹⁸OH

C = (structure)

(c) **A** = (structure)   **B** = (structure)   **C** = (structure)

CHAPTER 19

19.1 (c)   **19.2** (e)   **19.3** (b)

19.4 (a) **A** = (structure)   **B** = (structure)

CHAPTER 20

20.1 (d)   **20.2** (e)

20.3 (a) (2)   (b) (4)   (c) (3)

20.4 (a) **A** = HNO_3/H_2SO_4 **B** = $H_3C-\langle\bigcirc\rangle-NO_2$ **C** = $NaNO_2$, HCl

D = CuCN   **E** = $LiAlH_4$ **F** = $(CH_3)_2N-\langle\bigcirc\rangle$

(b) **A** = NaN_3 **B** = $\langle\bigcirc\rangle-\overset{\overset{\displaystyle O}{\|}}{C}-\overset{..}{\underset{..}{N}}-\overset{+}{N}\equiv N:$ **C** = $\langle\bigcirc\rangle-NH_2$

D = $Br-\langle\bigcirc\rangle-NH_2$ (with Br at top and bottom)   **E** = H_3PO_2

20.5 (a) (2)   (b) (2)   (c) (1)

CHAPTER 21

21.1 (a)   **21.2** (b)   **21.3** (b)   **21.4** (e)

21.5 (a) **A** = $Br-\langle\bigcirc\rangle-OH$ **B** = KNH_2, NH_3, -33°C

21.6 $Br-\langle\bigcirc\rangle-OH$ + CH_3Br

21.7 (a) (1)   (b) (1)

CHAPTER 22

22.1 (a) CH$_2$OH
$\quad\quad$ =O
H—OH
CH$_2$OH

(b) CHO
CHOH ⎫ OH on
CHOH ⎬ either
CHOH ⎭ side
H—OH
CH$_2$OH

(c) CHO
CHOH ⎫ OH on
CHOH ⎬ either
HO—H ⎭ side
CH$_2$OH

(d) CHO
(CHOH)$_n$
CH$_2$OH
n = 1,2,3...

(e)

(f)

22.2   C

22.3

22.4   (a)   (b)   (c)   (d)

22.5

22.6   (a)   (b)   (c)   Reducing

(d)   Active   (e)   Aldonic   (f)   Active   (g)   Aldaric   (h)   NaBH$_4$

(i)   Active

22.7   (a)   Galactose $\xrightarrow{\text{NaBH}_4}$ optically *inactive* alditol

(b)   HIO$_4$ oxidation $\longrightarrow$ different products:

Fructose $\longrightarrow$ 2 mol HCH + CO$_2$ + 3 HC–OH

Glucose $\longrightarrow$ 1 mol HCH + 5 HC–OH

22.8   (e)   **22.9**   (d)

CHAPTER 23

23.1 (a) $CH_3(CH_2)_{12}CO_2H$

(b) $CH_3(CH_2)_{12}\overset{O}{\overset{\|}{C}}-ONa$

(c) $\begin{array}{l} CH_2O\overset{O}{\overset{\|}{C}}(CH_2)_{12}CH_3 \\ | \\ CHO\overset{O}{\overset{\|}{C}}(CH_2)_{12}CH_3 \\ | \\ CH_2O\overset{O}{\overset{\|}{C}}(CH_2)_{12}CH_3 \end{array}$

(d) $\begin{array}{l} CH_2O\overset{O}{\overset{\|}{C}}(CH_2)_7CH=CH(CH_2)_5CH_3 \\ | \\ CHO\overset{O}{\overset{\|}{C}}(CH_2)_7CH=CH(CH_2)_5CH_3 \\ | \\ CH_2O\overset{O}{\overset{\|}{C}}(CH_2)_7CH=CH(CH_2)_5CH_3 \end{array}$

(e) $CH_3(CH_2)_{13}SO_3Na$

(f)

23.2 (a) I_2/OH^- (iodoform test)   (b) Br_2/CCl_4 (c) Ethynylestradiol only shows IR absorption at ~ 3300 cm^{-1}.

23.3 5α-Androstane

23.4 (a) $CH_3(CH_2)_4CH_2C\equiv CH$ (b) $CH_3(CH_2)_5C\equiv CNa$

(c) $CH_3(CH_2)_5C\equiv C(CH_2)_6CH_2Cl$ (d) KCN

(e) $CH_3(CH_2)_5C\equiv C(CH_2)_7CO_2H$ (f) H_2/Pd

23.5 Sesquiterpene

23.6

23.7 (e)

CHAPTER 24

24.1   (a)

(b)

(c)   CH₃CH₂CHCHCO₂⁻ with CH₃ and NH₂ substituents

24.2   Pro·Leu·Gly·Phe·Gly·Tyr

C

APPENDIX
Molecular Model Set Exercises

The exercises in this appendix are designed to help you gain an understanding of the three-dimensional nature of molecules. You are encouraged to perform these exercises with a model set as described.

These exercises should be performed as part of the study of the chapters shown below.

Chapter in Text	Accompanying Exercises
4	1, 3, 4, 5, 6, 8, 10, 11, 12, 14, 15, 16, 17, 18, 20, 21
5	2, 7, 9, 13, 24, 25, 26, 27
7	9, 19, 22, 28
2	31
3	23
22	29
24	30
12	31

The following molecular model set exercises were originally developed by Ronald Starkey.

Refer to the instruction booklet that accompanies your model set for details of molecular model assembly.

EXERCISE 1 (Chapter 4)

Assemble a molecular model of methane, CH_4. Note that the hydrogen atoms describe the apexes of a regular tetrahedron with the carbon atom at the center of the tetrahedron. Demonstrate by attempted superposition that two models of methane are identical.

Replace any one hydrogen atom on each of the two methane models with a halogen to form two molecules of CH_3X. Are the two structures identical? Does it make a difference which of the four hydrogen atoms on a methane molecular you replace? How many different configurations of CH_3X are possible?

Repeat the same considerations for two disubstituted methanes with two identical substituents (CH_2X_2), and then with two different substituents (CH_2XY). Two colors of atom-centers could be used for the two different substituents.

EXERCISE 2 (Chapter 5)

Construct a model of a trisubstituted methane molecule (CHXYZ). Four different colored atom-centers are attached to a central tetrahedral carbon atom center. Note that the carbon now has four different substituents. Compare this model with a second model of CHXYZ. Are the two structures identical (superposable)?

Interchange any two substituents on one of the carbon atoms. Are the two CHXYZ molecules identical now? Does the fact that interchange of any two substituents on the carbon interconverts the stereoisomers indicate that there are only two possible configurations of a tetrahedral carbon atom?

Compare the two models that were not identical. What is the relationship between them? Do they have a mirror-image relationship? That is, are they related as an object and its mirror image?

EXERCISE 3 (Chapter 4)

Make a model of ethane, CH_3CH_3. Does each of the carbon atoms retain a tetrahedral configuration? Can the carbon atoms be rotated with respect to each other without breaking the carbon-carbon bond?

Rotate about the carbon-carbon bond until the carbon-hydrogen bonds of one carbon atom are aligned with those of the other carbon atom. This is the eclipsed conformation. When the C–H bond of one carbon atom bisects the H–C–H angle of the other carbon atom the conformation is called staggered. Remember, conformations are arrangements of atoms in a molecule that can be interconverted by bond rotations.

In which of the two conformations of ethane you made are the hydrogen atoms of one carbon closer to those of the other carbon?

EXERCISE 4 (Chapter 4)

Prepare a second model of ethane. Replace one hydrogen, any one, on each ethane model with a substituent such as a halogen, to form two models of CH_3CH_2X. Are the structures identical? If not, can they be made identical by rotation about the C—C bond? With one of the models, demonstrate that there are three equivalent staggered conformations (see Exercise 3) of CH_3CH_2X. How many equivalent eclipsed conformations are possible?

EXERCISE 5 (Chapter 4)

Assemble a model of a 1,2-disubstituted ethane molecule, CH_2XCH_2X. Note how the orientation of and the distance between the X groups changes with rotation about the carbon-carbon bond. The arrangement in which the X substituents are at maximum separation is the *anti*-staggered conformation. The other staggered conformations are called *gauche*.

How many *gauche* conformations are possible? Are they energetically equivalent? Are they identical?

EXERCISE 6 (Chapter 4)

Construct two models of butane, $CH_3CH_2CH_2CH_3$. Note that the structures can be viewed as dimethyl substituted ethanes. Show that rotations about the C2,C3 bond of butane produce eclipsed, *anti*-staggered, and *gauche*-staggered conformations. Measure the distance between C1 and C4 in the conformations just mentioned. [The scale of the Darling Framework Molecular Model Set, for example, is: 2.0 inches in a model corresponds to approximately 1.0Å (0.1 nm) on a molecular scale.] In which eclipsed conformation are the C1 and C4 atoms closest to each other? How many eclipsed conformations are possible?

EXERCISE 7 (Chapter 5)

Using two models of butane, verify that the two hydrogen atoms on C2 are not stereochemically equivalent. Replacement of one hydrogen leads to a product that is not identical to that obtained by replacement of the other C2 hydrogen atom. Both replacement products have the same condensed formula, $CH_3CHXCH_2CH_3$. What is the relationship of the two products?

EXERCISE 8 (Chapter 4)

Make a model of hexane, $CH_3CH_2CH_2CH_2CH_2CH_3$. Extend the six-carbon chain as far as it will go. This puts C1 and C6 at maximum separation. Notice that this *straight-chain* structure maintains the tetrahedral bond angles at each carbon atom and therefore the carbon chain adopts a zigzag arrangement. Does this extended chain adopt staggered or eclipsed conformations of the hydrogen atoms? How could you describe the relationship of C1 and C4?

EXERCISE 9 (Chapter 7)

Prepare models of the four isomeric butenes, C_4H_8. Note that the restricted rotation about the double bond is responsible for the cis-trans stereoisomerisim. Verify this by observing that breaking the π bond of *cis*-2-butene allows rotation and thus conversion to *trans*-2-butene. Is any of the four isomeric butenes chiral (nonsuperposable with its mirror image)? Indicate pairs of butene isomers that are structural (constitutional) isomers. Indicate pairs that are diastereomers. How does the distance between the C1 and C4 atoms in *trans*-2-butene compare with that of the *anti* conformation of butane? Compare the C1 to C4 distance in *cis*-2-butene with that in the conformation of butane in which the methyls are eclipsed.

1-Butene   *cis*-2-Butene   *trans*-2-Butene   2-Methylpropene

EXERCISE 10 (Chapter 4)

Make a model of cyclopropane. Take care not to break your models due to the angle strain of the carbon-carbon bonds of the cyclopropane ring. It should be apparent that the ring carbon atoms must be coplanar. What is the relationship of the hydrogen atoms on adjacent carbon atoms? Are they staggered, eclipsed, or skewed?

EXERCISE 11 (Chapter 4)

A model of cyclobutane can be assembled in a conformation that has the four carbon atoms coplanar. How many eclipsed hydrogen atoms are there in the conformation? Torsional strain (strain caused by repulsions between the aligned electron pairs of eclipsed bonds) can be relieved at the expense of increased angle strain by a slight folding of the ring. The deviation of one ring carbon from the plane of the other three carbon atoms is about 25°. This folding compresses the C—C—C bond angle to about 88°. Rotate the ring carbon bonds of the planar conformation to obtain the folded conformation. Are the hydrogen atoms on adjacent carbon atoms eclipsed or skewed? Considering both structural and stereoisomeric forms, how many dimethylcyclobutane structures are possible? Do deviations of the ring from planarity have to be considered when determining the number of possible dimethyl structures?

EXERCISE 12 (Chapter 4)

Cyclopentane is a more flexible ring system than cyclobutane or cyclopropane. A model of cyclopentane in a conformation with all the ring carbon atoms coplanar exhibits minimal deviation of the C–C–C bond angles from the normal tetrahedral bond angle. How many eclipsed hydrogen interactions are there in this planar conformation? If one of the ring carbon atoms is pushed slightly above (or below) the plane of the other carbon atoms, a model of the envelope conformation is obtained. Does the envelope conformation relieve some of the torsional strain? How many eclipsed hydrogen interactions are there in the envelope conformation?

Cyclopentane

EXERCISE 13 (Chapter 5)

Make a model of 1,2-dimethylcyclopentane. How many stereoisomers are possible for this compound? Identify each of the possible structures as either cis or trans. Is it apparent that cis-trans isomerism is possible in this compound because of restricted rotation? Are any of the stereoisomers chiral? What are the relationships of the 1,2-dimethylcyclopentane stereoisomers?

EXERCISE 14 (Chapter 4)

Assemble the six-membered ring compound cyclohexane. Is the ring flat or puckered? Place the ring in a chair conformation and then in a boat conformation. Demonstrate that the chair and boat are indeed conformations of cyclohexane—that is, they may be inter-converted by rotations about the carbon-carbon bonds of the ring.

Chair form   Boat form

Note that in the chair conformation carbon atoms 2, 3, 5, and 6 are in the same plane and carbon atoms 1 and 4 are below and above the plane, respectively. In the boat conformation, carbon atoms 1 and 4 are both above (they could also both be below) the plane described by carbon atoms 2, 3, 5, and 6. Is it apparent why the boat is sometimes associated with the flexible form? Are the hydrogen atoms in the chair conformation staggered or eclipsed? Are any hydrogen atoms eclipsed in the boat conformation? Do carbon atoms 1 and 4 have an *anti* or *gauche* relationship in the chair conformation? (*Hint:* Look down the C2, C3 bond).

A twist conformation of cyclohexane may be obtained by slightly twisting carbon atoms 2 and 5 of the boat conformation as shown.

Boat form   Twist form

Note that the C2, C3 and the C5, C6 sigma bonds no longer retain their parallel orientation in the twist conformation. If the ring system is twisted too far, another boat conformation results. Compare the nonbonded (van der Waals repulsion) interactions and the torsional strain present in the boat, twist, and chair conformations of cyclohexane. Is it apparent why the relative order of thermodynamic stabilities is chair > twist > boat?

EXERCISE 15 (Chapter 4)

Construct a model of methylcyclohexane. How many chair conformations are possible? How does the orientation of the methyl group change in each chair conformation?

Identify carbon atoms in the chair conformation of methylcyclohexane that have in-tramolecular interactions corresponding to those found in the *gauche* and *anti*conforma-tions of butane. Which of the chair conformations has the greatest number of *gauche* interactions? How many more? If we assume, as in the case for butane, that the *anti* interaction is 3.8 kJ mol^{-1} more favorable than *gauche*, then what is the relative stability of the two chair conformations of methylcyclohexane? *Hint:* Identify the relative number of *gauche* interactions in the two conformations.

EXERCISE 16 (Chapter 4)

Compare models of the chair conformations of monosubstituted cyclohexanes in which the substituent alkyl groups are methyl, ethyl, isopropyl, and *tert*-butyl.

Rationalize the relative stability of axial and equatorial conformations of the alkyl group given in the table for each compound. The chair conformation with the alkyl group equatorial is more stable by the amount shown.

ALKYL GROUP	$\Delta G°$ (kJ mol^{-1}) EQUATORIAL $\rightleftharpoons$ AXIAL
CH_3	7.1
CH_2CH_3	7.5
$CH(CH_3)_2$	9.2
$C(CH_3)_3$	21 (approximate)

EXERCISE 17 (Chapter 4)

Make a model of 1,2-dimethylcyclohexane. Answer the questions posed in Exercise 13 with regard to 1,2-dimethylcyclohexane.

EXERCISE 18 (Chapter 4)

Compare models of the neutral and charged molecules shown next. Identify the structures that are isoelectronic, that is, those that have the same electronic structure. How do those structures that are isoelectronic compare in their molecular geometry?

CH_3CH_3	CH_3NH_2	CH_3OH
$CH_3CH_2^-$	$CH_3NH_3^+$	$CH_3OH_2^+$
CH_3NH^-		

EXERCISE 19 (Chapter 7)

Prepare a model of cyclohexene. Note that chair and boat conformations are no longer possible, as carbon atoms 1, 2, 3, and 6 lie in a plane. Are cis and trans stereoisomers possible for the double bond? Attempt to assemble a model of *trans*-cyclohexene. Can it be done? Are cis and trans stereoisomers possible for 2,3-dimethylcyclohexene? For 3,4-dimethylcyclohexene?

Cyclohexene

Assemble a model of *trans*-cyclooctene. Observe the twisting of the π-bond system. Would you expect the cis stereoisomer to be more stable than *trans*-cyclooctene? Is *cis*-cyclooctene chiral? Is *trans*-cyclooctene chiral?

EXERCISE 20 (Chapter 4)

Construct models of *cis*-decalin (*cis*-bicyclo[4.4.0]decane) and *trans*-decalin. Observe how it is possible to convert one conformation of *cis*-decalin in which both rings are in chair conformations to another all-chair conformation. This interconversion is not possible in the case of the *trans*-decalin isomer. Suggest a reason for the difference in the behavior of the cis and trans isomers. *Hint:* What would happen to carbon atoms 7 and 10 of *trans*-decalin if the other ring (indicated by carbon atoms numbered 1–6) is converted to the alternative chair conformation. Is the situation the same for *cis*-decalin?

trans-Decalin   *cis*-Decalin

EXERCISE 21 (Chapter 4)

Assemble a model of norbornane (bicyclo[2.2.1]heptane). Observe the two cyclopentane ring systems in the molecule. The structure may also be viewed as having a methylene (CH$_2$) bridge between carbon atoms 1 and 4 of cyclohexane. Describe the conformation of the cyclohexane ring system in norbornane. How many eclipsing interactions are present?

Norbornane

Using a model of twistane, identify the cyclohexane ring systems held in twist conformations. In adamantane, find the chair conformation cyclohexane systems. How many are present? Evaluate the torsional and angle strain in adamantane. Which of the three compounds in this exercise are chiral?

Twistane   Adamantane

EXERCISE 22 (Chapter 7)

An hypothesis known as Bredt's Rule states that a double bond to a bridgehead of a small-ring bridged bicyclic compound is not possible. The basis of this rule can be seen if you attempt to make a model of bicyclo[2.2.1]hept-1-ene, **A.** One approach to the assembly of this model is to try to bridge the number 1 and number 4 carbon atoms of cyclohexene with a methylene (CH_2) unit. Compare this bridging with the ease of installing a CH_2 bridge between the 1 and 4 carbon atoms of cyclohexane to form a model of norbornane (see Exercise 21). Explain the differences in ease of assembly of these two models.

A   **B**

Bridgehead double bonds can be accommodated in larger ring-bridged bicyclic compounds such as bicyclo[3.2.2]non-1-ene, **B.** Although this compound has been prepared in the laboratory, it is an extremely reactive alkene.

EXERCISE 23 (Chapter 14)

Not all cyclic structures with alternating double and single bonds are aromatic. Cyclooctatetraene shows none of the aromatic characteristics of benzene. From examination of molecular models of cyclooctatetraene and benzene, explain why there is π-electron delocalization in benzene but not in cyclooctatetraene. *Hint:* Can the carbon atoms of the eight-membered ring readily adopt a planar arrangement?

Benzene   Cyclooctatetraene

Note that benzene can be represented in several different ways with most molecular model sets. In this exercise, the Kekulé representation with alternating double and single bonds is appropriate. Alternative representations of benzene, such as a form depicting molecular orbital lobes, are shown in your model set instruction booklet.

EXERCISE 24 (Chapter 5)

Consider the $CH_3CHXCHYCH_3$ system. Assemble all possible stereoisomers of this structure. How many are there? Indicate the relationship among them. Are they all chiral? Repeat the analysis with the $CH_3CHXCHXCH_3$ system.

EXERCISE 25 (Chapter 5)

The $CH_3CHXCHXCH_3$ molecule can exist as the stereoisomers shown here. In the eclipsed conformation (meso) shown on the left, the molecule has a plane of symmetry that bisects the C2, C3 bond. This is a more energetic conformation than any of the three staggered conformations, but it is the only conformation of this configurational stereoisomer that has a plane of symmetry. Can you consider a molecule achiral if only one conformation, and in this case not even the most stable conformation, has a plane of symmetry? Are any of the staggered conformations achiral (superposable on their mirror image)? Make a model of the staggered conformation shown here and make another model that is the mirror image of it. Are these two structures different conformations of the same configurational stereoisomer (e.g., are they conformers that can be interconverted by bond rotations), or are they configurational stereoisomers? Based on your answer to the last question, suggest an explanation for the fact that the molecule is not optically active.

E   S

EXERCISE 26 (Chapter 5)

Not all molecular chirality is a result of a tetrahedral stereocenter, such as CHXYZ. Cumulated dienes (1,2-dienes or allenes) are capable of generating molecular chirality. Identify, using models, which of the following cumulated dienes are chiral.

A   B   C

Are the following compounds chiral? How are they structurally related to cumulated dienes?

D   E

Is the cumulated triene **F** chiral? Explain the presence or absence of molecular chirality. More than one stereoisomer is possible for triene **F**. What are the structures, and what is the relationship between those structures?

F

EXERCISE 27 (Chapters 5 and 14)

Substituted biphenyl systems can produce molecular chirality if the rotation about the bond connecting the two rings is restricted. Which of the three biphenyl compounds indicated here are chiral and would be expected to be optically active?

J. a = f = CH_3
   b = e = $\overset{+}{N}(CH_3)_3$

K. a = b = CH_3
   e = f = $\overset{+}{N}(CH_3)_3$

L. a = f = CH_3
   b = e = H

EXERCISE 28 (Chapter 8)

Assemble a simple model of ethyne (acetylene). The linear geometry of the molecule should be readily apparent. Now, use appropriate pieces of your model set to depict the σ and both the π bonds of the triple bond system using *sp* hybrid carbon atoms and pieces that represent orbitals. Based on attempts to assemble cycloalkynes, predict the smallest cycloalkyne that is stable.

EXERCISE 29 (Chapter 22)

Construct a model of β-D-glucopyranose. Note that in one of the chair conformations all the hydroxyl groups and the CH_2OH group are in an equatorial orientation. Convert the structure of β-D-glucopyranose to α-D-glucopyranose, to β-D-mannopyranose, and to β-D-galactopyranose. Indicate the number of large ring substituents (OH or CH_2OH) that are axial in the more favorable chair conformation of each of these sugars. Is it reasonable that the β-anomer is more stable than the α-anomer of D-glucopyranose?

Make a model of β-L-glucopyranose. What is the relationship between the D and L configurations? Which is more stable?

β-D-Glucopyranose   D-(+)-Glucose   D-(+)-Mannose   D-(+)-Galactose

EXERCISE 30 (Chapter 24)

Assemble a model of tripeptide **A** shown here. Restricted rotation of the C—N bond in the amide linkage results from resonance contribution of the nitrogen nonbonding electron pair. Note the planarity of the six atoms associated with the amide portions of the molecule caused by this resonance contribution. Which bonds along the peptide chain are free to rotate? The amide linkage can either be cisoid or transoid. How does the length (from the N-terminal nitrogen atom to the C-terminal carbon atom) of the tripeptide chain that is transoid compare with one that is cisoid? Which is more "linear"? Convert a model of tripeptide **A** in the transoid arrangement to a model of tripeptide **B**. Which tripeptide has a longer chain?

Tripeptide **A**	R = CH$_3$	(L-Alanine)
Tripeptide **B**	R = CH$_2$OH	(L-Serine)

EXERCISE 31 (Chapter 13)

Make models of the π molecular orbitals for the following compounds. Use the phase representation of each contributing atomic orbital shown in your molecular model set instruction booklet. Compare each model with π molecular orbital diagrams shown in the textbook.

(a) π_1 and π_2 of ethene (CH$_2$=CH$_2$)

(b) π_1 through π_4 of 1,3-butadiene (CH$_2$=CH—CH=CH$_2$)

(c) π_1, π_2, and π_3 of the allyl (propenyl) radical (CH$_2$=CH—CH$_2$)

EXERCISE 32

Use your model set to construct several of the interesting representative natural product structures shown here.

Progesterone   Caryophyllene   Longifolene

Morphine   Strychnine

MOLECULAR MODEL SET EXERCISES — SOLUTIONS

Solution 1   Replacement of any hydrogen atom of methane leads to the same monosubstituted product CH_3X. Therefore, there is only one configuration of a monosubstituted methane. There
is only one possible configuration for a disubstituted methane of either the CH_2X_2
or CH_2XY type.

Solution 2   Interchange of any two substituents converts the configuration of a tetrahedral stereocenter to that of its enantiomer. There are only two possible configurations. If the models are
not identical, they will have a mirror-image relationship.

Solution 3   The tetrahedral carbon atoms may be rotated without breaking the carbon-carbon bond.
There is no change in the carbon-carbon bond orbital overlap during rotation. The
eclipsed conformation places the hydrogen atoms closer together than they are in the staggered conformation.

Staggered   Eclipsed
conformation   conformation

Solution 4   All monosubstituted ethanes (CH_3CH_2X) may be made into identical structures by rotations about the C–C bond. The following structures are three energetically equivalent staggered conformations.

The three equivalent eclipsed conformations are

Solution 5 The two *gauche* conformations are energetically equivalent, but not identical (superposable) since they are conformational enantiomers. They bear a mirror-image relationship and are interconvertible by rotation about the carbon-carbon bond.

anti Conformation   *gauche* Conformation

Solution 6 There are three eclipsed conformations. The methyl groups (C1 and C4) are closest together in the methyl-methyl eclipsed conformation. The carbon-carbon internuclear distances between C1 and C4 are shown in the following table. The number of conformations of each type and the molecular distances in angstroms (Å) are shown.

CONFORMATION	NUMBER	DISTANCES (Å)
Eclipsed (CH_3, CH_3)	1	2.5
Gauche	2	2.8
Eclipsed (H, CH_3)	2	3.3
Anti	1	3.7

Solution 7 The enantiomers formed from replacement of the C2 hydrogen atoms of butane are

Solution 8 The extended chain assumes a staggered arrangement. The relationship of C1 and C4 is *anti*.

Solution 9   None of the isomeric butenes is chiral. They all have a plane of symmetry. All the isomeric butenes are related as constitutional (or structural) isomers except *cis*-2-butene and *trans*-2-butene, which are diastereomers.

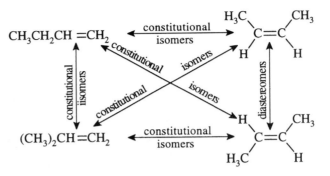

Molecular Model Set C1 to C4 distances:

COMPOUND	DISTANCES (Å)
cis-1-Butene	2.0
trans-2-Butene	3.7
Butane (*gauche*)	2.8
Butane (*anti*)	3.7

Solution 10   The hydrogen atoms are all eclipsed in cyclopropane.

Solution 11   All the hydrogen atoms are eclipsed in the planar conformation of cyclobutane. The folded ring system has skewed hydrogen interactions. There are six possible isomers of di-methylcyclobutane. Since the ring is not held in one particular folded conformation, deviations of the ring planarity need not be considered in determining the number of possible dimethyl structures.

Solution 12   In the planar conformation of cyclopentane, all five pairs of methylene hydrogen atoms are eclipsed. That produces 10 eclipsed hydrogen interactions. Some torsional strain is relieved in the envelope conformation since there are only six eclipsed hydrogen interactions.

Solution 13   The three configurational stereoisomers of 1,2-dimethylcyclopentane are shown here. Both trans stereoisomers are chiral, while the cis configuration is an achiral meso compound.

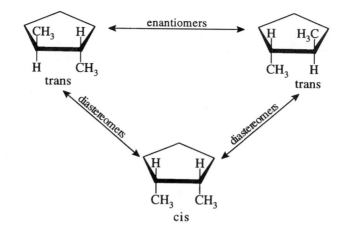

Solution 14 The puckered ring of the chair and the boat conformations can be interconverted by rotation about the carbon-carbon bonds. The chair is more rigid than the boat conformation. All hydrogen atoms in the chair conformation have a staggered arrangement. In the boat conformation, there are eclipsed relationships between the hydrogen atoms on C2 and C3, and also between those on C5 and C6. Carbon atoms that are 1,4 to each other in the chair conformation have a *gauche* relationship. An evaluation of the three conformations confirms the relative stability: chair > twist > boat. The boat conformation has considerable eclipsing strain and nonbonded (van der Waals repulsion) interactions, the twist conformation has slight eclipsing strain, and the chair conformation has a minimum of eclipsing and nonbonded interactions.

Solution 15 Interconversion of the two chair conformations of methylcyclohexane changes the methyl group from an axial to a less crowded equatorial orientation, or the methyl that is equatorial to the more crowded axial position.

Axial methyl   Equatorial methyl

The conformation with the axial methyl group has two *gauche* (1,3-diaxial) interactions that are not present in the equatorial methyl conformation. These *gauche* interactions are axial methyl to C3 and axial methyl to C5. The methyl to C3 and methyl to C5 relationships with methyl groups in an equatorial orientation are anti.

Solution 16 The $\Delta G°$ value reflects the relative energies of the two chair conformations for each structure. The crowding of the alkyl group in an axial orientation becomes greater as the bulk of the group increases. The increased size of the substituent has little effect on the steric interactions of the conformation that has the alkyl group equatorial. The *gauche* (1,3-diaxial) interactions are responsible for the increased strain for the axial conformation. Since the ethyl and isopropyl groups can rotate to minimize the nonbonded interactions, their effective size is less than their actual size. The *tert*-butyl group cannot relieve the steric interactions by rotation and thus there is a considerably greater difference in potential energy between the axial and equatorial conformations.

Solution 17 All four stereoisomers of 1,2-dimethylcyclohexane are chiral. The *cis*-1,2-dimethylcyclohexane conformations have equal energy and are readily interconverted, as shown here.

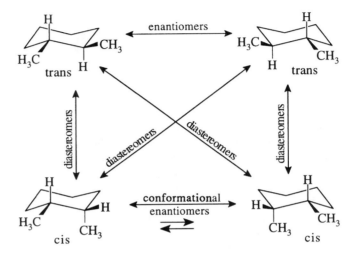

Solution 18 The structures that are isoelectronic have the same geometry. Isoelectronic structures are

$$CH_3CH_3 \qquad and \qquad CH_3NH_3{}^+$$
$$CH_3NH_2 \qquad CH_3CH_2{}^- \qquad and \qquad CH_3OH_2{}^+$$

Structure CH_3NH^- would be isoelectronic to CH_3OH.

Solution 19 Cis-trans stereoisomers are possible only for 3,4-dimethylcyclohexene. The ring size and geometry of the double bond prohibit a trans configuration of the double bond. Two configurational isomers (they are enantiomers) are possible for 2,3-dimethylcyclohexene.

cis-Cyclooctene is more stable because it has less strain than the *trans*-cyclooctene structure. The relative stability of cycloalkene stereoisomers in rings larger than cyclodecane generally favors trans. The *trans*-cyclooctene structure is chiral.

trans-Cyclooctene
(one enantiomer)

Solution 20 The ring fusion in *trans*-decalin is equatorial, equatorial. That is, one ring is attached to the other as 1,2-diequatorial substituents would be. Interconversion of the chair conformations of one ring (carbon atoms 1 through 6) in *trans*-decalin would require the other ring to adopt a 1,2-diaxial orientation. Carbon atoms 7 and 10 would both become axial substituents to the other ring. The four carbon atoms of the *substituent* ring (carbon atoms 7 through 10) cannot bridge the diaxial distance. In *cis*-decalin both conformations have an axial, equatorial ring fusion. Four carbon atoms can easily bridge the axial, equatorial distance.

Solution 21 The cyclohexane ring in norbornane is held in a boat conformation, and therefore has four hydrogen eclipsing interactions. All the six-membered ring systems in twistane are in twist conformations. All four of the six-membered ring systems in adamantane are chair conformations.

Solution 22 Bridging the 1 and 4 carbon atoms of cyclohexane is relatively easy since in the boat conformation the flagpole hydrogen atoms (on C1 and C4) are fairly close and their C–H bonds are directed toward one another. With cyclohexene, the geometry of the double bond and its inability to rotate freely make it impossible to bridge the C1, C4 distance with a single methylene group. Note, however, that a cyclohexene ring can accommodate a methylene bridge between C3 and C6. This bridged bicyclic system (bicyclo[2.2.1] hept-2-ene) does not have a bridgehead double bond.

Bicyclo[2.2.1]hept-2-ene

Solution 23 The 120° geometry of the double bond is ideal for incorporation into a planar six-membered ring, as the internal angle of a regular hexagon is 120°. Cyclooctatetraene cannot adopt a planar ring system without considerable angle strain. The eight-membered ring adopts a "tub" conformation that minimizes angle strain and does not allow significant *p*-orbital overlap other than that of the four double bonds in the system. Thus, the cyclooctatetraene has four isolated double bonds and is not a delocalized π-electron system.

Cyclooctatetraene (tub conformation)

Solution 24   In the $CH_3CHXCHYCH_3$ system, there are four stereoisomers, all of which are chiral.

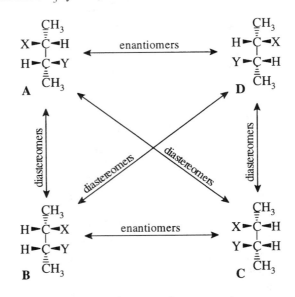

In the $CH_3CHXCHXCH_3$ system, there are three stereoisomers, two of which are chiral. The third stereoisomer (**G**) (shown on page 575) is an achiral meso structure.

Solution 25   If at least one conformation of a molecule in which free rotation is possible has a plane of symmetry, the molecule is achiral. For a molecule with the configurations specified, there are two achiral conformations: the eclipsed conformation **E** shown in the exercise and staggered conformation **F.**

A model of **F** is identical with its mirror image. It is achiral, although it does not have a plane of symmetry, due to the presence of a *center* of symmetry that is located between C2 and C3. A center of symmetry, like a plane of symmetry, is a reflection symmetry element. A center of symmetry involves reflection through a point; a plane of symmetry requires reflection about a plane. A model of the mirror image of **S** (structure **T**) is not identical to **S**, but is a conformational enantiomer of **S.** They can be made identical by rotation about the C2, C3 bond. Since **S** and **T** are conformational enantiomers, the two will be present in equal amounts in a solution of this configurational stereoisomer. Both conformation **S** and conformation **T** are chiral and therefore should rotate the plane of plane polarized light. Since they are enantiomeric, the rotations of light will be equal in magnitude but *opposite* in direction. The net result is a racemic form of conformational enantiomers, and thus optically inactive. A similar argument can be made for any other chiral conformation and this configuration of $CH_3CHXCHXCH_3$.

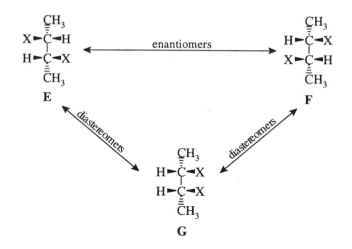

Solution 26   Structures B and C are chiral. Structure A has a plane of symmetry and is therefore achiral. Compounds D and E are both chiral. The relative orientation of the terminal groups in D and E is perpendicular, as is the case in the cumulated dienes.

Cumulated triene F is achiral. It has a plane of symmetry passing through all six carbon atoms. Structure F has a trans configuration. The cis diastereomer is the only other possible stereoisomer.

Solution 27   Structure **J** can be isolated as a chiral stereoisomer because of the large steric barrier to rotation about the bond connecting the rings. Biphenyl **K** has a plane of symmetry and is therefore achiral. The symmetry plane of **K** is shown here. Any chiral conformation of **L** can easily be converted to its enantiomer by rotation. It is only when a ≠ b and f ≠ e and rotation is restricted by bulky groups that chiral (optically active) stereoisomers can be isolated.

A plane of symmetry

Solution 28   A representation of the molecular orbitals in ethyne is given in Section 1.14 of the text. The smallest stable cycloalkyne is the nine-membered ring cyclononyne.

Solution 29   As shown here, the alternative chair conformation of β-D-glucopyranose has all large substituents in an axial orientation. The structures α-D-glucopyranose, β-D-mannopyranose, and β-D-galactopyranose all have one large axial substitutent in the most favorable conformation. β-L-Glucopyranose is the enantiomer (mirror image) of β-D-glucopyranose. Enantiomers are of equal thermodynamic stability.

β-D-Glucopyranose

α-D-Glucopyranose   β-D-Galactopyranose

β-D-Mannopyranose   β-L-Glucopyranose

Solution 30 The peptide chain bonds not free to rotate are indicated by the bold lines in the structure shown here. The transoid arrangement produces a more linear tripeptide chain. The length of the tripeptide chain does not change if you change the substituent **R** groups.

Solution 31 The models of the π molecular orbitals for ethene are shown in the Orbital Symmetry section of the Darling Framework Molecular Model Set instruction booklet. A representation of these orbitals can be found in the text in Section 1.13.

The π molecular orbitals for 1,3-butadiene are shown in the text in Section 13.7C. A model of the π molecular orbitals of 1,3-butadiene is also shown in the Orbital Symmetry section of the Darling Framework Molecular Model instruction booklet. The phases of the contributing atomic orbitals to the molecular orbitals of the allyl radical can be found in the text in Section 13.3A. The π molecular orbital of the allyl radical has a node at C2.

D
APPENDIX
Glossary

A

Absolute configuration (Section 5.14A): The actual arrangement of groups in a molecule. The absolute configuration of a molecule can be determined by X-ray analysis or by relating the configuration of a molecule, using reactions of known stereochemistry, to another molecule whose absolute configuration is known.

Absorption spectrum (Section 13.9A): A plot of the wavelength (λ) of a region of the spectrum versus the absorbance (A) at each wavelength. The absorbance at a particular wavelength (A_λ) is defined by the equation $A_\lambda = \log (I_R/I_S)$, where I_R is the intensity of the reference beam and I_S is the intensity of the sample beam.

Acetal (Section 16.7B): A functional group, consisting of a carbon bonded to alkoxy groups [i.e., $RCH(OR')_2$ or $R_2C(OR')_2$], derived by adding 2 molar equivalents of an alcohol to an aldehyde or ketone.

Acetylene (Sections 1.14, 7.11 and 7.12): A common name for ethyne. Also used as a general name for alkynes.

Acetylenic hydrogen atom (Sections 3.14, 4.18C and 7.12): A hydrogen atom attached to a carbon atom that is bonded to another carbon atom by a triple bond.

Achiral molecule (Section 5.2): A molecule that is superposable on its mirror image. Achiral molecules lack handedness and are incapable of existing as a pair of enantiomers.

Acid strength (Section 3.5): The strength of an acid is related to its acidity constant, K_a or to its pK_a. The larger the value of its K_a or the smaller the value of its pK_a, the stronger is the acid.

Acidity constant (Section 3.5A): An equilibrium constant related to the strength of an acid. For the reaction,

$$HA + H_2O \rightleftharpoons H_3O^+ + A^-$$

$$K_a = \frac{[H_3O^+][A^-]}{[HA]}$$

Activating group (Section 15.10): A group that when present on a benzene ring causes the ring to be more reactive in electrophilic substitution than benzene itself.

Activation Energy E_{act}; see **Energy of activation.**

Acyl group (Section 15.7): The general name for groups with the structure RCO— or ArCO—.

Acyl halide (Section 15.7): Also called an *acid halide*. A general name for compounds with the structure RCOX or ArCOX.

Acylation (Section 15.7): The introduction of an acyl group into a molecule.

Acylium ion (Section 15.7): The resonance-stabilized cation:

$$R-\overset{+}{C}=\overset{..}{O}: \longleftrightarrow R-C\equiv\overset{+}{O}:$$

Addition (chain-growth) polymer (Section 10.10): A polymer that results from a stepwise addition of monomers to a chain (usually through a chain reaction) with no loss of other atoms or molecules in the process.

Addition reaction (Section 8.1): A reaction that results in an increase in the number of groups attached to a pair of atoms joined by a double or triple bond. An addition reaction is the opposite of an elimination reaction.

Aglycone (Section 22.4): The alcohol obtained by hydrolysis of a glycoside.

Aldaric acid (Section 22.6C): An α,ω-dicarboxylic acid that results from oxidation of the aldehyde group and the terminal 1° alcohol group of an aldose.

Alditol (Section 22.7): The alcohol that results from the reduction of the aldehyde or keto group of an aldose or ketose.

Aldonic acid (Section 22.6B): A monocarboxylic acid that results from oxidation of the aldehyde group of an aldose.

Aliphatic compound (Section 14.1): A nonaromatic compound such as an alkane, cycloalkane, alkene, or alkyne.

Alkaloid (Special Topic F): A naturally occurring basic compound that contains an amino group. Most alkaloids have profound physiological effects.

Alkylation (Sections 4.18C and 15.7): The introduction of an alkyl group into a molecule.

Allyl group (Section 4.5): The $CH_2=CHCH_2-$ group.

Allylic substituent (Section 13.2): Refers to a substituent on a carbon atom adjacent to a carbon–carbon double bond.

Ambident nucleophile (Section 17.7C): Nucleophiles that are capable of reacting at two different nucleophilic sites.

Angle strain (Section 4.11): The increased potential energy of a molecule (usually a cyclic one) caused by deformation of a bond angle away from its lowest energy value.

Annulene (Section 14.7A): Monocyclic hydrocarbons that can be represented by structures having alternating single and double bonds. The ring size of an annulene is represented by a number in brackets, e.g., benzene is [6]annulene and cyclooctatetraene is [8]annulene.

Anomers (Section 22.2C): A term used in carbohydrate chemistry. Anomers are diastereomers that differ only in configuration at the acetal or hemiacetal carbon of a sugar in its cyclic form.

Anti addition (Section 8.7): An addition that places the parts of the adding reagent on opposite faces of the reactant.

Anti conformation (Section 4.9A): An anti conformation of butane, for example, has the methyl groups at an angle of 180° to each other:

Anti
conformation
of butane

Antiaromatic compound (Section 14.7D): A cyclic conjugated system whose π electron energy is greater than that of the corresponding open-chain compound.

Antibonding molecular orbital (antibonding MO) (Sections 1.11 and 1.13): A molecular orbital whose energy is higher than that of the isolated atomic orbitals from which it is constructed. Electrons in an antibonding molecular orbital destabilize the bond between the atoms that the orbital encompasses.

Anticodon (Section 25.5C): A sequence of three bases on transfer RNA (tRNA) that associates with a codon of messenger RNA (mRNA).

Aprotic solvent (Section 6.14C): A solvent whose molecules do not have a hydrogen atom attached to a strongly electronegative element (such as oxygen). For most purposes, this means that an aprotic solvent is one whose molecules lack an —OH group.

Arene (Section 15.1): A general name for an aromatic hydrocarbon.

Aromatic compound (Sections 14.1–14.8 and 14.11): A cyclic conjugated unsaturated molecule or ion that is stabilized by π electron delocalization. Aromatic compounds are characterized by having large resonance energies, by reacting by substitution rather than addition, and by deshielding of protons exterior to the ring in their ^{1}H NMR spectra caused by the presence of an induced ring current.

Aryl group (Section 15.1): The general name for a group obtained (on paper) by the removal of a hydrogen from a ring position of an aromatic hydrocarbon. Abbreviated Ar—.

Aryl halide (Section 6.1): An organic halide in which the halogen atom is attached to an aromatic ring, such as a benzene ring.

Atactic polymer (Special Topic A): A polymer in which the configuration at the stereocenters along the chain is random.

Atomic orbital (AO) (Section 1.10): A volume of space about the nucleus of an atom where there is a high probability of finding an electron. An atomic orbital can be described mathematically by its **wave function.** Atomic orbitals have characteristic quantum numbers; the *principle quantum number, n,* is related to the energy of the electron in an atomic orbital and can have the values 1, 2, 3, . . . The *azimuthal quantum number, l,* determines the angular momentum of the electron that results from its motion around the nucleus, and can have the values, 0, 1, 2, . . . , $(n − 1)$. The *magnetic quantum number, m,* determines the

orientation in space of the angular momentum and can have values from $+l$ to $−l$. The *spin quantum number, s,* measures the intrinsic angular momentum of an electron and can have the values of $+\frac{1}{2}$ and $−\frac{1}{2}$ only.

Aufbau principle (Section 1.10): A principle that guides us in assigning electrons to orbitals of an atom or molecule in its lowest energy state or ground state. The aufbau principle states that electrons are added so that orbitals of lowest energy are filled first.

Autoxidation (Section 10.11C): The reaction of an organic compound with oxygen to form a hydroperoxide.

Axial bond (Section 4.13): The six bonds of a cyclohexane ring (below) that are perpendicular to the general plane of the ring, and that alternate up

and down around the ring.

B

Base peak (Section 9.14): The most intense peak in a mass spectrum.

Base strength (Section 3.5): The strength of a base is inversely related to the strength of its conjugate acid; the weaker the conjugate acid, the stronger is the base. In other words, if the conjugate acid has a large pK_a, the base will be strong.

Benzenoid aromatic compound (Section 14.8A): An aromatic compound whose molecules have one or more benzene rings.

Benzyl group (Sections 2.5B and 15.15): The $C_6H_5CH_2$— group.

Benzylic substituent (Section 15.15): Refers to a substituent on a carbon atom adjacent to a benzene ring.

Benzyne (Section 21.11B): An unstable, highly reactive intermediate consisting of a benzene ring with an additional bond resulting from sideways overlap of sp^2 orbitals on adjacent atoms of the ring.

Betaine (Section 16.10): An electrically neutral molecule that has nonadjacent cationic and anionic sites and that does not possess a hydrogen atom bonded to the cationic site.

Bimolecular reaction (Section 6.6): A reaction whose rate-determining step involves two initially separate species.

Boat conformation (Section 4.12): A conformation of cyclohexane that resembles a boat and that has eclipsed bonds along its two sides:

It is of higher energy than the chair conformation.

Bond angle (Sections 1.12 and 1.16): The angle between two bonds originating at the same atom.

Bond dissociation energy, see **Homolytic bond dissociation energy.**

Bond length (Sections 1.11 and 1.14A): The equilibrium distance between two bonded atoms or groups.

Bond-line formula (Section 1.17D): A formula that shows the carbon skeleton of a molecule with lines. The number of hydrogen atoms necessary to fulfill each carbon's valence is assumed to be present but not written in. Other atoms (e.g., O, Cl, N) are written in.

Bonding molecular orbital (bonding MO) (Section 1.11): The energy of a bonding molecular orbital is lower than the energy of the isolated atomic orbitals from which it arises. When electrons occupy a bonding molecular orbital they help hold together the atoms that the molecular orbital encompasses.

Bromination (Section 8.6): A reaction in which bromine atoms are introduced into a molecule.

Bromohydrin (Section 8.8): A bromo alcohol.

Bromonium ion (Section 8.6A): An ion containing a positive bromine atom bonded to two carbon atoms.

Bronsted–Lowry theory of acids and bases (Section 3.2A): An acid is a substance that can donate (or lose) a proton; a base is a substance that can accept (or remove) a proton. The **conjugate acid** of a base is the molecule or ion that forms when a base accepts a proton. The **conjugate base** of an acid is the molecule or ion that forms when an acid loses its proton.

C

Carbanion (Section 3.3): A chemical species in which a carbon atom bears a formal negative charge.

Carbene (Section 8.9): An uncharged species in which a carbon atom is divalent. The species, : CH_2, called methylene, is a carbene.

Carbenoid (Section 8.9C): A carbene-like species. A species such as the reagent formed when diiodomethane reacts with a zinc–copper couple. This reagent, called the Simmons–Smith reagent, reacts with alkenes to add methylene to the double bond in a stereospecific way.

Carbocation (Section 3.3): A chemical species in which a trivalent carbon atom bears a formal positive charge.

Carbohydrate (Section 22.1A): A group of naturally occurring compounds that are usually defined as polyhydroxyaldehydes or polyhydroxyketones, or as substances that undergo hydrolysis to yield such compounds. In actuality, the aldehyde and ketone groups of carbohydrates are often present as hemiacetals and acetals. The name comes from the fact that many carbohydrates possess the empirical formula $C_x(H_2O)_y$.

Carbonyl group (Section 16.1): A functional group consisting of a carbon atom doubly bonded to an oxygen atom. The carbonyl group is found in aldehydes, ketones, esters, anhydrides, amides, acyl halides, and so on. Collectively these compounds are referred to as carbonyl compounds.

Cascade polymer (Special Topic B.5): A polymer produced from a multifunctional central core by progressively adding layers of repeating units.

CFC (see **Freon**): A chlorofluorocarbon.

Chain-growth polymer: See **Addition polymer.**

Chain reaction (Section 10.4): A reaction that proceeds by a se-

quential, stepwise mechanism, in which each step generates the reactive intermediate that causes the next step to occur. Chain reactions have *chain initiating steps, chain propagating steps, and chain terminating steps.*

Chair conformation (Section 4.12): The all-staggered conformation of cyclohexane that has no angle strain or torsional strain and is, therefore, the lowest energy conformation:

Chemical shift, δ (Sections 9.3C and 9.6): The position in an NMR spectrum, relative to a reference compound, at which a nucleus absorbs. The reference compound most often used is tetramethylsilane (TMS), and its absorption point is arbitrarily designated zero. The chemical shift of a given nucleus is proportional to the strength of the magnetic field of the spectrometer. The chemical shift in delta units, δ, is determined by dividing the observed shift from TMS in hertz multiplied by 10^6 by the operating frequency of the spectrometer in hertz.

Chiral molecule (Section 5.2): A molecule that is not superposable on its mirror image. Chiral molecules have handedness and are capable of existing as a pair of enantiomers.

Chirality (Section 5.2): The property of having handedness.

Chlorination (Sections 8.6 and 10.5): A reaction in which chlorine atoms are introduced into a molecule.

Chlorohydrin (Section 8.8): A chloro alcohol.

Cis–trans isomers (Sections 4.5 and 7.2): Diastereomers that differ in their stereochemistry at adjacent atoms of a double bond or on different atoms of a ring.

Codon (Section 25.5C): A sequence of three bases on messenger RNA (mRNA) that contains the genetic information for one amino acid. The codon associates, by hydrogen bonding, with an anticodon of a transfer RNA (tRNA) that carries the particular amino acid for protein synthesis on the ribosome.

Condensation (step-growth) polymer (Special Topic B): A polymer produced when bifunctional monomers (or potentially bifunctional monomers) react with each other through the intermolecular elimination of water or an alcohol. Polyesters, polyamides, and polyurethanes are all condensation polymers.

Condensation reaction (Section 17.4): A reaction in which molecules become joined through the intermolecular elimination of water or an alcohol.

Configuration (Section 5.6): The particular arrangement of atoms (or groups) in space that is characteristic of a given stereoisomer.

Conformation (Section 4.8): A particular temporary orientation of a molecule that results from rotations about its single bonds.

Conformational analysis (Section 4.9): An analysis of the energy changes that a molecule undergoes as its groups undergo rotation (sometimes only partial) about the single bonds that join them.

Conformer (Section 4.8): A particular staggered conformation of a molecule.

Conjugate acid (Section 3.2A): The molecule or ion that forms when a base accepts a proton.

Conjugate addition (Section 17.9): A form of nucleophilic addition to an α,β-unsaturated carbonyl compound in which the nucleophile adds to the β-carbon.

Conjugate base (Section 3.2A): The molecule or ion that forms when an acid loses its proton.

Conjugated system (Section 13.1): Molecules or ions that have an extended π system. A conjugated system has a p orbital on an atom adjacent to a multiple bond; the p orbital may be that of another multiple bond or that of a radical, carbocation, or carbanion.

Connectivity (Section 1.3): The sequence, or order, in which the atoms of a molecule are attached to each other.

Constitutional isomers (Section 1.3A): Compounds that have the same molecular formula but that differ in their connectivity (i.e., molecules that have the same molecular formula but have their atoms connected in different ways).

Copolymer (Special Topic A): A polymer synthesized by polymerizing two monomers.

Coupling constant, J_{ab} (Section 9.8): The separation in frequency units (hertz) of the peaks of a multiplet caused by spin–spin coupling between atoms a and b.

Covalent bond (Section 1.4B): The type of bond that results when atoms share electrons.

Cracking (Section 4.1C): A process used in the petroleum industry for breaking down the molecules of larger alkanes into smaller ones. Cracking may be accomplished with heat (thermal cracking), or with a catalyst (catalytic cracking).

Crown ether (Section 11.20A): Cyclic polyethers that have the ability to form complexes with metal ions. Crown ethers are named as x-crown-y where x is the total number of atoms in the ring and y is the number of oxygen atoms in the ring.

Cyanohydrin (Sections 16.9 and 18.3): A functional group consisting of a carbon atom bonded to a cyano group and to a hydroxyl group, i.e., RHC(OH)(CN) or R_2C(OH)(CN), derived by adding HCN to an aldehyde or ketone.

Cycloaddition (Section 13.11): A reaction, like the Diels–Alder reaction, in which two connected groups add to the end of a π system to generate a new ring.

D

D and L designations (Section 22.2B): A method for designating the configuration of monosaccharides and other similar compounds in which the reference compound is (+)- or (−)-glyceraldehyde. According to this system, (+)-glyceraldehyde is designated D-(+)-glyceraldehyde and (−)-glyceraldehyde is designated L-(−)-glyceraldehyde. Therefore, a monosaccharide whose highest numbered stereocenter has the same general configuration as D-(+)-glyceraldehyde is designated a D-sugar; one whose highest numbered stereocenter has the same general configuration as L-(+)-glyceraldehyde is designated an L-sugar.

Deactivating group (Section 15.10): A group that when present on a benzene ring causes the ring to be less reactive in electrophilic substitution than benzene itself.

Debromination (Section 7.9): The elimination of two atoms of bromine from a *vic*-dibromide, or, more generally, the loss of bromine from a molecule.

Debye unit (Section 2.3): The unit in which dipole moments are stated. One debye, D, equals 1×10^{-18} esu cm.

Decarboxylation (Section 18.11): A reaction whereby a carboxylic acid loses CO_2.

Degenerate orbitals (Section 1.10): Orbitals of equal energy. For example, the three $2p$ orbitals are degenerate.

Dehydration reaction (Section 7.7): An elimination that involves the loss of a molecule of water from the substrate.

Dehydrohalogenation (Section 6.16): An elimination reaction that results in the loss of HX from adjacent carbons of the substrate and the formation of a π bond.

Delocalization (Section 6.12B): The dispersal of electrons (or of electrical charge). Delocalization of charge always stabilizes a system.

Dextrorotatory (Section 5.7B): A compound that rotates plane-polarized light clockwise.

Diastereomers (Section 5.1): Stereoisomers that are not mirror images of each other.

Diastereoselective reaction (See **Stereoselective reaction** and Section 12.3)

Diastereotopic hydrogens (or ligands) (Section 9.7B): If replacement of each of two hydrogens (or ligands) by the same groups yields compounds that are diastereomers, the two hydrogen atoms (or ligands) are said to be diastereotopic.

Dielectric constant (Section 6.14D): A measure of a solvent's ability to insulate opposite charges from each other. The dielectric constant of a solvent roughly measures its polarity. Solvents with high dielectric constants are better solvents for ions than are solvents with low dielectric constants.

Dienophile (Section 13.11): The diene-seeking component of a Diels–Alder reaction.

Dipolar ion (Section 24.2C): The charge-separated form of an amino acid that results from the transfer of a proton from a carboxyl group to a basic group.

Dipole moment, μ (Section 2.3): A physical property associated with a polar molecule that can be measured experimentally. It is defined as the product of the charge in electrostatic units (esu) and the distance that separates them in centimeters: $\mu = e \times d$.

Dipole-dipole interaction (Section 2.14B): An interaction between molecules having permanent dipole moments.

Disaccharide (Section 22.1A): A carbohydrate that, on a molecular basis, undergoes hydrolytic cleavage to yield two molecules of a monosaccharide.

Distortionless enhanced polarization transfer (DEPT) spectra (Section 9.10E): A series of ^{13}C NMR spectra in which the signal for each type of carbon, C, CH, CH_2, and CH_3, is printed out separately. Data from DEPT spectra help identify the different types of carbon atoms in a ^{13}C NMR spectrum.

E

E–Z system (Section 7.2): A system for designating the stereochemistry of alkene diastereomers based on the priorities of groups in the Cahn–Ingold–Prelog convention.

E1 reaction (Section 6.18): A unimolecular elimination in which, in a slow, rate-determining step, a leaving group departs from the substrate to form a carbocation. The carbocation then in a fast step loses a proton with the resulting formation of a π bond.

E2 reaction (Section 6.17): A bimolecular 1,2 elimination in which, in a single step, a base removes a proton and a leaving group departs from the substrate, resulting in the formation of a π bond.

Eclipsed conformation (Section 4.8): A temporary orientation of groups around two atoms joined by a single bond such that the groups directly oppose each other.

An eclipsed
conformation

Electromagnetic spectrum (Section 9.2): The full range of energies propagated by wave fluctuations in an electromagnetic field.

Electron isodensity surface (Section 1.12B): An electron isodensity surface shows points in space that happen to have the same electron density. An electron isodensity surface can be calculated for any chosen value of electron density. A "high" electron isodensity surface (also called a "bond" electron density surface) shows the *core* of electron density around each atomic nucleus and regions where neighboring atoms share electrons (bonding regions). A "low" electron isodensity surface roughly shows the *outline* of a molecule's electron cloud. This surface gives information about molecular shape and volume, and usually looks the same as a van der Waals or space-filling model of the molecule. (Contributed by Alan Shusterman, Reed College, and Warren Hehre, Wavefunction, Inc.)

Electronegativity (Section 1.4A): A measure of the ability of an atom to attract electrons it is sharing with another and thereby polarize the bond.

Electrophile (Section 3.3 and 8.1): A Lewis acid, an electron-pair acceptor, an electron seeking reagent.

Electrophoresis (Section 25.6A): A technique for separating charged molecules based on their different mobilities in an electric field.

Electrostatic potential maps (Section 3.2C): Electrostatic potential maps are structures calculated by a computer that show the relative distribution of electron density at some surface of a molecule or ion. They are very useful for understanding interactions between molecules that are based on attraction of opposite charges. Usually we choose the van der Waals surface (approximately the outermost region of electron density) of a molecule to depict the electrostatic potential map because this is where the electron density of one molecule would first interact with another. In an electrostatic potential map, color trending toward red indicates a region with more negative charge, and color trending toward blue indicates a region with less negative charge (or more positive charge). An electrostatic potential map is generated by calculating the extent of charge interaction (electrostatic potential) between an imaginary positive charge and the electron density at a particular point or surface in a molecule. (Contributed by Alan Shusterman, Reed College, and Warren Hehre, Wavefunction, Inc.)

Elimination (Section 6.16): A reaction that results in the loss of two groups from the substrate and the formation of a π bond. The most common elimination is a 1,2 elimination or β elimination, in which the two groups are lost from adjacent atoms.

Empirical formula (Section 1.2B): A formula that expresses the relative proportions of atoms in a molecule as smallest whole numbers.

Enantiomeric excess or enantiomeric purity (Section 5.8B): A percentage calculated for a mixture of enantiomers by dividing the moles of one enantiomer minus the moles of the other enantiomer by the moles of both enantiomers and multiplying by 100. The enantiomeric excess equals the percentage optical purity.

Enantiomers (Section 5.1): Stereoisomers that are mirror images of each other.

Enantioselective reaction (See **Stereoselective reaction** and Section 5.9B)

Enantiotopic hydrogens (or **ligands**) (Section 9.7B): If replacement of each of two hydrogens (or ligands) by the same group yields compounds that are enantiomers, the two hydrogen atoms (or ligands) are said to be enantiotopic.

Endergonic reaction (Section 6.8): A reaction that proceeds with a positive free energy change.

Endothermic reaction (Section 3.8A): A reaction that absorbs heat. For an endothermic reaction $\Delta H°$ is positive.

Energy (Section 3.8): Energy is the capacity to do work.

Energy of activation, E_{act} (Section 6.8): A measure of the difference in potential energy between the reactants and the transition state of a reaction. It is related to, but not the same as, the free energy of activation, $\Delta G^{\ddagger}$.

Enolate anion (Section 17.1): The delocalized anion formed when an enol loses its hydroxylic proton or when the carbonyl tautomer that is in equilibrium with the enol loses an α proton.

Enthalpy change (Sections 3.8A and 3.9): Also called the heat of reaction. The *standard enthalpy change,* $\Delta H°$, is the change in enthalpy after a system in its standard state has undergone a transformation to another system, also in its standard state. For a reaction, $\Delta H°$ is a measure of the difference in the total bond energy of the reactants and products. It is one way of expressing the change in potential energy of molecules as they undergo reaction. The enthalpy change is related to the **free-energy change,** $\Delta G°$, and to the **entropy change,** $\Delta S°$, through the expression:

$$\Delta H° = \Delta G° + T\Delta S°$$

Entropy change (Section 3.9): The standard entropy change, $\Delta S°$, is the change in entropy between two systems in their standard states. Entropy changes have to do with changes in the relative order of a system. The more random a system is, the greater is its entropy. When a system becomes more disorderly its entropy change is positive.

Epoxide (Section 11.17). An oxirane. A three-membered ring containing one oxygen and two carbon atoms.

Equatorial bond (Section 4.13): The six bonds of a cyclohexane ring that lie generally around the "equator" of the molecule:

Equilibrium constant (Section 3.5A): A constant that expresses the position of an equilibrium. The equilibrium constant is calculated by multiplying the molar concentrations of the products together and then dividing this number by the number obtained by multiplying together the molar concentrations of the reactants.

Equilibrium control, see Thermodynamic control.

Essential oil (Section 23.3): A volatile odoriferous compound obtained by steam distillation of plant material.

Exergonic reaction (Section 6.8): A reaction that proceeds with a negative free-energy change.

Exothermic reaction (Section 3.8A): A reaction that evolves heat. For an exothermic reaction, $\Delta H°$ is negative.

F

Fat (Section 23.2): A triacylglycerol. The triester of glycerol with carboxylic acids.

Fatty acid (Section 23.2): A long-chained carboxylic acid (usually with an even number of carbon atoms) that is isolated by the hydrolysis of a fat.

Fischer projection formula (Sections 5.12 and 22.2C): A two-dimensional formula for representing the configuration of a chiral molecule. By convention, Fischer projection formulas are written with the main carbon chain extending from top to bottom with all groups eclipsed. Vertical lines represent bonds that project behind the plane of the page (or that lie in it). Horizontal lines represent bonds that project out of the plane of the page.

Fischer   Wedge-dashed
projection   wedge formula

Fluorination (Section 10.5C): A reaction in which fluorine atoms are introduced into a molecule.

Formal charge (Section 1.7): The difference between the number of electrons assigned to an atom in a molecule and the number of electrons it has in its outer shell in its elemental state. Formal charge can be calculated using the formula: $F = Z - S/2 - U$, where F is the formal charge, Z is the group number of the atom (i.e., the number of electrons the atom has in its outer shell in its elemental state), S equals the number of electrons the atom is sharing with other atoms, and U is the number of unshared electrons the atom possesses.

Free energy of activation, $\Delta G^{\ddagger}$, (Section 6.8): The difference in free energy between the transition state and the reactants.

Free-energy change (Section 3.9): The *standard free-energy change*, $\Delta G°$, is the change in free energy between two systems in their standard states. At constant temperature, $\Delta G° = \Delta H° - T\Delta S° = -RT \ln K_{eq}$, where $\Delta H°$ is the standard enthalpy change, $\Delta S°$ is the standard entropy change, and K_{eq} is the equilibrium constant. A negative value of $\Delta G°$ for a reaction means that the formation of products is favored when the reaction reaches equilibrium.

Free energy diagram (Section 6.8): A plot of free energy changes that take place during a reaction versus the reaction coordinate. It displays free energy changes as a function of changes in bond orders and distances as reactants proceed through the transition state to become products.

Freon (Section 10.11E): A chlorofluorocarbon or CFC.

Frequency (abbreviated ν) (Section 2.16 and 9.2): The number of full cycles of a wave that pass a given point in each second.

Functional group (Section 2.5): The particular group of atoms in a molecule that primarily determines how the molecule reacts.

Functional group interconversion (Section 6.15): A process that converts one functional group into another.

Furanose (Section 22.2C): A sugar in which the cyclic acetal or hemiacetal ring is five membered.

G

Gauche conformation (Section 4.9A): A gauche conformation of butane, for example, has the methyl groups at an angle of 60° to each other:

**A gauche
conformation
of butane**

Geminal (gem) substituents (Section 7.9): Substituents that are on the same atom.

Glycol (Sections 4.3F and 8.10): A diol.

Glycoside (Section 22.4): A cyclic mixed acetal of a sugar with an alcohol.

Grignard reagent (Section 12.6B): An organomagnesium halide, usually written RMgX.

Ground state (Section 1.12): The lowest electronic energy state of an atom or molecule.

H

Halogenation (Section 10.3): A reaction in which halogen atoms are introduced into a molecule.

Halohydrin (Section 8.8): A halo alcohol.

Halonium ion (Section 8.6A): An ion containing a positive halogen atom bonded to two carbon atoms.

Hammond–Leffler postulate (Section 6.14A): A postulate stating that the structure and geometry of the transition state of a given step will show a greater resemblance to the reactants or products of that step depending on which is closer to the transition state in energy. This means that the transition state of an endothermic step will resemble the products of that step more than the reactants, whereas the transition state of an exothermic step will resemble the reactants of that step more than the products.

Heat of combustion (Section 4.10A): The standard enthalpy change for the complete combustion of 1 mol of a compound.

Heat of hydrogenation (Section 7.3A): The standard enthalpy change that accompanies the hydrogenation of 1 mol of a compound to form a particular product.

Heisenberg uncertainty principle (Section 1.11): A fundamental principle that states that both the position and momentum of an electron (or of any object) cannot be exactly measured simultaneously.

Hemiacetal (Section 16.7A): A functional group, consisting of a carbon atom bonded to an alkoxy group and to a hydroxyl group, [i.e., $RCH(OH)(OR')$ or $R_2C(OH)(OR')$]. Hemiacetals are synthesized by adding one molar equivalent of an alcohol to an aldehyde or a ketone.

Hemiketal (Section 16.7A): Also called a hemiacetal. A functional group, consisting of a carbon atom bonded to an alkoxy group and to a hydroxyl group, [i.e., $R_2C(OH)(OR')$]. Hemiketals are synthesized by adding one molar equivalent of an alcohol to a ketone.

Hertz (abbreviated Hz) (Section 9.2): Now used instead of the equivalent cycles per second as a measure of the frequency of a wave.

Heterocyclic compound (Section 14.9): A compound whose molecules have a ring containing an element other than carbon.

Heterolysis (Section 3.1A): The cleavage of a covalent bond so that one fragment departs with both of the electrons of the covalent bond that joined them. Heterolysis of a bond normally produces positive and negative ions.

Hofmann rule (Sections 7.6B and 20.13A). When an elimination yields the alkene with the less substituted double bond, it is said to follow the Hofmann rule.

HOMO (Sections 3.2C and 13.9B): The highest occupied molecular orbital.

Homologous series (Section 4.7): A series of compounds in which each member differs from the next member by a constant unit.

Homolysis (Section 3.1A): The cleavage of a covalent bond so that each fragment departs with one of the electrons of the covalent bond that joined them.

Homolytic bond dissociation energy, $DH°$, (Section 10.2): The enthalpy change that accompanies the homolytic cleavage of a covalent bond.

Hückel's rule (Section 14.7): A rule stating that planar monocyclic rings with ($4n + 2$) delocalized π electrons (i.e., with 2, 6, 10, 14, . . . , delocalized π electrons) will be aromatic.

Hund's rule (Section 1.10): A rule used in applying the **aufbau principle.** When orbitals are of equal energy (i.e., when they are **degenerate**), electrons are added to each orbital with their spins unpaired, until each degenerate orbital contains one electron. Then electrons are added to the orbitals so that the spins are paired.

Hybridization of atomic orbitals (Section 1.12): A mathematical (and theoretical) mixing of two or more atomic orbitals to give the same number of new orbitals, called *hybrid orbitals,* each of which has some of the character of the original atomic orbitals.

Hydration (Sections 8.5 and 11.4): The addition of water to a molecule, such as the addition of water to an alkene to form an alcohol.

Hydroboration (Section 11.6): The addition of a boron hydride (either BH_3 or an alkylborane) to a multiple bond.

Hydrogen bond (Section 2.14C): A strong dipole–dipole interaction ($4–38$ kJ mol^{-1}) that occurs between hydrogen atoms bonded to small strongly electronegative atoms (O, N, or F) and the nonbonding electron pairs on other such electronegative atoms.

Hydrogenation (Sections 4.18A and 7.13–7.15): A reaction in which hydrogen adds to a double or triple bond. Hydrogenation is often accomplished through the use of a metal catalyst such as platinum, palladium, rhodium, or ruthenium.

Hydrophilic group (Section 2.14E): A polar group that seeks an aqueous environment.

Hydrophobic group (also called a **lipophilic group**) (Sections 2.14E and 11.20): A nonpolar group that avoids an aqueous surrounding and seeks a nonpolar environment.

Hydroxylation (Sections 8.10 and 11.19): The addition of hydroxyl groups to each carbon or atom of a double bond.

I

Index of hydrogen deficiency (Section 7.16): The index of hydrogen deficiency (or IHD) equals the number of pairs of hydrogen atoms that must be subtracted from the molecular formula of the corresponding alkane to give the molecular formula of the compound under consideration.

Inductive effect (Sections 3.7B and 15.11B): An intrinsic electron-attracting or -releasing effect that results from a nearby dipole in the molecule and that is transmitted through space and through the bonds of a molecule.

Infrared (IR) spectroscopy (Section 2.16): A type of optical spectroscopy that measures the absorption of infrared radiation. Infrared spectroscopy provides structural information about functional groups present in the compound being analyzed.

Intermediate (Sections 3.1, 6.10, and 6.11): A transient species that exists between reactants and products in a state corresponding to a local energy minimum on a potential energy diagram.

Iodination (Section 10.5C): A reaction in which iodine atoms are introduced into a molecule.

Ion (Sections 1.4A and 6.3): A chemical species that bears an electrical charge.

Ion–dipole interaction (Section 2.14E): The interaction of an ion with a permanent dipole. Such interactions (resulting in solvation) occur between ions and the molecules of polar solvents.

Ionic bond (Section 1.4A): A bond formed by the transfer of electrons from one atom to another resulting in the creation of oppositely charged ions.

Ionic reaction (Sections 3.1A and 6.3): A reaction involving ions as reactants, intermediates, or products. Ionic reactions occur through the heterolysis of covalent bonds.

Isoelectric point (Section 24.2C): The pH at which the number of positive and negative charges on an amino acid or protein are equal.

Isomers (Sections 1.3A and 5.1): Different molecules that have the same molecular formula.

Isoprene unit (Section 23.3): A name for the structural unit:

found in all terpenes.

Isotactic polymer (Special Topic A): A polymer in which the configuration at each stereocenter along the chain is the same.

K

Kekulé structure (Sections 2.2D and 14.4): A structure in which lines are used to represent bonds. The Kekulé structure for benzene is a hexagon of carbon atoms with alternating single and double bonds around the ring, and with one hydrogen atom attached to each carbon.

Ketal (Section 16.7B): Properly called an acetal. A functional group, consisting of a carbon bonded to alkoxyl groups, [i.e., $R_2C(OR')_2$], derived by adding two molar equivalents of an alcohol to a ketone.

Kinetic control (Section 13.10A): A principle stating that when the ratio of products of a reaction is determined by relative rates of reaction, the most abundant product will be the one that is formed fastest.

Kinetic energy (Section 3.8): Energy that results from the motion of an object. Kinetic energy $(KE) = \frac{1}{2}mv^2$, where m is the mass of the object, and v is its velocity.

Kinetics (Section 6.6): A term that refers to rates of reactions.

L

Lactam (Section 18.8I): A cyclic amide.

Lactone (Section 18.7C): A cyclic ester.

Leaving group (Section 6.5): The substituent with an unshared electron pair that departs from the substrate in a nucleophilic substitution reaction.

Leveling effect of a solvent (Section 3.14): An effect that restricts the use of certain solvents with strong acids and bases. In principle, no acid stronger than the conjugate acid of a particular solvent can exist to an appreciable extent in that solvent, and no base stronger than the conjugate base of the solvent can exist to an appreciable extent in that solvent.

Levorotatory (Section 5.7B): A compound that rotates plane-polarized light in a counterclockwise direction.

Lewis structure (or **electron-dot structure**) (Section 1.4B): A representation of a molecule showing electron pairs as a pair of dots or as a dash.

Lewis theory of acids and bases (Section 3.2B): An acid is an electron pair acceptor, and a base is an electron pair donor.

Lipid (Section 23.1): A substance of biological origin that is soluble in nonpolar solvents. Lipids include fatty acids, triacylglycerols (fats and oils), steroids, prostaglandins, terpenes and terpenoids, and waxes.

Lipophilic group (or hydrophopic group) (Sections 2.14E and 11.20): A nonpolar group that avoids an aqueous surrounding and seeks a nonpolar environment.

LUMO (Sections 3.2C and 13.9B): The lowest unoccupied molecular orbital.

M

Macromolecule (Section 10.10): A very large molecule.

Magnetic resonance imaging (Section 9.11B): A technique based on NMR spectroscopy that is used in medicine.

Markovnikov's rule (Section 8.2): A rule for predicting the regiochemistry of electrophilic additions to alkenes and alkynes that can be stated in various ways. As originally stated (in 1870) by Vladimir Markovnikov, the rule provides that "if an unsymmetrical alkene combines with a hydrogen halide, the halide ion adds to the carbon with the fewer hydrogen atoms." More commonly the rule has been stated in reverse: that in the addition of HX to an alkene or alkyne the hydrogen atom adds to the carbon atom that already has the greater number of hydrogen atoms. A modern expression of Markovnikov's rule is: *In the ionic addition of an unsymmetrical reagent to a multiple bond, the positive portion of the reagent (the electrophile) attaches itself to a carbon atom of the reagent in the way that leads to the formation of the more stable intermediate carbocation.*

Mass spectrometry (Section 9.12): A technique, useful in structure elucidation, that is based on generating ions of a molecule in a magnetic field, then instrumentally determining the mass/charge ratio and relative amounts of the ions that result.

Mechanism, see **Reaction mechanism.**

Meso compound (Section 5.11A): An optically inactive compound whose molecules are achiral even though they contain tetrahedral atoms with four different attached groups.

Mesylate (Section 11.10): A methanesulfonate ester.

Methylene (Section 8.9A): The carbene with the formula CH_2.

Methylene group (Section 6.2): The $-CH_2-$ group.

Micelle (Section 23.2C): A spherical cluster of ions in aqueous solution (such as those from a soap) in which the nonpolar groups are in the interior and the ionic (or polar) groups are at the surface.

Molar absorptivity (abbreviated ϵ) (Section 13.9A): A proportionality constant that relates the observed absorbance (A) at a particular wavelength (λ) to the molar concentration of the sample (C) and the length (l) (in centimeters) of the path of the light beam through the sample cell:

$$\epsilon = A/C \times l$$

Molecular formula (Section 1.2B): A formula that gives the total number of each kind of atom in a molecule. The molecular formula is a whole number multiple of the empirical formula. For example the molecular formula for benzene is C_6H_6; the empirical formula is CH.

Molecular ion (Section 9.13): The radical cation produced in a mass spectrometer when one electron is dislodged from the parent molecule.

Molecular orbital (MO) (Section 1.11): Orbitals that encompass more than one atom of a molecule. When atomic orbitals combine to form molecular orbitals, the number of molecular orbitals that results always equals the number of atomic orbitals that combine.

Molecularity (Section 6.6): The number of species involved in a single step of a reaction (usually the rate determining step).

Monomer (Section 10.10): The simple starting compound from which a polymer is made. For example, the polymer polyethylene is made from the monomer, ethylene.

Monosaccharide (Section 22.1A): The simplest type of carbohydrate, one that does not undergo hydrolytic cleavage to a simpler carbohydrate.

Mutarotation (Section 22.3): The spontaneous change that takes place in the optical rotation of α and β anomers of a sugar when they are dissolved in water. The optical rotations of the sugars change until they reach the same value.

N

Neighboring group participation (Problem 6.42): The effect on the course or rate of a reaction brought about by another group near the functional group undergoing reaction.

Newman projection formula (Section 4.8): A means of representing the spatial relationships of groups attached to two atoms of a molecule. In writing a Newman projection formula we imagine ourselves viewing the molecule from one end directly along the bond axis joining the two atoms. Bonds that are attached to the front atom are shown as radiating from the center of a circle; those attached to the rear atom are shown as radiating from the edge of the circle:

Front atom

Rear atom

Nitrogen rule (Section 9.15A): A rule that states that if the mass of the molecular ion in a mass spectrum is an even number, the parent compound contains an even number of nitrogen atoms, and conversely.

Node (Section 1.9): A place where a wave function (ψ) is equal to zero. The greater the number of nodes in an orbital, the greater is the energy of the orbital.

Nonbenzenoid aromatic compound (Section 14.8B): An aromatic compound, such as azulene, that does not contain benzene rings.

Nuclear magnetic resonance (NMR) spectroscopy (Section 9.3): A spectroscopic method for measuring the absorption of radiofrequency radiation by certain nuclei when the nuclei are in a strong magnetic field. The most important NMR spectra for organic chemists are ^{1}H NMR spectra and ^{13}C NMR spectra. These two types of spectra provide structural information about the carbon framework of the molecule, and about the number and environment of hydrogen atoms attached to each carbon atom.

Nucleic acids (Sections 25.1 and 25.2): Biological polymers of nucleotides. DNA and RNA are, respectively, nucleic acids that preserve and transcribe hereditary information within cells.

Nucleophile (Section 3.3): A Lewis base, an electron pair donor that seeks a positive center in a molecule.

Nucleophilic substitution reaction (Section 6.3): A reaction initiated by a nucleophile (a species with an unshared electron pair) in which the nucleophile reacts with a substrate to replace a substituent (called the leaving group), that departs with an unshared electron pair.

Nucleophilicity (Section 6.14B): The relative reactivity of a nucleophile in an S_N2 reaction as measured by relative rates of reaction.

Nucleoside (Section 25.2): A five-carbon monosaccharide bonded at the 1' position to a purine or pyrimidine.

Nucleotide (Section 25.2): A five-carbon monosaccharide bonded at the 1' position to a purine or pyrimidine and at the 3' or 5' position to a phosphate group.

O

Olefin (Section 7.1): An old name for an alkene.

Optical purity (Section 5.8B): A percentage calculated for a mixture of enantiomers by dividing the observed specific rotation for the mixture by the specific rotation of the pure enantiomer and multiplying by 100. The optical purity equals the enantiomeric purity or enantiomeric excess.

Optically active compound (Section 5.7): A compound that rotates the plane of polarization of plane polarized light.

Orbital (Section 1.10): A volume of space in which there is a high probability of finding an electron. Orbitals are described mathematically by the squaring of wave functions, and each orbital has a characteristic energy. An orbital can hold two electrons when their spins are paired.

Organometallic compound (Section 12.5): A compound that contains a carbon–metal bond.

Oxidation (Section 12.2): A reaction that increases the oxidation state of atoms in a molecule or ion. For an organic substrate, oxidation usually involves increasing its oxygen content or decreasing its hydrogen content. Oxidation also accompanies any reaction in which a less electronegative substituent is replaced by a more electronegative one.

Oxonium ion (Section 3.12): A chemical species with an oxygen atom that bears a formal positive charge.

Oxymercuration (Section 11.5): The addition of —OH and —HgO$_2$CR to a multiple bond.

Ozonolysis (Section 8.11A): The cleavage of a multiple bond that makes use of the reagent O$_3$, called ozone. The reaction leads to the formation of a cyclic compound called an *ozonide,* which is then reduced to carbonyl compounds by treatment with zinc and acetic acid.

P

p orbitals (Section 1.10): A set of three degenerate (equal energy) atomic orbitals shaped like two tangent spheres with a nodal plane at the nucleus. For p orbitals, the principal quantum number, n (see **atomic orbital**), is 2; the azimuthal quantum number, l, = 1; and the magnetic quantum numbers, m are +1, 0, or −1.

Paraffin (Section 4.17): An old name for an alkane.

Pauli exclusion principle (Section 1.10): A principle that states that no two electrons of an atom or molecule may have the same set of four quantum numbers. It means that only two electrons can occupy the same orbital, and then only when their spin quantum numbers are opposite. When this is true, we say that the spins of the electrons are paired.

Periplanar (Section 7.6C): A conformation in which vicinal groups lie in the same plane.

Peroxide (Section 10.1A): A compound with an oxygen-oxygen single bond.

Peroxy acid (Section 11.17): An acid with the general formula RCO_3H, containing an oxygen–oxygen single bond.

Phase sign (Section 1.9): Signs, either + or −, that are characteristic of all equations that describe the amplitudes of waves.

Phase-transfer catalyst (Section 11.20): A reagent that transports an ion from an aqueous phase into a nonpolar phase where reaction takes place more rapidly. Tetraalkylammonium ions and crown ethers are phase-transfer catalysts.

Phospholipid (Section 23.6): Compounds that are structurally derived from *phosphatidic acids*. Phosphatidic acids are derivatives of glycerol in which two hydroxyl groups are joined to fatty acids, and one terminal hydroxyl group is joined in an ester linkage to phosphoric acid. In a phospholipid the phosphate group of the phosphatidic acid is joined in ester linkage to a nitrogen-containing compound such as choline, 2-aminoethanol, or L-serine.

Pi (π) bond (Section 1.13): A bond formed when electrons occupy a bonding π molecular orbital (i.e., the lower energy molecular orbital that results from overlap of parallel p orbitals on adjacent atoms).

Pi (π) molecular orbital (Section 1.13): A molecular orbital formed when parallel p orbitals on adjacent atoms overlap. Pi molecular orbitals may be *bonding* (p lobes of the same phase sign overlap) or *antibonding* (p orbitals of opposite phase sign overlap).

pK_a (Section 3.5): The pK_a is the negative logarithm of the acidity constant, K_a. $pK_a = -\log K_a$.

Plane of symmetry (Section 5.5): An imaginary plane that bisects a molecule in a way such that the two halves of the molecule are mirror images of each other. Any molecule with a plane of symmetry will be achiral.

Plane-polarized light (Section 5.7A): Ordinary light in which the oscillations of the electrical field occur only in one plane.

Polar covalent bond (Section 2.3): A covalent bond in which the electrons are not equally shared because of differing electronegativities of the bonded atoms.

Polar molecule (Section 2.4): A molecule with a dipole moment.

Polarimeter (Section 5.7B): A device used for measuring optical activity.

Polarizability (Section 6.14C): The susceptibility of the electron cloud of an uncharged molecule to distortion by the influence of an electric charge.

Polymer (Section 10.10): A large molecule made up of many repeating subunits. For example, the polymer polyethylene is made up of the repeating subunit $-(CH_2CH_2)_n-$.

Polysaccharide (Section 22.1A): A carbohydrate that, on a molecular basis, undergoes hydrolytic cleavage to yield many molecules of a monosaccharide.

Potential energy (Section 3.8): Potential energy is stored energy; it exists when attractive or repulsive forces exist between objects.

Primary carbon (Section 2.6): A carbon atom that has only one other carbon atom attached to it.

Primary structure (Section 24.5): The covalent structure of a polypeptide or protein. This structure is determined, in large part, by determining the sequence of amino acids in the protein.

Prochiral (Section 12.3): A group is prochiral if replacement of one of two identical groups at a tetrahedral atom, or if addition of a group to a trigonal planar atom, leads to a new stereocenter. At a tetrahedral atom where there are two identical groups, the identical groups can be designated pro-R and pro-S depending on what configuration would result when it is imagined that each is replaced by a group of next higher priority (but not higher than another existing group).

Protecting group (Sections 12.10, 16.7C, and 24.7A): A group that is introduced into a molecule to protect a sensitive group from reaction while a reaction is carried out at some other location in the molecule. Later, the protecting group is removed.

Protein (Section 24.1): A large biological polymer of α-amino acids joined by amide linkages.

Protic solvent (Sections 3.11 and 6.14C): A solvent whose molecules have a hydrogen atom attached to a strongly electronegative element such as oxygen or nitrogen. Molecules of a protic solvent can therefore form hydrogen bonds to unshared electron pairs of oxygen or nitrogen atoms of solute molecules or ions, thereby stabilizing them. Water, methanol, ethanol, formic acid, and acetic acid are typical protic solvents.

Proton decoupling (Section 9.10B): An electronic technique used in ^{13}C NMR spectroscopy that allows decoupling of spin–spin interactions between ^{13}C nuclei and ^1H nuclei. In spectra obtained in this mode of operation all carbon resonances appear as singlets.

Proton off-resonance decoupling (Section 9.10D): An electronic technique used in ^{13}C NMR spectroscopy that allows one-bond couplings between ^{13}C nuclei and ^1H nuclei. In spectra obtained in this mode of operation, CH_3 groups appear as quartets, CH_2 groups appear as triplets, CH groups appear as doublets, and carbon atoms with no attached hydrogen atoms appear as singlets.

Psi function (ψ function or wave function) (Section 1.9): A mathematical expression derived from *quantum mechanics* corresponding to an energy state for an electron. The square of the ψ function, ψ^2, gives the probability of finding the electron in a particular location in space.

Pyranose (Section 22.2C): A sugar in which the cyclic acetal or hemiacetal ring is six membered.

R

R (Section 2.5A): A symbol used to designate an alkyl group. Oftentimes it is taken to symbolize any organic group.

(R–S) System (Section 5.6): A method for designating the configuration of tetrahedral stereocenters.

Racemic form (racemate or racemic mixture) (Section 5.8A): An equimolar mixture of enantiomers. A racemic form is optically inactive.

Racemization (Section 6.13A): A reaction that transforms an optically active compound into a racemic form is said to proceed with racemization. Racemization takes place whenever a reaction causes chiral molecules to be converted to an achiral intermediate.

Radical (or **free radical**) (Section 3.1A): An uncharged chemical species that contains an unpaired electron.

Radical reaction (Section 10.1): A reaction involving radicals. Homolysis of covalent bonds occurs in radical reactions.

Radicofunctional nomenclature (Section 4.3E): A system for naming compounds that uses two or more words to describe the compound. The final word corresponds to the functional group present, the preceding words, usually listed in alphabetical order, describe the remainder of the molecule. Examples are: methyl alcohol, ethyl methyl ether, and ethyl bromide.

Rate control, see **Kinetic control.**

Rate-determining step (Section 6.10A): if a reaction takes place in a series of steps, and if the first step is intrinsically slower than all of the others, then the rate of the overall reaction will be the same as (will be determined by) the rate of this slow step.

Reaction coordinate (Section 6.8): The abscissa in a potential energy diagram that represents the progress of the reaction. It represents the changes in bond orders and bond distances that must take place as reactants are converted to products.

Reaction mechanism (Section 3.1): A step-by-step description of the events that are postulated to take place at the molecular level as reactants are converted to products. A mechanism will include a description of all intermediates and transition states. Any mechanism proposed for a reaction must be consistent with all experimental data obtained for the reaction.

Rearrangement (Sections 3.1 and 7.8A): A reaction that results in a product with a different carbon skeleton from the reactant. The type of rearrangement called a 1,2 shift involves the migration of an organic group (with its electrons) from one atom to the atom next to it.

Reducing sugar (Section 22.6A): Sugars that reduce Tollens' or Benedict's reagents. All sugars that contain hemiacetal or hemiketal groups (and therefore are in equilibrium with aldehydes or α-hydroxyketones) are reducing sugars. Sugars in which only acetal or ketal groups are present are nonreducing sugars.

Reduction (Section 12.2): A reaction that lowers the oxidation state of atoms in a molecule or ion. Reduction of an organic compound usually involves increasing its hydrogen content or decreasing its oxygen content. Reduction also accompanies any reaction that results in replacement of a more electronegative substituent by a less electronegative one.

Regioselective reaction (Section 8.2C): A reaction that yields only one (or a predominance of one) constitutional isomer as the product when two or more constitutional isomers are possible products.

Relative configuration (Section 5.14A): The relationship between the configurations of two chiral molecules. Molecules are said to have the same relative configuration when similar or identical groups in each occupy the same position in space. The configurations of molecules can be related to each other through reactions of known stereochemistry, for example, through reactions that cause no bonds to a stereocenter to be broken.

Resolution (Sections 5.15 and 20.3E): The process by which the enantiomers of a racemic form are separated.

Resonance effect (Sections 3.10A, 13.5 and 15.11B): An effect by which a substituent exerts either an electron-releasing or -withdrawing effect through the pi system of the molecule.

Resonance energy (Section 14.5): An energy of stabilization that represents the difference in energy between the actual compound and that calculated for a single resonance structure. The resonance energy arises from delocalization of electrons in a conjugated system.

Resonance structures (or **resonance contributors**) (Sections 1.8 and 13.5): Lewis structures that differ from one another only in the position of their electrons. A single resonance structure will not adequately represent a molecule. The molecule is better represented as a *hybrid* of all the resonance structures.

Retrosynthetic analysis (Section 4.20A): A method for planning syntheses that involves reasoning backward from the target molecule through various levels of precursors and thus finally to the starting materials.

Ring flip (Sections 4.12 and 4.13): The change in a cyclohexane ring (resulting from partial bond rotations) that converts one ring conformation to another. A chair–chair ring flip converts any equatorial substitutent to an axial substituent and vice versa.

Ring strain (Section 4.11): The increased potential energy of the cyclic form of a molecule (usually measured by heats of combustion) when compared to its acyclic form.

S

s orbital (Section 1.10): A spherical atomic orbital. For s orbitals the azimuthal quantum number $l = 0$ (see atomic orbital).

Saponification (Section 18.7B): Base-promoted hydrolysis of an ester.

Saturated compound (Section 2.2): A compound that does not contain any multiple bonds.

Secondary carbon (Section 2.6): A carbon atom that has two other carbon atoms attached to it.

Secondary structure (Section 24.8): The local conformation of a polypeptide backbone. These local conformations are specified in terms of regular folding patterns such as pleated sheets, α-helixes, and turns.

Shielding and deshielding (Section 9.5): Effects observed in NMR spectra caused by the circulation of sigma and pi electrons within the molecule. Shielding causes signals to appear at higher magnetic fields (upfield), deshielding causes signals to appear at lower magnetic fields (downfield).

Sigma (σ) bond (Section 1.12): A single bond. A bond formed when electrons occupy the bonding σ orbital formed by the end-on overlap of atomic orbitals (or hybrid orbitals) on adjacent atoms. In a sigma bond the electron density has circular symmetry when viewed along the bond axis.

Sigma (σ) orbital (Section 1.12): A molecular orbital formed by end-on overlap of orbitals (or lobes of orbitals) on adjacent atoms. Sigma orbitals may be *bonding* (orbitals or lobes of the same phase sign overlap) or *antibonding* (orbitals or lobes of opposite phase sign overlap).

S$_N$1 reaction (Sections 6.10 and 6.14): Literally, substitution nucleophilic unimolecular. A multistep nucleophilic substitution in which the leaving group departs in a unimolecular step before the attack of the nucleophile. The rate equation is first order in substrate but zero order in the attacking nucleophile.

S$_N$2 reaction (Sections 6.6, 6.7 and 6.14): Literally, substitution nucleophilic bimolecular. A bimolecular nucleophilic substitution reaction that takes place in a single step in which a nucleophile attacks a carbon bearing a leaving group from the backside, causing an inversion of configuration at this carbon and displacement of the leaving group.

Solvent effect (Section 6.14D): An effect on relative rates of reaction caused by the solvent. For example, the use of a polar solvent will increase the rate of reaction of an alkyl halide in an S$_N$1 reaction.

Solvolysis (Section 6.13B): Literally, cleavage by the solvent. A nucleophilic substitution reaction in which the nucleophile is a molecule of the solvent.

sp orbital (Section 1.14): A hybrid orbital that is derived by mathematically combining one s atomic orbital and one p atomic orbital. Two sp hybrid orbitals are obtained by this process, and they are oriented in opposite directions with an angle of 180° between them.

sp^2 orbital (Section 1.13): A hybrid orbital that is derived by mathematically combining one s atomic orbital and two p atomic orbitals. Three sp^2 hybrid orbitals are obtained by this process, and they are directed toward the corners of an equilateral triangle with angles of 120° between them.

sp^3 orbital (Section 1.12): A hybrid orbital that is derived by mathematically combining one s atomic orbital and three p atomic orbitals. Four sp^3 hybrid orbitals are obtained by this process, and they are directed toward the corners of a regular tetrahedron with angles of 109.5° between them.

Specific rotation (Section 5.7C): A physical constant calculated from the observed rotation of a compound using the following equation:

$$[\alpha]_D = \frac{\alpha}{c \times l}$$

where α is the observed rotation using the D line of a sodium lamp, c is the concentration of the solution in grams per milliliter or the density of a neat liquid in g mL^{-1}, and l is the length of the tube in decimeters.

Spin decoupling (Section 9.9): An effect that causes spin–spin splitting not to be observed in NMR spectra.

Spin–spin splitting (Section 9.8): An effect observed in NMR spectra. Spin–spin splittings result in a signal appearing as a multiplet (i.e., doublet, triplet, quartet, etc.) and are caused by magnetic couplings of the nucleus being observed with nuclei of nearby atoms.

Staggered conformation (Section 4.8): A temporary orientation of groups around two atoms joined by a single bond such that the bonds of the back atom exactly bisect the angles formed by the bonds of the front atom in a Newman projection formula:

A staggered
conformation

Step-growth polymer: See Condensation polymer.

Stereocenter (Section 5.2): An atom bearing groups of such nature that an interchange of any two groups will produce a stereoisomer.

Stereochemistry (Section 5.4): Chemical studies that take into account the spatial aspects of molecules.

Stereoisomers (Sections 1.13B, 5.1, and 5.2): Compounds with the same molecular formula that differ *only* in the arrangement of their atoms in space. Stereoisomers have the same connectivity and, therefore, are not constitutional isomers. Stereoisomers are classified further as being either enantiomers or diastereomers.

Stereoselective reaction (Sections 5.9B, 8.15, and 12.3): In reactions where stereocenters are altered or created, a stereoselective reaction produces a preponderance of one stereoisomer. Furthermore, a stereoselective reaction can be either enantioselective, in which case the reaction produces a preponderance of one enantiomer, or diastereoselective, in which case the reaction produces a preponderance of one diastereomer.

Stereospecific reaction (Section 8.7A): A reaction in which a particular stereoisomeric form of the reactant reacts in such a way that it leads to a specific stereoisomeric form of the product.

Steric effect (Section 6.14A): An effect on relative reaction rates caused by the space-filling properties of those parts of a molecule attached at or near the reacting site.

Steric hindrance (Section 6.14A): An effect on relative reaction rates caused when the spatial arrangement of atoms or groups at or near the reacting site hinders or retards a reaction.

Steroid (Section 23.4): Steroids are lipids that are derived from the following perhydrocyclopentanophenanthrene ring system:

Structural formula (Sections 1.3A and 1.17): A formula that shows how the atoms of a molecule are attached to each other.

Substituent effect (Sections 3.7B and 15.10): An effect on the rate of reaction (or on the equilibrium constant) caused by the replacement of a hydrogen atom by another atom or group. Substituent effects include those effects caused by the size of the atom or group, called steric effects, and those effects caused by the ability of the group to release or withdraw electrons, called electronic effects. Electronic effects are further classified as being inductive effects or resonance effects.

Substitution reaction (Sections 6.3 and 10.3): A reaction in which one group replaces another in a molecule.

Substitutive nomenclature (Section 4.3F): A system for naming compounds in which each atom or group, called a substituent, is cited as a prefix or suffix to a parent compound. In the IUPAC system only one group may be cited as a suffix. Locants (usually numbers) are used to tell where the group occurs.

Substrate (Section 6.3): The molecule or ion that undergoes reaction.

Sugar (Section 22.1A): A carbohydrate.

Syn addition (Section 7.14A): An addition that places both parts of the adding reagent on the same face of the reactant.

Syndiotactic polymer (Special Topic A): A polymer in which the configuration at the stereocenters along the chain alternate regularly: (*R*), (*S*), (*R*), (*S*), etc.

Synthon (Section 8.15): The fragments that result (on paper) from the disconnection of a bond. The actual reagent that will, in a synthetic step, provide the synthon is called the *synthetic equivalent.*

T

Tautomers (Section 17.2): Constitutional isomers that are easily interconverted. Keto and enol tautomers, for example, are rapidly interconverted in the presence of acids and bases.

Terpene (Section 23.3): Terpenes are lipids that have a structure that can be derived on paper by linking isoprene units.

Tertiary carbon (Section 2.6): A carbon atom that has three other carbon atoms attached to it.

Tertiary structure (Section 24.8B): The three-dimensional shape of a protein that arises from folding of its polypeptide chains superimposed on its α helixes and pleated sheets.

Thermodynamic control (Section 13.10A): A principle stating that the ratio of products of a reaction that reaches equilibrium is determined by the relative stabilities of the products (as measured by their standard free energies, $\Delta G°$). The most abundant product will be the one that is the most stable.

Torsional barrier (Section 4.8): The barrier to rotation of groups joined by a single bond caused by repulsions between the aligned electron pairs in the eclipsed form.

Torsional strain (Section 4.11): The strain associated with an eclipsed conformation of a molecule; it is caused by repulsions between the aligned electron pairs of the eclipsed bonds.

Tosylate (Section 11.10): A *p*-toluenesulfonate ester.

Transition state (Sections 6.8 and 6.9): A state on a potential energy diagram corresponding to an energy maximum (i.e., characterized by having higher potential energy than immediately adjacent states). The term transition state is also used to refer to the species that occurs at this state of maximum potential energy; another term used for this species is *the activated complex.*

U

Ultraviolet–Visible (UV–Vis) spectroscopy (Section 13.9): A type of optical spectroscopy that measures the absorption of light in the visible and ultraviolet regions of the spectrum. UV–Visible spectra primarily provide structural information about the kind and extent of conjugation of multiple bonds in the compound being analyzed.

Unimolecular reaction (Section 6.10): A reaction whose rate-determining step involves only one species.

Unsaturated compound (Section 2.2): A compound that contains multiple bonds.

V

van der Waals force (or **London force**) (Section 2.14D and 4.7): Weak forces that act between nonpolar molecules or between parts of the same molecule. Bringing two groups (or molecules) together first results in an attractive force between them because a temporary unsymmetrical distribution of electrons in one group induces an opposite polarity in the other. When groups are brought closer than their *van der Waals radii,* the force between them becomes repulsive because their electron clouds begin to interpenetrate each other.

Vicinal (*vic*) **substituents** (Section 7.9): Substituents that are on adjacent atoms.

Vinyl group (Section 4.5): The CH_2=CH— group.

Vinylic halide (Section 6.1): An organic halide in which the halogen atom is attached to a carbon atom of a double bond.

Vinylic substituent (Section 6.1): Refers to a substituent on a carbon atom that participates in a carbon–carbon double bond.

W

Wave function (or ψ function) (Section 1.9): A mathematical expression derived from *quantum mechanics* corresponding to an energy state for an electron, i.e., for an orbital. The square of the ψ function, $ψ^2$, gives the probability of finding the electron in a particular place in space.

Wavelength (abbreviated λ) (Sections 2.16 and 9.2): The distance between consecutive crests (or troughs) of a wave.

Wavenumber (Section 2.16): A way to express the frequency of a wave. The wavenumber is the number of waves per centimeter, expressed as cm^{-1}.

Y

Ylide (Section 16.10): An electrically neutral molecule that has a negative carbon with an unshared electron pair adjacent to a positive heteroatom.

Z

Zaitsev's rule (Section 7.6A): A rule stating that an elimination will give as the major product the most stable alkene (i.e., the alkene with the most highly substituted double bond).

Zwitterion (see **Dipolar ion**): Another name for a dipolar ion.